HIGH PERFORMANCE CONCRETE: PROPERTIES AND APPLICATIONS

HIGH PERFORMANCE CONCRETE: PROPERTIES AND APPLICATIONS

Edited by

S P Shah
*Walter P Murphy Professor of Civil Engineering, and
Director of NSF Center for Science and Technology
of Advanced Cement Based Materials
North Western University, Evanston, IL, USA*

S H Ahmad
*Professor of Civil Engineering
North Carolina State University
Raleigh, NC, USA*

McGraw-Hill, Inc.
New York   St. Louis   San Francisco   Montreal   Toronto

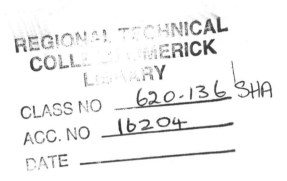
First published in Great Britain 1994

Library of Congress Cataloging-in-Publication Data

High performance concrete : properties and applications / edited by
  S. P. Shah, S. H. Ahmad.
      p.   cm.
  Includes bibliographical references and index.
  ISBN 0-07-056974-6
  1. Concrete.   2. Concrete construction.   I. Shah, S. P.
(Surendra P.)   II. Ahmad, S. H. (Shuaib Haroon), 1953–   .
  TA439.H55   1994
  620.1'36—dc20
                                                        93-40885
                                                            CIP

Printed and bound in Great Britain.

Contents

viii   *Contents*

Preface

High performance concretes (HPC) represents a rather recent development in concrete materials technology. HPC is not a commodity but a range of products, each specifically designed to satisfy in the most effective way the performance requirements for the intended application.

Concrete has large number of properties or attributes. These attributes can be grouped into three general categories: (1) attributes which benefit the construction process; (2) enhanced mechanical properties; (3) enhanced non-mechanical properties such as durability etc. For hardened concrete, strength and durability are the two most important attributes.

In the last three or four years, several national-scale research programs have been established to study various aspects of high performance concretes. These include the two in the US: Center for Science and Technology for Advanced Cement-Based Materials (ACBM), Strategic Highway Research Program (SHRP); The Canadian Network of Centers of Excellence (NCE) Program on High Performance Concrete; the Royal Norwegian Council for Scientific and Industrial Research Program; the Swedish National Program on HPC; the French National Program called 'New Ways for Concrete' and the Japanese New Concrete Program. As the results from these programs start to be disseminated and digested in the concrete industry, the concrete technology will experience a significant advancement.

Historically, more attention has been given to the strength attribute and concrete performance has been specified and evaluated in terms of compressive strength — the higher the compressive strength, the better the expected performance. However, experience has shown that durability considerations become more important for structures exposed to hostile environments (e.g. marine structures and sanitary structures) and for structures such as bridges and pavements which are designed for longer service life. The SHRP program on High Performance Concrete has defined HPC for highway applications in terms of strength and durability attributes and water-cementious materials ratio. HPC is defined as concrete that meets the following criteria:

- It shall have one of the following strength characteristics:
    4-hour strength $\geqslant$2500 psi (17.5 MPa)
    24-hour strength $\geqslant$5000 psi (35 MPa)
    28-day strength $\geqslant$10,000 psi (70 MPa)

- It shall have a durability factor >80% after 300 cycles of freezing and thawing
- It shall have a water-cementitious materials ratio ≤0.35

During the last decade, developments in mineral and chemical admixtures have made it possible to produce concretes with relatively much higher strengths than was thought possible. Presently concretes with strengths of 14,000 to 16,000 psi (98 to 112 MPa) are being commercially produced and used in the construction industry in USA. Other countries such as England, Canada, Norway, Sweden, France, Italy, Japan, Hong Kong and South Korea are aggressively employing the high strength concrete technology in their construction practice.

The aim of this book is to summarize the developments of the last decade in the area of materials development for producing higher strength concrete, production methods, mechanical properties evaluation, non mechanical properties such as durability, and the implication of material properties on the structural design and performance. The use of higher strength concretes in the construction industry has steadily increased during the last decade, and therefore two chapters have been devoted to summarize the applications of higher strength concretes. Experts from USA, Canada, France, Norway, Spain and Japan have contributed individual chapters so as to give the book a broad perspective of the prevailing state-of-the-art in different parts of the world. The book is intended for the academics, engineers, consultants, contractors and researchers.

The eleven chapters in the book are arranged so that the reader can be selective. Chapter 1 provides the background for the selection of materials and proportions. This chapter also provides information on quality control aspects for concretes with higher strengths.

Chapter 2 addresses the short term mechanical properties such as compressive strength, modulus of elasticity, tensile strength etc. The long term mechanical properties such as creep, shrinkage and temperature effects are discussed in Chapter 3. Chapter 4 provides the information on the bond and fatigue characteristics. The important aspect of the durability and its implication for the performance of concrete are discussed in Chapter 5.

The fracture mechanics approach to the understanding of the structural response is outlined in Chapter 6. The behavior of the structural members such as beam, columns and slabs is detailed in Chapter 7. The ductility issues of the structural members and the structural ductility is presented in Chapter 8.

Chapter 9 addresses structural design considerations and the structural applications with special emphasis on high-rise buildings and bridges. This chapter also summarizes the special construction considerations needed for these concretes. Chapter 10 is dedicated to high strength lightweight aggregate concrete and its applications. The last chapter is devoted to the applications of HPC in Japan and South East Asia.

List of Contributors

P Acker
Head, Division: 'Betons et Ciments pour Ouvrages d'Art', Laboratoire
Central des Ponts et Chaussees, Paris, France

S H Ahmad
Professor, Department of Civil Engineering, North Carolina State
University, Raleigh, NC, USA

P N Balaguru
Professor, Civil Engineering Department, Rutgers University, Piscataway,
NJ, USA

T W Bremner
Professor of Civil Engineering, University of New Brunswick, Fredericton,
Canada

F de Larrard
Senior Scientist, Division: 'Betons et Ciments pour Ouvrages d'Art',
Laboratoire Central des Ponts et Chaussees, Paris, France

A S Ezeldin
Assistant Professor, Department of Civil, Environmental and Coastal
Engineering, Stevens Institute of Technology, Hoboken, NJ, USA

R Gettu
Senior Researcher, Technical University of Catalunya, Barcelona, Spain

S K Ghosh
Director, Engineered Structures and Codes, Portland Cement
Association, Skokie, IL, USA

O E Gjørv
Professor, Division of Building Materials, Norwegian Institute of
Technology – NTH, Trondheim – NTH, Norway

T A Holm
Vice President of Engineering, Solite Corporation, PO Box 27211, Richmond, VA, USA

R Le Roy
Research Engineer, Division: 'Betons et Ciments pour Ouvrages d'Art', Laboratoire Central des Ponts et Chaussees, Paris, France

S Mindess
Professor, Department of Civil Engineering, University of British Columbia, Vancouver, Canada

S Nagataki
Professor of Civil Engineering, Tokyo Institute of Technology, O-okayama, Meguru-ku, Tokyo 152, Japan

A H Nilson (Professor)
162 Round Pound Road, HC-60, Box 162, Medomak, Maine, USA (formerly of Cornell University)

H G Russell
Vice President, Construction Technology Laboratories Inc., 5420 Old Orchard Road, Skokie, IL, USA

M Saatcioglu
Associate Professor of Civil Engineering, University of Ottawa, Canada

E Sakai
Manager, Special Cement Additives Division, Denki Kagaku Kogyo Co. Ltd, Yuraku-cho, Chiyoda-ku, Tokyo 100, Japan

S P Shah
Walter P Murphy Professor of Civil Engineering; Director, NSF Center for Science and Technology of Advanced Cement-Based Materials; and Director, Center for Concrete and Geomaterials, Northwestern University, Evanston, IL, USA

1 Materials selection, proportioning and quality control

S Mindess

1.1   Introduction

High performance concretes (HPC) are concretes with properties or attributes which satisfy the performance criteria. Generally concretes with higher strengths and attributes superior to conventional concretes are desirable in the construction industry. For the purposes of this book, HPC is defined in terms of strength and durability. The researchers of Strategic Highway Research Program SHRP-C-205 on High Performance Concrete[1] defined the high performance concretes for pavement applications in terms of strength, durability attributes and water-cementitious materials ratio as follows:

- It shall have one of the following strength characteristics:
    4-hour compressive strength ⩾2500 psi (17.5 MPa) termed as very early strength concrete (VES), or
    24-hour compressive strength ⩾5000 psi (35 MPa) termed as high early strength concrete (HES), or
    28-day compressive strength ⩾10,000 psi (70 MPa) termed as very high strength concrete (VHS).
- It shall have a durability factor greater than 80% after 300 cycles of freezing and thawing.
- It shall have a water-cementitious materials ratio ⩽0.35.

High strength concrete (HSC) could be considered as high performance if other attributes are satisfactory in terms of its intended application. Generally concretes with higher strengths exhibit superiority of other attributes. In North American practice, high strength concrete is usually considered to be a concrete with a 28-day compressive strength of at least 6000 psi (42 MPa). In a recent CEB-FIP State-of-the-Art Report on High

Strength Concrete[2] it is defined as concrete having a minimum 28-day compressive strength of 8700 psi (60 MPa). Clearly then, the definition of 'high strength concrete' is relative; it depends upon both the period of time in question, and the location.

The proportioning (or mix design) of normal strength concretes is based primarily on the w/c ratio 'law' first proposed by Abrams in 1918. At least for concretes with strengths up to 6000 psi (42 MPa), it is implicitly assumed that almost any normal-weight aggregates will be stronger than the hardened cement paste. There is thus no explicit consideration of aggregate strength (or elastic modulus) in the commonly used mix design procedures, such as those proposed by the American Concrete Institute.[3] Similarly, the interfacial regions (or the cement-aggregate bond) are also not explicitly addressed. Rather, it is assumed that the strength of the hardened cement paste will be the limiting factor controlling the concrete strength.

For high strength concretes, however, *all* of the components of the concrete mixture are pushed to their critical limits. High strength concretes may be modelled as three-phase composite materials, the three phases being (i) the hardened cement paste (hcp); (ii) the aggregate; and (iii) the interfacial zone between the hardened cement paste and the aggregate. These three phases must all be optimized, which means that each must be considered explicitly in the design process. In addition, as has been pointed out by Mindess and Young,[4]

> 'it is necessary to pay careful attention to all aspects of concrete production (i.e. selection of materials, mix design, handling and placing). It cannot be emphasized too strongly that quality control is an essential part of the production of high-strength concrete and requires full cooperation among the materials or ready-mixed supplier, the engineer, and the contractor'.

In essence then, the proportioning of high strength concrete mixtures consists of three interrelated steps: (1) selection of suitable ingredients – cement, supplementary cementing materials, aggregates, water and chemical admixtures, (2) determination of the relative quantities of these materials in order to produce, as economically as possible, a concrete that has the desired rheological properties, strength and durability, (3) careful quality control of every phase of the concrete-making process.

1.2   Selection of materials

As indicated above, it is necessary to get the maximum performance out of all of the materials involved in producing high strength concrete. For convenience, the various materials are discussed separately below. However, it must be remembered that prediction with any certainty as to how they will behave when combined in a concrete mixture is not feasible. Particularly when attempting to make high strength concrete, any material incompatibilities will be highly detrimental to the finished product. Thus, the culmination of any mix design process must be the extensive testing of trial mixes.

High strength concrete will normally contain not only portland cement, aggregate and water, but also superplasticizers and supplementary cementing materials. It is possible to achieve compressive strengths of up to 14,000 psi (98 MPa) using fly ash or ground granulated blast furnace slag as the supplementary cementing material. However, to achieve strengths in excess of 14,000 psi (100 MPa), the use of silica fume has been found to be essential, and it is frequently used for concretes in the strength range of 9000–14,000 psi (63–98 MPa) as well.

Portland cement

There are two different requirements that any cement must meet: (i) it must develop the appropriate strength; and (ii) it must exhibit the appropriate rheological behaviour.

High strength concretes have been produced successfully using cements meeting the ASTM Standard Specification C150 for Types I, II and III portland cements. Unfortunately, ASTM C150 is very imprecise in its chemical and physical requirements, and so cements which meet these rather loose specifications can vary quite widely in their fineness and chemical composition. Consequently, cements of nominally the same type will have quite different rheological and strength characteristics, particularly when used in combination with chemical admixtures and supplementary cementing materials. Therefore, when choosing portland cements for use in high strength concrete, it is necessary to look carefully at the cement fineness and chemistry.

Fineness

Increasing the fineness of the portland cement will, on the one hand, increase the early strength of the concrete, since the higher surface area in contact with water will lead to a more rapid hydration. On the other hand, too high a fineness may lead to rheological problems, as the greater amount of reaction at early ages, in particular the formation of ettringite, will lead to a higher rate of slump loss. Early work by Perenchio[5] indicated that fine cements produced higher early age concrete strengths, though at later ages differences in fineness were not significant. Most cements now used to produce high strength concrete have Blaine finenesses that are in the range of 1467 to 1957 ft²/lb (300 to 400 m²/kg), though when Type III (high early strength) cements are used, the finenesses are in the range of 2201 ft²/lb (450 m²/kg).

Chemical composition of the cement

The previously cited work of Perenchio[5] indicates that cements with higher C_3A contents leads to higher strengths. However, subsequent work[6] has shown that high C_3A contents generally leads to rapid loss of flow in the fresh concrete, and as a result high C_3A contents should be avoided in cements used for high strength concrete. Aitcin[7] has shown that the C_3A should be primarily in its cubic, rather than its orthorhombic, form. Further, Aitcin[7] suggests that attention must be paid not only to the total amount of SO_3 in the cement, but also to the amount of soluble sulfates. Thus, the degree of sulfurization of the clinker is an important parameter.

In addition to commercially available cements conforming to ASTM Types I, II and III, a number of cements have been formulated specifically for high strength concrete. For instance, in Norway, Norcem Cement has developed two special cements for high strength concrete, in addition to their ordinary portland cement. The characteristics of these cements are given in Table 1.1.[8] Note that for the two special cements (SP30-4A and SP30-4A MOD), the C_3A contents were held to 5.5%.

Table 1.1   Composition of special cements for high strength concrete (developed by Norcem Cement[8])

	*SP30**	*SP30-4A*	*SP30-4A MOD*
C_2S (%)	18	28	28
C_3S (%)	55	50	50
C_3A (%)	8	5.5	5.5
C_4AF (%)	9	9	9
MgO (%)	3	1.5–2.0	1.5–2.0
SO_3 (%)	3.3	2–3	2–3
Na_2O equivalent (%)	1.1	0.6	0.6
Blaine fineness (m^2/kg)	300	310	400
heat of hydration (kcal/kg)	71	56	70
setting time (min): initial	120	140	120
final	180	200	170

* Ordinary portland cement, for comparison
1 m^2/kg = 4.89 ft^2/lb

Supplementary cementing materials

As indicated above, most modern high strength concretes contain at least one supplementary cementing material: fly ash, blast-furnace slag, or silica fume. Very often, the fly ash or slag is used in conjunction with silica fume. These materials are all specified in the Canadian CSA Standard A23.5.[9] In the United States, fly ash is specified in ASTM C618,[10] and blast furnace slag in ASTM C989[11]; there is, as yet, no U.S. standard for silica fume. These materials are described in detail in *Supplementary Cementing Materials for Concrete*.[12]

Using a somewhat different approach, a high silica modulus portland

Table 1.2 Bogue composition and other properties
of HTS cement (after Aitcin *et al.*[13])

C_2S (%)	22
C_3S (%)	62
C_3A (%)	3.6
C_4AF (%)	6.9
Na_2O equivalent (%)	0.38
lime saturation factor	92.7
silica modulus	4.8
Blaine fineness, m^2/kg	320

$1\ m^2/kg = 4.89\ ft^2/lb$

cement (referred to as HTS, or Haute Teneur en Silica, or high silica
content) was developed,[13] with the composition shown in Table 1.2. Note
that, compared to more conventional cements (such as the SP-30 of
Table 1.1), there is a very high total silicate content (84%), and C_3A
content of only 3.6%. The cement is rather coarsely ground (Blaine
fineness of 1565 ft^2/lb (320 m^2/kg)). It is made from a clinker composed of
small alite and belite crystals, and minute C_3A crystals. It is capable of
producing concretes with excellent 28-day compressive strengths, as indicated in Table 1.3, when used in conjunction with 10% silica fume.

Table 1.3 28 day compressive strengths of concrete
made with HTS cement and 10% silica fume[13]

w/c	f_c' (MPa)
0.31	74
0.23	106
0.20	115
0.17	124

1 ksi = 6.89 MPa
1 MPa = 0.145 ksi

Silica fume

It is possible to make high strength concrete without silica fume, at
compressive strengths of up to about 14,000 psi (98 MPa). Beyond that
strength level, however, silica fume becomes essential, and even at lower
strengths 9000–14,000 psi (63–98 MPa), it is easier to make HSC with silica
fume than without it. Thus, when it is available at a reasonable price, it
should generally be a component of the HSC mix.

Silica fume[14] is a waste by-product of the production of silicon and
silicon alloys, and is thus not a very well-defined material. Consequently, it
is important to characterize any new source of silica fume, by determining
the specific surface area by nitrogen adsorption, and the silica, alkali and
carbon contents. In addition, it is desirable to minimize the content of

Table 1.4   Some Canadian specifications for silica fume (taken from CSA Standard A23.5[9])

Chemical requirements	
SiO_2, min (%)	85
SO_3, max (%)	1.0
Loss in ignition, max (%)	6.0
Physical requirements	
Accelerated pozzolanic activity index, min, (%) of control	85
Fineness, max, (%) retained on 45 μm sieve	10
Soundness – autoclave expansion or contraction (%)	0.2
Relative density, max variation from average (%)	5
Fineness, max variation from average (%)	5
Optional physical requirements	
Increase of drying shrinkage, max (%) of control	0.03
Reactivity with cement alkalis: min reduction (%)	80

crystalline material. The acceptance limits for silica fume, taken from CSA-A23.5[9] are given in Table 1.4.

Silica fume is available in several forms. In its bulk form, its unit weight is in the range of 118 to 147.5 pcf (200–250 kg/m^3), which makes handling difficult. More commonly now, silica fume is available in a densified form, in which the bulk densities are about twice as great as those of the bulk form (i.e. 400–500 kg/m^3). In general, this makes it easier to handle. In addition, silica fume is available in slurry form (often in conjunction with superplasticizers in the liquid phase), with a solids content of about 50%. This form of silica fume requires special equipment for its use. Finally, silica fume is available already blended with portland cement (at percentages of the total mass of cementitious material in the range of 6.7 to 9.3%) in Canada, France and Iceland. In spite of this apparently wide selection, however, in any one location the choice of silica fumes will be very limited, and one must use what is locally available.

Fly ash

Fly ash has, of course, been used very extensively in concrete for many years. Fly ashes are, unfortunately, much more variable than silica fumes in both their physical and chemical characteristics. Any fly ash which works well in ordinary concrete mixes is likely to work well in high strength concrete as well. However, most fly ashes will result in strengths of not much more than 10,000 psi (70 MPa), though there have been a few reports of high strength concretes with strengths of up to 14,000 psi (98 MPa) in which fly ash has been used. For higher strengths, silica fume must be used in conjunction with the fly ash, though this practice has not been common in the past.

In general, for high strength concrete applications, fly ash is used at dosage rates of about 15% of the cement content. Because of the variability of the fly ash produced even from a single plant, however, quality control is particularly important. This involves determinations of the Blaine specific surface area, as well as the chemical composition (in particular the contents of SiO_2, Al_2O_3, Fe_2O_3, CaO, alkali, carbon and sulfates). And, as with silica fume, it is important to check the degree of crystallinity; the more glassy the fly ash, the better.

Blast furnace slag

In North America, slag is not as widely available as in Europe, and hence there is not much information available as to its performance in high strength concrete. However, the indications are that, as with fly ash, slags that perform well in ordinary concrete are suitable for use in high strength concrete, at dosage rates between 15% and 30%. The lower dosage rates should be used in the winter, so that the concrete develops strength rapidly enough for efficient form removal. For very high strengths, in excess of 14,000 psi (98 MPa), it will likely be necessary to use the slag in conjunction with silica fume.

The chemical composition of slags does not generally vary very much. Therefore, routine quality control is generally confined to Blaine specific surface area tests, and X-ray diffraction studies to check on the degree of crystallinity (which should be low).

Limitation on the use of silica fume, fly ash or slag

There appear to be no particular deleterious effects when silica fume is used in concrete. However, the use of fly ash and slag may lead to some problems:

(i) The *early strength development* of mixes in which some of the portland cement has been replaced by slag or fly ash is less rapid than that when only portland cement is used, and this may adversely affect the time at which the forms can be stripped, particularly at low temperatures. One way of dealing with this problem is by further reductions in the w/c ratio, through the use of even more superplasticizer. Clearly, this is not economically very attractive; if high *early* strength is needed, it may well be necessary to reduce the fly ash or slag content.

(ii) The existing test data are rather ambiguous with regard to the *free-thaw durability* of high strength concrete made with supplementary cementitious materials. This is true both for air-entrained and non-air-entrained mixes. Therefore, until more data are available, designers should be cautious when using high strength concrete in an environment in which the concrete will be subjected to many freeze-thaw cycles in a saturated state.

(iii)   At the substitution levels used (15–30%), fly ash or slag will have very little effect on the *maximum temperature development* in mass concrete pours.

Superplasticizers

In modern concrete practice, it is essentially impossible to make high strength concrete at adequate workability in the field without the use of superplasticizers. Unfortunately, different superplasticizers will behave quite differently with different cements (even cements of nominally the same type). This is due in part to the variability in the minor components of the cement (which are not generally specified), and in part to the fact that the acceptance standards for superplasticizers themselves are not very tightly written. Thus, some cements will simply be found to be incompatible with certain superplasticizers.

There are, basically, three principal types of superplasticizer: (i) *ligno-sulfonate-based*; (ii) polycondensate of formaldehyde and melamine sulfonate (often referred to simply as *melamine sulfonate*; and (iii) polycondensate of formaldehyde and naphthalene sulfonate, (often referred to as *naphthalene sulfonate*).

In addition, a variety of other molecules might be mixed in with these basic formulations. It may thus be very difficult to determine the precise chemical composition of most superplasticizers; certainly manufacturers try to keep their formulations as closely guarded secrets.

It should be noted that much of what we know about superplasticizers comes from tests carried out on normal strength concretes, at relatively low superplasticizer contents. This does not necessarily reflect their performance at very low w/c ratios and very high superplasticizer addition rates.

Lignosulfonate-based superplasticizers

In high strength concrete, lignosulfonate superplasticizers are generally used in conjunction with either melamine or naphthalene superplasticizers. They tend not to be efficient enough for the economic production of very high strength concretes on their own. Sometimes, lignosulfonates are used for initial slump control, with the melamines or naphthalenes used subsequently for slump control in the field.

Melamine sulfonate superplasticizers

Until recently, only one melamine superplastizer was available (tradename Melment), but now other melamine-based superplasticizers are likely to become commercially available.

Melamine superplasticizers are clear liquids, containing about 22% solid particles; they are generally in the form of their sodium salt. These superplasticizers have been used for many years now with good results, and so they remain popular with high strength concrete producers.

Naphthalene sulfonate superplasticizers

Naphthelene superplasticizers have been in use longer than any of the others, and are available under a greater number of brand names. They are available as both a powder and a brown liquid; in the liquid form they typically have a solids content of about 40%. They are generally available as either calcium salts, or more commonly, sodium salts. (Calcium salts should be used in case where a potentially alkali-reactive aggregate is to be used.)

The particular advantages of naphthalene superplasticizers, apart from their being slightly less expensive than the other types, appears to be that they make it easier to control the rheological properties of high strength concrete, because of their slight retarding action.

Superplasticizer dosage

There is no *a priori* way of determining the required superplasticizer dosage; it must be determined, in the end, by some sort of trial and error procedure. Basically, if strength is the primary criterion, then one should work with the lowest w/c ratio possible, and thus the highest superplasticizer dosage rate. However, if the rheological properties of the high strength concrete are very important, then the highest w/c ratio consistent with the required strength should be used, with the superplasticizer dosage then adjusted to get the desired workability. In general, of course, some intermediate position must be found, so that the combination of strength and rheological properties can be optimized. Typical superplasticizer dosages for a number of high strength concrete mixes are given below, in Tables 1.5 to 1.10.

Table 1.5 Mix proportions for Interfirst Plaza, Dallas (adapted from Cook[15])

	1 cm max size aggregate	*25 cm max size aggregate*
water (kg/m^3)	166	148
cement, Type I (kg/m^3)	360	357
fly ash, Class C (kg/m^3)	150	149
coarse aggregate (kg/m^3)	1052	1183
fine aggregate (kg/m^3)	683	604
water reducer L/m^3	1.01	1.01
superplasticizer L/m^3	2.54	2.52
w/cementitious ratio	0.33	0.29
f_c' 28-day (MPa)- moist cured	79.5	85.8
f_c' 91-day (MPa)- moist cured	89.0	92.4

1 lb/yd^3 = 0.59 kg/m^3	or	1 kg/m^3 = 1.69 pcf
1 in. = 25.4 mm	or	1 in. = 0.0393 mm

Table 1.6 High strength concrete mix design guidelines (after Peterman and Carrasquillo[16])

	H-H-00	H-H-01	H-H-10	H-H-11
water (kg/m^3)	195	143	173	134
cement (kg/m^3)	558	474	391	335
fly ash (kg/m^3)	–	–	167	144
coarse agg./fine agg. ratio	2.0	2.0	2.0	2.0
superplasticizer	–	yes*	–	yes*
w/cementitious ratio	0.34	0.30	0.31	0.27
f_c' 56-day (MPa)	66	72	69	76

* Use highest dosage of superplasticizer which will not lead to segregation or excessive retardation.

1 lb/yd^3 = 0.59 kg/m^3 or   1 kg/m^3 = 1.69 pcf
1 in.   = 25.4 mm   or   1 in.   = 0.0393 mm

Table 1.7 Mix proportions for high strength concrete at Pacific First Center, Seattle (adapted from Randall and Foot[17])

water (kg/m^3)	131
cement – Type II (kg/m^3)	534
fly ash – Type F (kg/m^3)	59
silica fume (kg/m^3)	40
coarse aggregate – 1 cm max. size (kg/m^3)	1069
fine aggregate – F.M. = 3.2 (kg/m^3)	623
water reducer I (L/m^3)	1.77
water reducer II (L/m^3)	7.39
w/cementitious ratio	0.21
f_c' 56-day (MPa)	124

1 lb/yd^3 = 0.59 kg/m^3 or   1 kg/m^3 = 1.69 pcf
1 in.   = 25.4 mm   or   1 in.   = 0.0393 mm

Table 1.8 Five examples of commercially produced high strength concrete mix designs (after Aitcin, Shirlaw and Fines[18])

water (kg/m^3)	195	165	135	145	130
cement (kg/m^3)	505	451	500	315	513
fly ash (kg/m^3)	60	–	–	–	–
slag (kg/m^3)	–	–	–	137	–
silica fume (kg/m^3)	–	–	30	36	43
coarse aggregate (kg/m^3)	1030	1030	1110	1130	1080
fine aggregate (kg/m^3)	630	745	700	745	685
water reducer (L/m^3)	0.975	–	–	0.9	–
retarder (L/m^3)	–	–	4.5	1.8	–
superplasticizer (L/m^3)	–	11.25	14	5.9	15.7
w/cementitious ratio	0.35	0.37	0.27	0.31	0.25
f_c' 28-day (MPa)	64.8	79.8	42.5	83.4	119
f_c' 91-day (MPa)	78.6	87.0	106.5	93.4	145

1 lb/yd^3 = 0.59 kg/m^3 or   1 kg/m^3 = 1.69 pcf
1 in.   = 25.4 mm   or   1 in.   = 0.0393 mm

Table 1.9 High strength mixtures in the Chicago area (adapted from Burg and Ost[19])

	Mix number				
	1	*2*	*3*	*4*	*5*
water (kg/m^3)	158	160	155	144	151
cement (kg/m^3)	564	475	487	564	475
fly ash (kg/m^3)	–	59	–	–	104
silica fume (kg/m^3)	–	24	47	89	74
coarse aggregate, SSD 12 mm max size	1068	1068	1068	1068	1068
fine aggregate, SSD (kg/m^3)	647	659	676	593	593
superplasticizer – Type F (L/m^3)	11.61	11.61	11.22	20.12	16.45
retarder – Type D (L/m^3)	1.12	1.04	0.97	1.47	1.51
w/cementitious ratio	0.281	0.287	0.291	0.220	0.231
f_c' 28-day (MPa) – moist cured	78.6	88.5	91.9	118.9	107.0
f_c' 56-day (MPa) – moist cured	81.4	97.3	94.2	121.2	112.0
f_c' 91-day (MPa) – moist cured	86.5	100.4	96.0	131.8	119.3

Table 1.10 Mix design for a high strength concrete designed for a low heat of hydration (adapted from Burg and Ost[19])

water (kg/m^3)	141
cement – Type I (kg/m^3)	327
fly ash – Type F (kg/m^3)	87
silica fume (kg/m^3)	27
coarse aggregate – 25 mm max. size (kg/m^3)	121
fine aggregate (kg/m^3)	742
superplasticizer, ASTM Type F (L/m^3)	6.31
superplasticizer, ASTM Type G (L/m^3)	3.25
water/cementitious ratio	0.32
f_c' 28-day (MPa) – moist cured	3.1
f_c' 91-day (MPa) – moist cured	88.6

Retarders

At one time retarders were recommended for some high strength concrete applications, to minimize the problem of over rapid slump loss. However, it is difficult to maintain a compatibility between the retarder and the superplasticizer, i.e. to minimize slump loss without excessively reducing early strength gain. In modern practice, retarders are recommended only as a last resort; the rheology is better controlled by the use of the appropriate supplementary cementing materials described above.

Aggregates

The aggregate properties that are most important with regard to high strength concrete are: particle shape, particle size distribution, mechanical properties of the aggregate particles, and possible chemical reactions between the aggregate and the paste which may affect the bond. Unlike their use in ordinary concrete, where we rarely consider the strength of the aggregates, in high strength concrete the aggregates may well become the strength limiting factor. Also, since it is necessary to maintain a low w/c ratio to achieve high strength, the aggregate grading must be very tightly controlled.

Coarse aggregate

It goes without saying that, for high strength concrete, the coarse aggregate particles themselves must be strong. A number of different rock types have been used to make high strength concrete; these include limestone, dolomite, granite, andesite, diabase, and so on. It has been suggested[1] that in most cases the aggregate strength itself is not usually the limiting factor in high strength concrete; rather, it is the strength of the cement–aggregate bond. As with ordinary concretes, however, aggregates that may be susceptible to alkali–aggregate reaction, or to D-cracking, should be avoided if at all possible, even though the low w/c ratios used will tend to reduce the severity of these types of reaction.

From both strength and rheological considerations, the coarse aggregate particles should be roughly equi-dimensional; either crushed rock or natural gravels, particularly if they are of glacial origin, are suitable. Flat or elongated particles must be avoided at all costs. They are inherently weak, and lead to harsh mixes. In addition, it is important to ensure that the aggregate is clean, since a layer of silt or clay will reduce the cement–aggregate bond strength, in addition to increasing the water demand. Finally, the aggregates should not be highly polished (as is sometimes the case with river-run gravels), because this too will reduce the cement–aggregate bond.

Not much work has been carried out on the effects of aggregate mineralogy on the properties of high strength concrete. However, a detailed study by Aitcin and Mehta,[20] involving four apparently hard strong aggregates (diabase, limestone, granite, natural siliceous gravel) revealed that the granite and the gravel yielded much lower strengths and E-values than the other two aggregates. These effects appeared to be related both to aggregate strength and to the strength of the cement–aggregate transition zone. Cook[15] has also pointed out the effect of the modulus of elasticity of the aggregate on that of the concrete. However, much work remains to be done to relate the mechanical and mineralogical properties of the aggregate to those of the resulting high strength concrete.

It is commonly assumed that a smaller maximum size of coarse aggregate will lead to higher strengths,[1,2,5,6,21] largely because smaller sizes will improve the workability of the concrete. However, this is not necessarily the case. While Mehta and Aitcin[6] recommend a maximum size of 10–12 mm, they report that 20–25 mm maximum size may be used for high strength concrete. On the other hand, using South African materials, Addis[22] found that the strength of his high strength concrete increased as the maximum size of aggregate increased from 13.2 to 26.5 mm. This, then, is another area which requires further study.

Fine aggregate

The fine aggregate should consist of smooth rounded particles,[2] to reduce the water demand. Normally, the fine aggregate grading should conform to the limits established by the American Concrete Institute[3] for normal strength concrete. However, it is recommended that the gradings should lie on the coarser side of these limits; a fineness modulus of 3.0 or greater is recommended,[1,6] both to decrease the water requirements and to improve the workability of these paste-rich mixes. Of course, the sand too must be free of silt or clay particles.

1.3   Mix proportions for high strength concrete

Only a few formal mix design methods for high strength concrete have been developed to date.[7,22,23] Most commonly, purely empirical procedures based on trial mixtures are used. For instance, according to the Canadian Portland Cement Association, 'the trial mix approach is best for selecting proportions for high-strength concrete'.[24] In other cases, mix design 'recipes' are provided for different classes of high strength concrete; an example of this approach is given by Peterman and Carrasquillo.[16]

In this section, it is not the intention to provide a canonical mix proportioning method. Much work remains to be done before any mix proportioning method for high strength concrete becomes as universally accepted, at least in North America, as has the *ACI Standard 211.1*[3] for normal strength concretes. Rather, the principles on which such a mix design method should be based will be discussed, and some general guidelines (and a number of empirically derived mixes drawn from the literature) will be presented.

Proportions of materials

Water/cementitious ratio

For normal strength concretes, mix proportioning is based to a large extent on the w/c ratio 'law'. For these concretes, in which the aggregate strength is generally much greater than the paste strength, the w/c ratio does indeed determine the strength of the concrete for any given set of raw materials. For high strength concretes, however, in which the aggregate strength, or the strength of the cement–aggregate bond, are often the strength-controlling factors, the role of the w/c ratio is less clear. To be sure, it is necessary to use very low w/c ratios to manufacture high strength concrete. However, the relationship between w/c ratio and concrete strength is not as straightforward as it is for normal strength concretes.

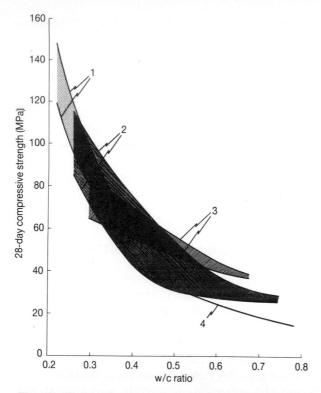

Fig. 1.1   Compressive strength versus w/c material ratio: (1) after Aitcin[7]; (2) after Fiorato[25]; (3) after Cook[15]; (4) normal strength concrete from CPCA[24]

Figure 1.1 shows a series of w/cementitious material vs compressive strength curves for high strength concrete. The sets of curves numbered 1, 2 and 3 show the strength range that might be expected for a given w/cementitious ratio. (Curve 1 is from Aitcin[7]; curve 2 is from Fiorato[25]; curve 3 is from Cook.[15]) For comparison, the w/c ratio vs strength curve for normal strength concrete is shown as curve 4.[24] Figure 1.2 shows a similar series of w/cementitious vs strength curves obtained by other investigators. Curve 1 is from Addis and Alexander,[23] who used high early strength cement. Curve 2 is from Hattori.[25] Curves 3 and 4 are from Suzuki[27]; curve 3 is for ordinary portland cement, and curve 4 for high early strength cement.

Several conclusions may be drawn from Figs. 1.1 and 1.2. First, while strength clearly increases as the w/cementitious ratio decreases, there is a considerable scatter of the results, which must be due to variations in the materials, used in the different investigations. Second, and more important, the range of strengths for a given w/cementitious ratio increases as the w/cementitious ratio decrease. If one looks at all of the curves in Figs. 1.1 and 1.2, at a w/cementitious ratio of 0.45, the range in strength is from 5400 psi (37 MPa) to 9500 psi (66 MPa); at a ratio of 0.26, the range is from 11,300 psi (78 MPa) to 17,400 psi (120 MPa). Therefore, the

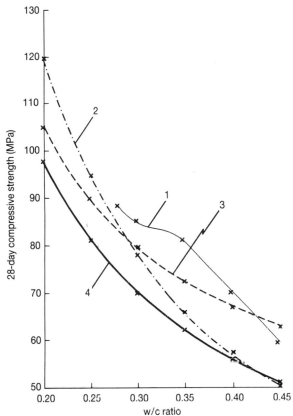

Fig. 1.2   Compressive strength versus w/c material ratio: (1) high early strength cement, after Addis and Alexander[23]; (2) after Hattori[26]; (3) ordinary Portland cement, after Suzuki[27]; (4) high early strength cement after Suzuki[24]

w/cementitious ratio by itself is not a very good predictor of compressive strength. The w/cementitious vs strength relationship must thus be determined for any given set of raw materials.

Cementitious materials content

For normal strength concretes, cement contents are typically in the range of 590 to 930 pcf (350 to 550 kg/m^3). For high strength concretes, however, the content of cementitious materials (cement, fly ash, slag, silica fume) is higher, ranging from about 845 to 1090 pcf (500 to 650 kg/m^3). The quantity of supplementary cementing materials may vary considerably, depending upon workability, economy and heat of hydration considerations.

Supplementary cementing materials

As indicated earlier, it is possible to make high strength concrete without

using fly ash, slag or silica fume. For higher strengths, however, supplementary cementing materials are generally necessary. In particular, the
use of silica fume is required for strengths much in excess of 14,000 psi
(98 MPa). In any event, the use of silica fume (which is now readily
available in most areas) makes the production of high strength concrete
much easier; it is generally added at rates of 5% to 10% of the total
cementitious materials.

Superplasticizers

With very careful mix design and aggregate grading, it is possible to
achieve strengths of about 14,000 psi (98 MPa) without superplasticizers.
However, as they are readily available they are now almost universally
used in high strength concrete, since they make it much easier to achieve
adequate workability at very low w/cementitious ratios.

Ratio of coarse to fine aggregate

For normal strength concretes, the ratio of coarse to fine aggregate (for a
0.55 in., 14 mm max size of aggregate) is in the range of 0.9 to 1.4.[24]
However, for high strength concrete, the coarse/fine ratio is much higher.
For instance, Peterman and Carrasquillo[16] recommend a coarse/fine ratio
of 2.0. And, as seen in Tables 1.5 to 1.10, coarse/fine ratios used in practice
vary in the range of 1.5 to 1.8.

Examples of high strength concrete mixes

As stated earlier, there is yet no generally agreed upon method of mix
proportioning. Mix designs for high strength concrete have, hitherto, been
developed empirically, depending on the raw material available in any
location. In this section, a number of typical mix designs, drawn from the
recent literature, will be presented.

Table 1.5 shows the mix proportions for Interfirst Plaza, Dallas,[15] in
which the concrete achieved compressive strength of about 11,500 psi
(80 MPa). Table 1.6 gives high strength concrete mix design guidelines
originally developed for the Texas State Department of Highways and
Public Transportation.[16] The expected 56-day strengths for these four
mixes range from 9500 to 11,000 psi (66 to 76 MPa). It should be noted that
the mix designs in Tables 1.5 and 1.6 do not involve the use of silica fume.
Table 1.7 shows the mix proportions of Pacific First Center, Seattle[17] in
which the concrete reached a 56-day compressive strength of 18,000 psi
(126 MPa).

Table 1.8 gives a series of mix designs for a number of high strength
concrete projects,[18] while Table 1.9 describes high strength concrete mixes
commercially available in Chicago.[19] In Tables 1.7, 1.8 and 1.9, it should
be noted that the higher strength mixes all contained silica fume. Finally,

Table 1.10 presents a mix design for a high strength, low heat of hydration concrete.[19]

From Tables 1.5 to 1.10, it may be seen that the mix designs, even for concretes of approximately the same strength, vary considerably. This reflects the differences in the quality of *all* of the raw materials available for each specific mix. So, while these examples may serve as a general guideline for the production of high strength concrete, copying a mix design used in one location is unlikely to produce the same concrete properties in another area.

In the end, as with conventional concrete, mix design will require the production of a number of trial mixes, though the examples given above may provide reasonable guidance for the first trial batch. In particular, it is essential first to ensure that the available raw materials are *capable* of producing the desired strengths, and that there are no incompatibilities between the cements, the admixture(s) and the supplementary cementing materials. With materials for which there is not much field experience, it may be necessary to try different brands of cement, different brands of superplasticizers, and different sources of fly ash, slag, or silica fume, in order to optimize both the materials and the concrete mixture. This sounds like a lot of work, and in general it is. At present, there is simply no straightforward procedure for proportioning a high strength concrete mixture with unfamiliar materials.

1.4   Quality control and testing

Conventional normal strength concrete is a relatively forgiving material; it can tolerate small changes in materials, mix proportions or curing conditions without large changes in its mechanical properties. However, high strength concrete, in which all of the components of the mix are working at their limits, is not at all a forgiving material. Thus, to ensure the quality of high strength concrete, every aspect of the concrete production must be monitored, from the uniformity of the raw materials to proper batching and mixing procedures, to proper transportation, placement, vibration and curing, through to proper testing of the hardened concrete.

The quality control procedures, such as the types of test on both the fresh and hardened concretes, the frequency of testing, and interpretation of test results are essentially the same as those for ordinary concrete. However, Cook[15] has presented data which indicate that for his high strength concrete, the compressive strength results were *not* normally distributed, and the standard deviation for a given mix was not independent of test age and strength level. This led him to conclude that the 'quality control techniques used for low to moderate strength concretes may not necessarily be appropriate for very high strength concretes.' To this date, however, separate quality control/quality assurance procedures for high strength concrete have not been developed.

The remainder of this section deals primarily with the determination of the compressive strength, f_c', since this is the basis on which high strength concrete is designed and specified.

Age at test

Traditionally, the acceptance standards for concrete involve strength determinations at an age of 28 days. Although there is, of course, nothing magical about this particular test age, it has been used universally as the reference time at which concrete strengths are reported. However, for high strength concretes, it has become common to determine compressive strengths at 56 days, or even 90 days. The justification for this is that concrete in structures will rarely, if ever, be loaded to anything approaching its design strength in less than 3 months, given the pace of construction. The increase in strength between 28 and 56 or 90 days can be considerable (10% to 20%), and this can lead to economies in construction. It is thus perfectly reasonable to measure strengths at later ages, and to specify the concrete strength in terms of these longer curing times.

There are, however, two drawbacks to this approach. First, it can be misleading to compare the compressive strengths of normal and high strength concretes, if these are measured at different times. Of more importance, there is a certain margin of safety when concrete strengths are measured at 28 days, since the concrete will generally be substantially stronger when it finally has to carry its design loads, perhaps at the age of one year for a typical high-rise concrete building. If strengths are specified at later ages, this margin is reduced (by an unknown amount), and hence there is an implicit reduction in the factor of safety. And, of course, finding higher strengths at later test ages does not in any way imply that the concrete has somehow become 'better' than a concrete whose strength was measured in the conventional way at 28 days.

Curing conditions

In general, the highest concrete strengths will be obtained with specimens continuously moist cured (at 100% relative humidity) until the time of testing. Unfortunately, the available data on this point are ambiguous. Carrasquillo, Nilson and Slate[28] found that high strength concrete, moist-cured for 7 days and then allowed to dry at 50% relative humidity till 28 days showed a strength loss of about 10% when compared to continuously moist-cured specimens. However, in subsequent work, Carrasquillo and Carrasquillo[29] found that up to an age of 15 days, specimens treated with a curing compound and allowed to cure in the field under ambient conditions yielded slightly higher strengths than moist-cured specimens. At 28 days, moist-cured specimens and field-cured specimens (with or without curing compounds) yielded approximately the same results. Only at later ages (56 and 91 days) did the strengths of the moist-cured specimens surpass those of the field-cured specimens treated with a curing compound. Similarly, for the mixes shown in Table 1.9, Burg and Ost[19] found that, when specimens that had been moist cured for 28 days were then subjected to air curing, their strengths at 91 days *exceeded* those of continuously moist-cured specimens; however, by 426 days, the continuously moist-cured specimens were from about 3% to 10% higher in strength than the air-cured ones.

On the other hand, several investigators have reported that, as long as a week or so of moist curing is provided, subsequent curing under ambient conditions is not particularly detrimental to strength development. Peterman and Carrasquillo[16] have stated that 'the 28-day compressive strength of high strength concrete which has been cured under ideal conditions for 7 days after casting is not seriously affected by curing in hot or dry conditions from 7 to 28 days after casting.'

Finally, contrary results were reported by Moreno[30] who indicated that air-cured specimens were about 10% stronger than moist-cured specimens at all ages up to 91 days.

Type of mold for casting cylindrical specimens

ASTM C470: *Molds for Forming Concrete Test Cylinders Vertically*, describes the requirements for both reusable and single-use molds, and ASTM C31: *Making and Curing Concrete Test Specimens on the Field* permits both types of mold to be used. However, it has long been known that different molds conforming to ASTM C470 will result in specimens with different measured strengths. This is true for both normal strength and high strength concretes. In general, more flexible molds will yield lower strengths than very rigid molds, because the deformation of the flexible molds during rodding or vibration leads to less efficient compaction than when using rigid molds. The experimental data largely bear this out. It should be noted that, whatever the mold materials, the molds must be properly sealed to prevent leakage of the mix water. If any significant leakage does occurs, the apparent strength will generally increase, because of the lower effective w/c ratio, and increased densification of the specimens.

For the standard 6 × 12 in. (150 × 300 mm) molds, Carrasquillo and Carrasquillo[29] found that steel molds gave strengths about 5% higher than plastic molds, while Hester[31] found about a 10% difference. Similar results were reported by Howard and Leatham.[32] Peterman and Carrasquillo[16] reported that steel molds gave strengths about 10% higher than those obtained with cardboard molds, and Hester[31] showed that steel molds gave strengths about 6% higher than tin molds.

On the other hand, Cook[15] reported that 'good success was experienced on the use of single-use rigid plastic molds', while Aitcin[33] reports increasing use of rigid, reusable plastic molds. In addition, Carrasquillo and Carrasquillo[29] have reported that for the smaller 4 × 8 in. (100 × 200 mm) molds, there were no strength differences between steel, plastic or cardboard molds.

In view of the above results, it would be prudent to use rigid steel molds whenever practicable, particularly for concrete strengths in excess of about 14,000 psi (98 MPa), at least until more test data become available for the smaller molds.

Specimen size

For most materials, including concrete, it has generally been observed that the smaller the test specimen, the higher the strength. For high strength concrete, however, though this effect is often observed, there are contradictory results reported in the literature. The results of a number of studies are compared in Table 1.11. It may be seen that the observed strength ratios of 4×8 in. (100×200 mm) cylinders to 6×12 in. (150×300 mm)cylinders range from about 1.1 to 0.93. These contradictory results may be due to differences in testing procedures amongst the various investigators.

It must be noted that while for a given set of materials and test procedures, it may be possible to increase the apparent concrete strength by decreasing the specimen size, this does not in any way change the strength of the concrete in the structure. One particular specimen size does not give 'truer' results than any other. Thus, one should be careful to specify a particular specimen size for a given project, rather than leaving it as a matter of choice.

Specimen end conditions

According to ASTM C39: *Compressive Strength of Cylindrical Concrete Specimens*, the ends of the test specimens must be plane within 0.002 in. (0.05 mm). This may be achieved either by capping the ends (usually with a sulfur mortar) or by sawing or grinding. Unfortunately, different end

Table 1.11   Effect of specimen size on the compressive strength of high strength concrete

Investigator	f_c' (100×200 mm cylinder) / f_c' (150×300 mm cylinder)
Peterman and Carrasquillo[16]	~1.1
Carrasquillo, Slate and Nilson[34]	~1.1
Howard and Leatham[32]	~1.08
Cook[15]	~1.05
Burg and Ost[19]	~1.01
Aitcin[33]	ambiguous results
Moreno[30]	
83 MPa concrete	~1.0
119 MPa concrete	~0.93
Carrasquillo and Carrasquillo[29]	~0.93

conditions can lead to different measured strengths, and so the end preparation for testing high strength concrete specimens should be specified explicitly for any given project.

The most common method for preparing the ends of normal strength concrete is to use sulfur caps; for high strength concrete, high strength sulfur mortars are commercially available. However, if the strength of the cap is less than the strength of the concrete, the compressive load will not be transmitted uniformly to the specimen ends, leading to invalid results. Thus, for high strength concrete, in addition to high strength capping compounds, a number of other end preparation techniques are being investigated. These include grinding the specimen ends, or using unbonded systems, consisting of a pad constrained in a confining ring which fits over the specimen ends.

Most compressive strength tests on high strength concrete are still carried out using a high strength capping compound. The materials available in North America will achieve compressive strengths of 12,000 psi to 13,000 psi (84 MPa to 91 MPa) when tested as 2 in. (50 mm) cubes.[33] Peterman and Carrasquillo[21] recommend the use of such capping compounds, since they give higher concrete strengths than ordinary capping compounds. Cook[16] has used such compounds for concrete strengths up to 10,000 psi (70 MPa), while Moreno[30] considers them to be satisfactory at strengths up to 17,000 psi (119 MPa).

Burg and Ost[19] report that a high strength capping material may be used with concrete strengths of up to 15,000 psi (105 MPa); beyond that, the mode of failure of the cylinders changed from the normal cone failure of a columnar one. They recommend grinding of the cylinder ends for strengths beyond 15,000 psi (105 MPa). Similarly, Aitcin[33] has reported that above about 17,000 psi (119 MPa), the high strength capping material is pulverized as the specimens fail, which might well affect the measured strength. He too recommends grinding of the specimen ends for very high strength concretes. (It might be noted that end grinders for concrete cylinders are now commercially available. In 1992, the cost of such a machine was approximately US$12,000.)

Because of the uncertainty with high strength capping compounds, and the costs and time involved in end grinding, a considerable amount of research has been carried out on unbonded capping systems. These consist of metal restraining caps into which elastomeric inserts are placed; the assemblies then fit over the ends of the cylinder. As the elastomeric inserts deteriorate with repeated use, they are replaced from time to time.

Richardson[35] used a system of neoprene inserts in aluminium caps for testing normal strength concretes in the range of 3000 psi to 6000 psi (21 MPa to 42 MPa). He found that below 4000 psi (28 MPa), the neoprene pads gave somewhat lower strengths than conventional sulfur caps, while above 4000 psi (28 MPa) they gave somewhat higher strengths. Overall, however, the mean compressive strengths were not significantly different between the two systems.

Carrasquillo and Carrasquillo[29] compared a high strength sulfur capping compound to an unbonded system consisting of a polyurethane pad in an aluminium restraining ring. They found that up to about 10,000 psi (70 MPa), the unbonded system gave strengths that were 97% of those obtained with the capping compound. Beyond 10,000 psi (70 MPa), however, the unbonded system gave much higher strengths; they hypothesized that this might be due to greater end restraint of the cylinders with such a system. In subsequent work,[36] they found that up to 10,000 psi (70 MPa), polyurethane pads in an aluminium cap gave results within 5% of those achieved with high strength sulfur caps, while up to 11,000 psi (77 MPa), neoprene pads in steel caps gave results within 3% of those obtained with the sulfur end caps. However, they concluded that the use of either unbonded system was questionable; substantial differences in test results were obtained when two sets of restraining caps (from the same manufacturer) were used.

To improve the results obtained with unbonded systems, Boulay[37] developed a system in which, instead of elastomeric inserts, a mixture of dry sand and wax is used. It was found[38] that the sand mixture gave results which were intermediate between those obtained with ground ends or with sulfur mortar caps.

In summary, then, below about 14,000 psi (98 MPa), a thin, high strength sulfur mortar cap may be used successfully. Beyond that strength level, it would appear that grinding specimen ends is currently the only way to ensure valid test results.

Testing machine characteristics

In general, for normal strength concrete, the characteristics of the testing machine itself are assumed to have little or no effect on the peak load. However, for very high strength concretes the machine may well have some effect on the response of the specimen to load. From a review of the literature, Hester[31] concluded that the longitudinal stiffness of the testing machine will not affect the maximum load, and this view is shared also by Aitcin.[33] However, if the machine is not stiff enough, the specimens may fail explosively. and, of course, a very stiff machine (with servo-controls) is required if one wishes to determine the post-peak response of the concrete. On the other hand, Hester[31] also reports that if the machine is not stiff enough *laterally*, compressive strengths may be adversely affected.

One must also be concerned about the capacity of the testing machine when testing very high strength concretes. Aitcin[33] calculated the required machine capacities for different strength levels and specimen sizes, using the common assumption that the failure load should not exceed ~2/3 of the machine capacity. Some of his results are reproduced in Table 1.12. Relatively few commercial laboratories are equipped to test high strength concrete, since a common capacity of commercial testing machine is 292,500 lbs (1.3 MN). To test a 6×12 in. (150×300 mm) cylinder of

Table 1.12 Machine capacity required for high strength concrete[33]

	Failure load		Machine capacity	
Specimen size	$f_c' = 100$ MPa	$f_c' = 150$ MPa	$f_c' = 100$ MPa	$f_c' = 150$ MPa
100×200 mm	0.785 MN	1.18 MN	1.2 MN	1.75 MN
150×300 mm	1.76 MN	2.65 MN	2.65 MN	4.0 MN

Note: 1 MN = 225,000 lbf

21,400 psi (150 MPa) concrete requires a 900,000 lb (4.0 MN) testing machine, and relatively few machines of this size are available in commercial laboratories. This then, is probably the driving force behind the move to the smaller 4×8 in. (100×200 mm) cylinders.

Effect of loading platens

Again, for ordinary concrete, the effects of the spherically seated bearing blocks (platens) are not explicitly considered, as long as they meet the requirements of ASTM C39: *Compressive Strength of Cylindrical Concrete Specimens*. However, recent work at the Construction Technology Laboratories in Skokie, Illinois[39] has shown that, for high strength concrete, even this cannot be ignored. Spherical bearing blocks which deform in such a way that the stresses are higher around the periphery of the specimen than at the centre, yield higher compressive strengths than blocks which deform so that the highest stresses are at the centre of the specimen, and fall off towards the edges (i.e. a 'concave' rather than a 'convex' stress distribution). Measured differences can be as high as 15% for concretes with compressive strengths greater than 16,000 psi (112 MPa).

1.5 Conclusions

In conclusion, then, it has been shown that the production of high strength concrete requires careful attention to details. It also requires close cooperation between the owner, the engineer, the suppliers and producers of the raw materials, the contractor, and the testing laboratory.[32] Perhaps most important, we must remember that the well-known 'laws' and 'rules-of-thumb' that apply to normal strength concrete may well not apply to high strength concrete, which is a distinctly different material. Nonetheless, we now know enough about high strength concrete to be able to produce it consistently, not only in the laboratory, but also in the field. It is to be hoped that codes of practice and testing standards catch up with the high strength concrete technology, so that the use of this exciting new material can continue to increase.

Acknowledgements

This work was supported by the Canadian Network of Centres of Excellence on High-Performance Concrete.

References

1   SHRP-C/FR-91-103 (1991) *High performance concretes, a state of the art report*. Strategic Highway Research Program, National Research Council, Washington, DC.
2   FIP/CEB (1990) *High strength concrete, state of the art report*. Bulletin d'Information No. 197.
3   ACI Standard 211.1 (1989) *Recommended practice for selecting proportions for normal weight concrete*. American Concrete Institute, Detroit.
4   Mindess, S. and Young, J.F. (1981) *Concrete*. Prentice Hall Inc., Englewood Cliffs.
5   Perenchio, W.F. (1973) An evaluation of some of the factors involved in producing very high-strength concrete. *Research and Development Bulletin*, No. RD014-01T, Portland Cement Association, Skokie.
6   Mehta, P.K. and Aitcin, P.-C. (1990) Microstructural basis of selection of materials and mix proportions for high-strength concrete, in *Second International Symposium on High-Strength Concrete, SP-121*. American Concrete Institute, Detroit, 265–86.
7   Aitcin, P.-C. (1992) private communication
8   Ronneburg, H. and Sandvik, M. (1990) High Strength Concrete for North Sea Platforms, *Concrete International*, **12**, 1, 29–34
9   CSA Standard A23.5-M86 (1986) *Supplementary cementing materials*. Canadian Standards Association, Rexdale, Ontario.
10   ASTM C618 *Standard specification for fly ash and raw or calcined natural pozzolan for use as a mineral admixture in portland cement concrete*. American Society for Testing and Materials, Philadelphia, PA.
11   ASTM C989 *Standard specification for ground iron blast-furnace slag for use in concrete and mortars*. American Society for Testing and Materials, Philadelphia, PA.
12   Malhotra, V.M. (ed) (1987) *Supplementary cementing materials for concrete*. Minister of Supply and Services, Canada.
13   Aitcin, P.-C., Sarkar, S.L., Ranc, R. and Levy, C. (1991) A High Silica Modulus Cement for High-Performance Concrete, in S. Mindess (ed.), *Advances in cementitious materials*. Ceramic Transactions **16**, The American Ceramic Society Inc., 102–21.
14   Malhotra, V.M., Ramachandran, V.S., Feldman, R.F. and Aitcin, P.-C. (1987) *Condensed silica fume in concrete*. CRC Press Inc., Boca Ratan, Florida.
15   Cook, J.E. (1989) 10,000 psi Concrete. *Concrete International*, **11**, 10, 67–75.
16   Peterman, M.B. and Carrasquillo, R.L. (1986) *Production of high strength concrete*. Noyes Publications, Park Ridge.
17   Randall, V.R. and Foot, K.B. (1989) High strength concrete for Pacific First Center. *Concrete International: Design and Construction*, **11**, 4, 14–16.
18   Aitcin, P.-C., Shirlaw, M. and Fines, E. (1992) High performance concrete: removing the myths, in *Concrescere*, Newsletter of the High-Performance Concrete Network of Centres of Excellence (Canada), 6, March.
19   Burg, R.G. and Ost, B.W. (1992) *Engineering properties of commercially available high-strength concretes*. Research and Development Bulletin RD104T, Portland Cement Association, Skokie.

20 Aitcin, P.-C. and Mehta, P.K. (1990) Effect of coarse aggregate type or mechanical properties of high strength concrete. *ACI Materials Journal*, American Concrete Institute, Detroit, **87**, 2, 103–107.

21 ACI Committee 363 (1984) *State-of-the-art report on high strength concrete* (ACI 363R-84). American Concrete Institute, Detroit.

22 Addis, B.H. (1992) *Properties of High Strength Concrete Made with South African Materials*, Ph.D. Thesis, University of the Witwatersrand, Johannesburg, South Africa.

23 Addis, B.J. and Alexander, M.G. (1990) A method of proportioning trial mixes for high-strength concrete, in ACI Sp-121, *High strength concrete, Second International Symposium*, American Concrete Institute, Detroit, 287–308.

24 Canadian Portland Cement Association (1991) *Design and control of concrete*. Edition CPCA, Ottawa.

25 Fiorato, A.E. (1989) PCA research on high-strength concrete. *Concrete International*, **11**, 4, 44–50.

26 Hattori, K. (1979) Experiences with mighty superplasticizer in Japan, in ACI SP-62, *Superplasticizers in concrete*, American Concrete Institute, Detroit, 37–66.

27 Suzuki, T. (1987) Experimental studies on high-strength superplasticized concrete, in *Utilization of high strength concrete, Symposium proceedings*. Stavanger, Norway: Tapis Publishers, Trondheim, 53–4.

28 Carrasquillo, R.C., Nilson, A.H. and Slate, F.O. (1981) Properties of high strength concrete subject to short-term loads. *Journal of American Concrete Institute*, **78**, 3, 171–8.

29 Carrasquillo, P.M. and Carrasquillo, R.L. (1988). Evaluation of the use of current concrete practice in the production of high-strength concrete. *ACI Materials Journal*, **85**, 1, 49–54.

30 Moreno, J. (1990) 225 W. Wacker Drive. *Concrete International*, **12**, 1, 35–9.

31 Hester, W.T. (1980) Field testing high-strength concretes: a critical review of the state-of-the-art. *Concrete International*, **2**, 12, 27–38.

32 Howard, N.L. and Leatham, D.M. (1989) The production and delivery of high-strength concrete. *Concrete International*, **11**, 4, 26–30.

33 Aitcin, P.-C. (1989) Les essais sue les betons a tres hautes performances, in *Annales de L'Institut Technique du Batiment et des Travaux Publics*, No. 473. Mars-Avril. Serie: Beton 263, 167–9.

34 Carrasquillo, R.L., Slate, F.O. and Nilson, A.H. (1981) Microcracking and behaviour of high strength concrete subjected to short term loading. *American Concrete Institute Journal*, **78**, 3, 179–86.

35 Richardson, D.N. (1990) Effects of testing variables on the comparison of neoprene pad and sulfur mortar-capped concrete test cylinders. *ACI Material Journal*, **87**, 5, 489–95.

36 Carrasquillo, P.M. and Carrasquillo, R.L. (1988) Effect of using unbonded capping systems on the compressive strength of concrete cylinders. *ACI Materials Journal*, **85**, 3, 141–7.

37 Boulay, C. (1989) La boite a sable, pour bien ecraser les betons a hautes performances. *Bulletin de Liaison des Laboratoires des Ponts et Chausses*, Nov/Dec.

38 Boulay, C., Belloc, A., Torrenti, J.M. and De Larrard, F. (1989) *Mise au point d'un nouveau mode operatoire d'essai de compression pour les betons a haute performances*. Internal report, Laboratoire Central des Ponts et Chaussees, Paris, December.

39 *CTL Review* (1992) Construction Technology Laboratories, Inc., Skokie, Illinois, **15**, 2.

2 Short term mechanical properties

S H Ahmad

2.1 Introduction

Chapter 1 discussed the production of concrete and the effects of a large number of constituent materials – cement, water, fine aggregate, coarse aggregate (crushed stone or gravel), air and other admixtures on the production process. Some quality control issues were also addressed. In the present chapter, the mechanical properties of hardened concrete under short term conditions or loadings are discussed.

Concrete must be proportioned and produced to carry imposed loads, resist deterioration and be dimensionally stable. The quality of concrete is characterized by its mechanical properties and ability to resist deterioration. The mechanical properties of concrete can be broadly classified as short-term (essentially instantaneous) and long-term properties. Short-term properties include strength in compression, tension, modulus of elasticity and bond characteristics. The long-term properties include creep, shrinkage, behavior under fatigue, and durability characteristics such as porosity, permeability, freeze-thaw resistance and abrasion resistance. The creep and shrinkage characteristics are discussed in Chapter 3, the behavior under fatigue and the bond characteristics is addressed in Chapter 4. The important aspect of durability is presented in Chapter 5.

While information on high performance concretes (HPC) as defined in Chapter 1 is scarce, there is a substantial body of information on the mechanical properties of high strength concrete and additional information is being developed rapidly. One class of high performance concretes are the early strength concretes. The mechanical properties of these types of high performance concretes are being investigated under the Strategic Highway Research Program SHRP C-205 which is in progress at North Carolina State University. Since high performance concretes typically have low water/cementitious materials (w/c) ratios and high paste contents, characteristics will in many cases be similar to those of high strength concrete. Much of the discussion in this chapter will therefore concentrate on high strength concretes.

A significant difference in behavior between the early strength and the high strength concretes is in the relationship of compressive strength to mechanical properties. Strength gain in compression is typically much faster than strength gain in aggregate–paste bond, for instance. This will lead to relative differences in elastic modulus and tensile strength of early strength concretes and high strength concretes, expressed as a function of compressive strength. The relationships of mechanical properties to 28-day compressive strength developed in other studies cannot necessarily be expected to apply to early strength concretes. The information developed under the SHRP program will be useful to fill this knowledge gap.

2.2   Strength

The strength of concrete is perhaps the most important overall measure of quality, although other characteristics may also be critical. Strength is an important indicator of quality because strength is directly related to the structure of hardened cement paste. Although strength is not a direct measure of concrete durability or dimensional stability, it has a strong relationship to the w/c ratio of the concrete. The w/c ratio, in turn, influences durability, dimensional stability and other properties of the concrete by controlling porosity. Concrete compressive strength, in particular, is widely used in specifying, controlling and evaluating concrete quality.

The strength of concrete depends on a number of factors including the properties and proportions of the constituent materials, degree of hydration, rate of loading, method of testing and specimen geometry.

The properties of the constituent materials which affect the strength are the quality of fine and coarse aggregate, the cement paste and the paste–aggregate bond characteristics (properties of the interfacial, or transition, zone). These, in turn, depend on the macro- and microscopic structural features including total porosity, pore size and shape, pore distribution and morphology of the hydration products, plus the bond between individual solid components. A simplified view of the factors affecting the strength of concrete is shown in Fig. 2.1.

Testing conditions including age, rate of loading, method of testing, and specimen geometry significantly influence the measured strength. The strength of saturated specimens can be 15% to 20% lower than that of dry specimens. Under impact loading, strength may be as much as 25% to 35% higher than under a normal rate of loading (10 to 20 microstrains per second). Cube specimens generally exhibit 20% to 25% higher strengths than cylindrical specimens. Larger specimens exhibit lower average strengths.

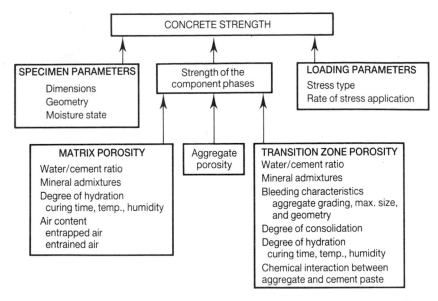

Fig. 2.1   An oversimplified view of the factors influencing strength of plain concrete[53]

Constituent materials and mix proportions

Concrete composition limits the ultimate strength which can be obtained and significantly affects the levels of strength attained at early ages. A more complete discussion of the effects of constituent materials and mix proportions is given in Chapter 1. However, a review of the two dominant constituent materials on strength is useful at this point. Coarse aggregate and paste characteristics are typically considered to control maximum concrete strength.

Coarse aggregate

The important parameters of coarse aggregate are its shape, texture and the maximum size. Since the aggregate is generally stronger than the paste, its strength is not a major factor for normal strength concrete, or in early strength concrete. However, the aggregate strength becomes important in the case of higher-strength concrete or lightweight aggregate concrete. Surface texture and mineralogy affect the bond between the aggregates and the paste and the stress level at which microcracking begins. The surface texture, therefore, may also affect the modulus of elasticity, the shape of the stress-strain curve and, to a lesser degree, the compressive strength of concrete. Since bond strength increases at a slower rate than compressive strength, these effects will be more pronounced in early strength concretes. Tensile strengths may be very sensitive to differences in aggregate surface texture and surface area per unit volume.

The effect of different types of coarse aggregate on concrete strength has been reported in numerous articles. A recent paper[12] reports results of four different types of coarse aggregates in a very high strength concrete mixture (w/c = 0.27). The results showed that the compressive strength was significantly influenced by the mineralogical characteristics of the aggregates. Crushed aggregates from fine-grained diabase and limestone gave the best results. Concretes made from a smooth river gravel and from crushed granite that contained inclusions of a soft mineral were found to be relatively weaker in strength.

The use of larger maximum size of aggregate affects the strength in several ways. Since larger aggregates have less specific surface area, the bond strength between aggregates and paste is lower, thus reducing the compressive strength. Larger aggregate results in a smaller volume of paste thereby providing more restraint to volume changes of the paste. This may induce additional stresses in the paste, creating microcracks prior to application of load, which may be a critical factor in very high strength concretes.

The effect of the coarse aggregate size on concrete strength was discussed by Cook *et al.*[22] Two sizes of aggregates were investigated: a 3/8 in. (10 mm) and a 1 in. (25 mm) limestone. A superplasticizer was used in all the mixes. In general, the smallest size of the coarse aggregate produces the highest strength for a given w/c ratio, see Figs 2.2–2.6. It may be noted that compressive strengths in excess of 10,000 psi (70 MPa) can be produced using a 1 in. (25 mm) maximum size aggregate when the mixture is properly proportioned.

Although these studies[12,22] provide useful data and insight, much more research is needed on the effects of aggregate mineral properties and

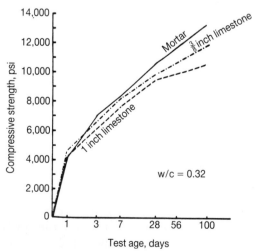

Fig. 2.2   Effect of aggregate type on strength at different ages for a constant w/c materials ratio without superplasticizer[22]

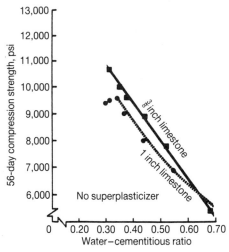

Fig. 2.3 Effect of aggregate type on 56 day strength for concrete for different w/c materials ratio[22]

particle shape on the strength and durability of higher strength concrete. This was recognized as one of the research needs by the ACI 363 Committee.[3]

Paste characteristics

The most important parameter affecting concrete strength is the w/c ratio,

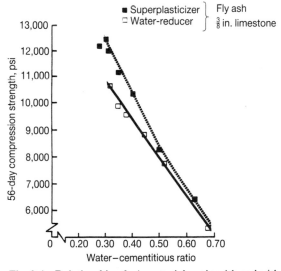

Fig. 2.4 Relationship of w/c materials ratio with and without a high range water-reducing admixture for coarse aggregate size not exceeding $\frac{3}{8}$ in. (10 mm)[22]

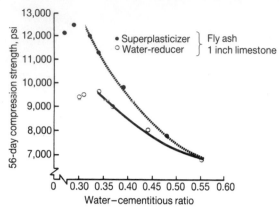

Fig. 2.5 Relationship of w/c materials ratio with and without a high range water-reducing admixture for coarse aggregate size not exceeding 1 in. (25.4 mm)[22]

sometimes referred to as the w/b (binder) ratio. Even though the strength of concrete is dependent largely on the capillary porosity or gel/space ratio, these are not easy quantities to measure or predict. The capillary porosity of a properly compacted concrete is determined by the w/c ratio and degree of hydration. The effect of w/c ratio on the compressive strength is shown in Fig. 2.7. The practical use of very low w/c ratio concretes has been made possible by use of both conventional and high range water reducers, which permit production of workable concrete with very low water contents.

Supplementary cementitious materials (fly ash, slag and silica fume) have been effective additions in the production of high strength concrete. Although fly ash is probably the most common mineral admixture, on a volume basis, silica fume (ultra-fine amorphous silica, derived from the production of silicon or ferrosilica alloys) in particular, used in combina-

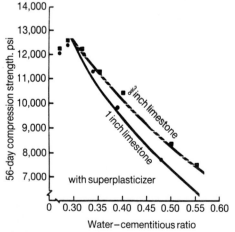

Fig. 2.6 Effect of aggregate type on strength at different ages for a constant w/c materials ratio, with superplasticizer[22]

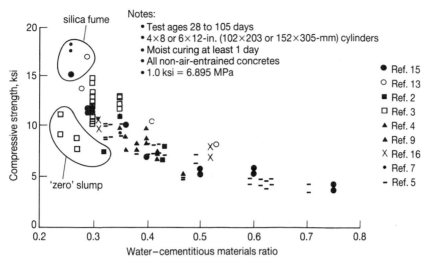

Fig. 2.7   Summary of strength data as a function of w/c materials ratio[29]

tion with high-range water reducers, has increased achievable strength levels dramatically (Fig. 2.7).[10,51,52]

The effect of condensed silica fume on the strength of concrete was reported in a very comprehensive study.[28] The beneficial effect of using up to 16% (by weight of cement) condensed silica on the compressive strength is shown in Fig. 2.8. The data indicate that to achieve 10,000 psi (70 MPa) 28 day $4 \times 4 \times 4$ in. ($100 \times 100 \times 100$ mm) cube strength, the w/c ratio

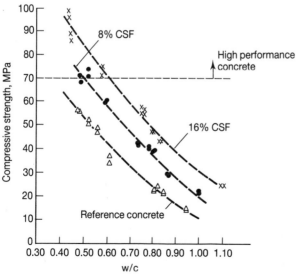

Fig. 2.8   28-day compressive strength versus w/c materials ratio for concrete with different condensed silica fume contents[28]

required is about 0.35 if no silica fume is used; however, with 8% silica fume, the w/c needed is about 0.50, and with 16% silica fume content the w/c ratio requirement increases to about 0.65. This indicates that higher compressive strength can be achieved very easily with high silica fume content at relatively higher w/c ratios.

The efficiency of silica fume in producing concrete of higher strength depends on water/cement + silica fume ratio, dosage of silica fume, age and curing conditions. Yogenendram *et al.*[85] investigated the efficiency of silica fume at lower w/c ratio. Their results indicated that the efficiency is much lower at w/c ratio of 0.28 as compared to the efficiency at w/c ratio of 0.48.

The performance of chemical admixtures is influenced by the particular cement and other cementitious materials. Combinations which have been shown to be effective in many cases may not work in all situations, due to adverse cement and admixture interaction (see Fig. 2.9). Substantial testing should be conducted with any new combination of cements, and mineral or chemical admixtures prior to large scale use.

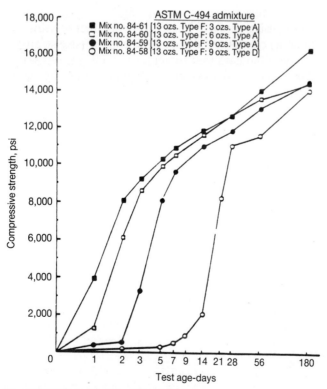

Fig. 2.9 Effect of varying dosage rates of normal retarding water-reducing admixtures on the strength development of concrete[22]

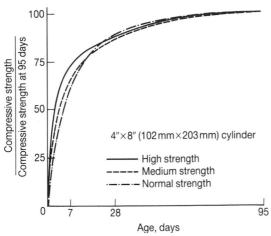

Fig. 2.10   Normalized strength gain with age for limestone concretes moist-cured until testing[16]

Strength development and curing temperature

The strength development with time is a function of the constituent materials and curing techniques. An adequate amount of moisture is necessary to ensure that hydration is sufficient to reduce the porosity to a level necessary to attain the desired strength. Although cement paste will never completely hydrate in practice, the aim of curing is to ensure sufficient hydration. In pastes with lower w/c ratios, self-desiccation can occur during hydration and thus prevent further hydration unless water is supplied externally.

The strength development with time up to 95 days for normal, medium and high strength concretes utilizing limestone aggregate sand moist cured until testing are shown in Fig. 2.10. The results indicate a higher rate of strength gain for higher strength concrete at early ages. At later ages the difference is not significant. The compressive strength development of 9000 psi, 11,000 psi, and 14,000 psi (62 MPa, 76 MPa, 97 MPa) concretes up to a period of 400 days is shown in Fig. 2.11. The results shown in the figure are for mixes containing cement only or cement and fly ash, with some mixes using high range water-reducing agents. The data indicate that for moist-cured specimens, strengths at 56 days are about 10% greater than 28 day strengths. Strengths at 90 days are about 15% greater than 28 day strengths. While it is inappropriate to generalize from such results, they do indicate the potential for strength gain at later ages.

In a recent study[45] at North Carolina State University (NCSU), concretes utilizing a number of different aggregates and mineral admixtures, with strengths from 7000 psi to 12,000 psi (48 MPa to 83 MPa) at 28 days and from 10,000 psi to almost 18,000 psi (69 MPa to 124 MPa) at one year were tested. On examining the absolute strength gain against the percentage strength gain with time, it was concluded that there appears to be no

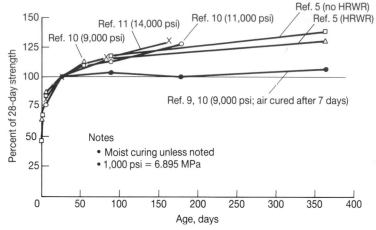

Fig. 2.11   Compressive strength development for concretes with and without high range water reducers[29]

single, constant factor which can be used to predict later strengths accurately from early strengths except in a very general sense. This is no doubt due to the contributions of not only the ultimate strength of the aggregate and the mortar, but to the strength of the transition zone. The transition zone strength, or interfacial bond strength of the mortar to the aggregate, of concretes of higher strengths, is typically affected by the binder composition as well as the ultimate strength of the mortar. Results for splitting tensile strength and modulus of rupture were similar.

The effect of condensed silica fume (CSF) on concrete strength development at 20 °C generally takes place from about 3 to 28 days after mixing. Johansen[40] measured strength up to 3 years and concluded that there was little effect of CSF on either the strength gain between 28 days and 1 year or between 1 and 3 years for water-stored specimens.

The effect of cement types on the strength development is presented in Table 2.1. At ordinary temperatures, for different types of portland and

Table 2.1   Approximate relative strength of concrete as affected by cement type

Type of portland cement		Compressive strength (percent of strength of Type I or normal portland cement concrete)			
ASTM	Description	1 day	7 days	28 days	90 days
I	Normal or general purpose	100	100	100	100
II	Moderate heat of hydration and moderate sulfate resisting	75	85	90	100
III	High early strength	190	120	110	100
IV	Low heat of hydration	55	65	75	100
V	Sulfate resisting	65	75	85	100

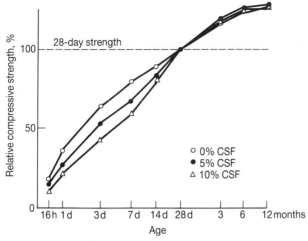

Fig. 2.12 Compressive strength development of concrete cured at 20 °C with different dosages of condensed silica fume[48]

blended cements, the degree of hydration at 90 days and above is usually similar; therefore, the influence of cement composition on the porosity of the matrix and strength is primarily a concern at early ages. The effect of condensed silica fume on the strength development of concretes with four different types of cement was investigated by Maage and Hammer.[48] The four cement types were ordinary portland cement, 10% and 25% pulverized fuel ash (fly ash) blends, and a 15% slag blend. Concrete mixes without CSF and with 0%, 5%, and 10% CSF were made at 5 °C, 20 °C and 35 °C and maintained at these temperatures in water for up to one year. The compressive strengths were measured from 16 hours up to a period of one year. Mixes in three strength classes were made: 2000 psi, 3500 psi and 6500 psi (15 MPa, 25 MPa and 45 MPa). Figure 2.12 shows the compressive strength development of concrete water-cured at 20 °C, with various CSF dosages and utilizing different cement types. In the figure each curve represents a mean value for four cement types, and relative compressive strength of 100% represents 28 day strength for each mix type. From the figure, it can be seen that at 20 °C curing, regardless of the cement type, the CSF had the same influence on the strength–age relationship. Figures 2.13 and 2.14 show relative strength development at 5 °C with and without 10% CSF for the four cement types, and similar data for 35 °C curing are shown in Figs. 2.15 and 2.16. At 5 °C curing, the blended cement lags behind ordinary portland cement concrete (OPC) up to 28 days; with 10% CSF the lag increases which indicates that the pozzolanic reactions have not contributed much to the strength in the 28 day period. At 35 °C the CSF mix is more strongly accelerated (in comparison with 20 °C curing) than the reference mixes, particularly between the first and the seventh day.

Curing at elevated temperatures has a greater accelerating effect on condensed silica fume (CSF) concrete than on control concrete.

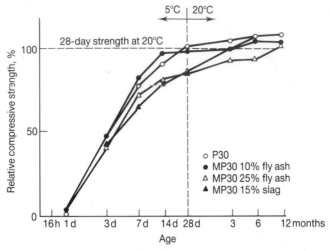

Fig. 2.13　Development of compressive strength in reference concrete cured in water at 5 °C for 28 days then at 20 °C. 100% represents 28-day strength at 20 °C for each cement type[48]

Evidence[28] indicates that a curing temperature of roughly 50 °C is necessary for CSF concrete to equal one day strength of an equivalent control mix. Curing at temperatures below 20 °C retards strength development more for CSF concrete than for control concrete. CSF makes it possible to design low-heat concrete over a wide range of strength levels.[28] Therefore the condensed silica fume concrete is more sensitive to curing temperature than ordinary portland cement concrete. The effect of curing on the condensed silica fume and fly ash concrete was studied in a recent investigation,[68] in which concrete was exposed to six different curing

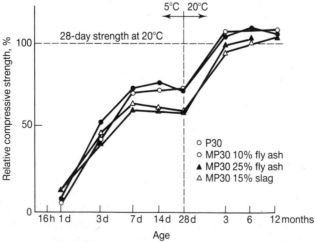

Fig. 2.14　Development of compressive strength in concrete containing 10% condensed silica fume and cured in water at 5 °C for 28 days then at 20 °C. 100% represents 28-day strength at 20 °C for each cement type[48]

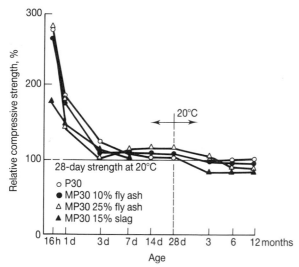

Fig. 2.15   Development of compressive strength in reference concrete cured in water at 35 °C for 28 days then at 20 °C. 100% represents 28-day strength at 20 °C for each cement type[48]

conditions. It was concluded that concrete cured at 20 °C continuously in water(reference) exhibited increasing strengths at all ages; concrete cured in water for 3 days before exposure to 50% RH showed higher initial strength, but the strength decreased after 2–4 months with respect to the reference; and concrete exposed to 50% RH showed lower strength after 28 days of curing than that cured in water.

Curing techniques have significant effects on the strength. The key concerns in curing, especially for concrete of higher strength, are maintaining adequate moisture and temperatures to permit continued cement hydration. Water curing of higher strength concrete is highly recommended[2] due to its low w/c ratio. At w/c ratio below 0.40, the

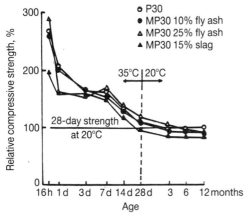

Fig. 2.16   Development of compressive strength in concrete containing 10% condensed silica fume and water-cured at 35 °C for 28 days then at 20 °C. 100% represents 28-day strength at 20 °C for each cement type[48]

Table 2.2   Effect of drying on compressive strength[16]

			Strength attained after drying*			
			Strength attained when moist cured until test age			
Moist cured, days	Drying period, days	Test age, days	Compressive strength f_c'		Modulus of rupture f_r'	
			Normal strength	High strength	Normal strength	High strength
0–7	8–28	28	0.98	0.91	0.83	0.74
0–7	8–28	28	0.94	0.89	0.86	0.74
0–7	8–28	28	0.95	0.88	0.88	0.74
0–28	29–95	95	0.99	0.95	0.97	0.91
0–28	29–95	95	1.01	0.96	0.96	0.93
0–28	29–95	95	0.99	0.96	0.99	0.91

Normal strength: f_c' = 3330 psi at 28 days; f_c' = 3750 psi at 95 days
High strength: f_c' = 10,210 psi at 28 days; f_c' = 11,560 psi at 95 days
* Average of three tests

ultimate degree of hydration is significantly reduced if free water is not provided. The effects of two different curing conditions on concrete strength were investigated.[16] The two conditions were moist curing for seven days followed by drying at 50% relative humidity until testing at 28 days, and moist curing the 28 days followed by drying at 50% relative humidity until testing at 95 days. Higher strength concrete showed a larger reduction in compressive strength when allowed to dry before completion of curing. The results are shown in Table 2.2. It has been reported that the strength is higher with moist curing as compared to field curing.[19]

Compressive strength

Conventionally, in the USA, concrete properties such as elastic modulus, tensile or flexural strength, shear strength, stress-strain relationships and bond strength are usually expressed in terms of uniaxial compressive strength of 6×12 in. (150×300 mm) cylinders, moist cured to 28 days. Compressive strength is the common basis for design for most structures, other than pavements, and even then is the common method of routine quality testing. The terms 'strength' and 'compressive strength' are used virtually interchangeably. The discussion above generally applies equally well to all measures of strength, although most results and conclusions were based either primarily or exclusively on compressive strength results.

Maximum, practically achievable, compressive strengths have increased steadily in the last decade. Presently, 28 day strengths of up to 12,000 psi (84 MPa) are routinely obtainable. The trend for the future has been examined in a recent ACI Committee 363 article[3] which identified development of concrete with compressive strength in excess of 20,000 psi (138 MPa) as one of the research needs.

Testing variables have a considerable influence on the measured compressive strength. The major testing variables are: mold type, specimen size, end conditions and rate of loading. The sensitivity of measured compressive strength to testing variables varies with level of compressive strength.

Since the compressive strength of early strength concretes are at conventional levels, conventional testing procedures can be used for the most part, although curing during the first several hours can affect test results dramatically. Testing of very high strength concretes is much more demanding. However, in all concretes, not just high performance concrete, competent testing is critical.

The effect of mold type on strength was reported in a recent paper by Carrasquillo and Carrasquillo.[18] Their results indicated that use of 6×12 in. (150×300 mm) plastic molds gave strengths lower than steel molds, and use of 4×8 in. (102×203 mm) plastic molds gave negligible difference with steel molds. They concluded that steel molds should be used for concrete with compressive strengths up to 15,000 psi (103 MPa). It seems appropriate that steel molds should also be used for concrete of higher strengths. The specimen size effect on the strength is shown in Fig. 2.17, which shows the relationship between the compressive strength of 4×8 in. (102×203 mm) cylinders and 6×12 in. (150×300 mm) cylinders. The figure indicates that 4×8 in. (102×203 mm) cylinders exhibit approximately 5% higher strengths than 6×12 in. (150×300 mm) cylinders. Similar results were also obtained in a recent study at North Carolina

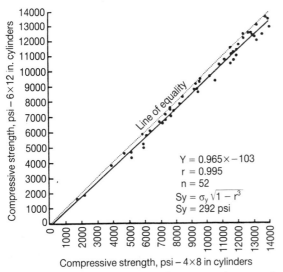

Fig. 2.17 Relationship between the compressive strength of 4×8 in. (102×203 mm) cylinders and 6×12 in (152×304 mm) cylinders[22]

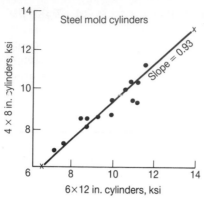

Fig. 2.18 Compressive strength of concrete cylinders cast in 4 × 8 in. (102 × 203 mm) steel molds versus 6 × 12 in. (152 × 304 mm) steel molds[18]

State University.[45] A contradictory result[19] is reported, however, which indicates that the compressive strength of 4 × 8 in. (102 × 203 mm) cylinders is slightly lower than 6 × 12 in. (150 × 300 mm) cylinders, see Fig. 2.18. The strength gain for 17,000 psi (117 MPa) concrete as shown by 6 × 12 in. (150 × 300 mm) and 4 × 8 in. (102 × 203 mm) cylinders has been reported by Moreno[54] and the results are shown in Fig. 2.19. His study also showed that the specimen size effect on the compressive strength is negligible on the basis of 29 tests, see Table 2.3. Another study[16] concluded that the ratio of 6 × 12 in. (150 × 300 mm) cylinder to 4 × 8 in. (102 × 203 mm) cylinder was close to 0.90 regardless of the strength of concrete for the ranges tested between 3000 and 11,000 psi (21 and 76 MPa).

The relationship between the compressive strength of 6 × 12 in. (150 × 300 mm) and cores from a column was studied for concrete with a strength of 10,000 psi (69 MPa), see Table 2.4. It was concluded that the 85% criterion specified in the ACI Building Code (ACI 318–89)[1] would be

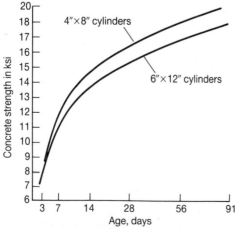

Fig. 2.19 Compressive strength development for 17,000 psi (117 MPa) concrete[54]

Table 2.3 Comparison of compressive strength test results of 12,000 psi (83 MPa) concrete at 56 days as obtained by 6 × 12 in. (152 × 304 mm) and 4 × 8 in. (102 × 203 mm) cylinders[54]

Cylinder size	6 × 12 in.	4 × 8 in.
Mean	13,444	13,546
Standard deviation	463	515
Coefficient of variation, percent	3.4	3.8
Number of tests	29	29

applicable to high strength concrete. This study also confirms the belief that job cured specimens do not give accurate measurements of the in-place strength. The reason for lower strength in the middle portion of the columns is probably due to temperature rise, i.e. 100 °F (38 °C) for high strength mixtures. In a recent study,[11] it was shown that the strength of 4 in. (100 mm) cores taken from a mock column at two and a half years after casting was nearly identical to that of specimens cured for 28 days in lime-saturated water at room temperature. The strength of the concrete tested was 12,300 psi (85 MPa).

The effect of end conditions on the compressive strength of concrete is summarized in a recent paper.[18] More than five hundred 6 × 12 in. (150 × 300 mm) cylinders from concretes having compressive strengths from 2500 to 16,500 psi (17 MPa to 114 MPa) were tested with either unbonded caps (two types) or sulfur mortar caps. It was concluded that use of unbonded caps (with a restraining ring and elastomeric insert) could provide a cleaner, safer and more cost-effective alternative to sulfur mortar for capping concrete cylinders. For concretes between 4000 and 10,000 psi (28 and 69 MPa), the use of polyurethane inserts with aluminium restraining rings in testing concrete cylinders yielded average test results within 5% of those obtained using sulfur mortar. For concrete strengths below 11,000 psi (76 MPa), the use of neoprene inserts with steel restraining rings in testing concrete cylinders yielded average test results within 3% of those obtained using sulfur mortar. For higher strength concrete, the use of either unbonded capping system is questionable. Substantial differences in compressive strength test results were obtained when two sets of restraining rings obtained from the same manufacturer were used. It was recommended that prior to acceptance, each set should be tested for correlation to results obtained from cylinders capped according to ASTM C617 for all strength levels of concrete for which the unbonded caps are to be used. (Equipment now exists for parallel grinding the ends of concrete cylinders prior to compression testing, thereby eliminating the need for any type of end cap.)

Measured compressive strength increases with higher rates of loading. This trend has been reported in a number of studies[14,15,26,38,49,83] for concrete with strengths in the range of 2000 to 5500 psi (14 to 49 MPa). However, only one study[8] has reported the effect of strain rate on concretes with compressive strengths in excess of 6000 psi (41 MPa). Based

Table 2.4 Column core strengths versus 6 × 12 in. (152 × 304 mm) cylinders*[22]

Test age Maximum size stone	7-Days		28-Days		56-Days		180-Days		1 Year	
	⅛ in	1 in	⅛ in	1 in	⅛ in	1 in	⅛ in	1 in	⅛ in	1 in
Compressive strength of 6 × 12 in cylinders, psi										
Field-cured	8,596	8,139	10,177	10,204	10,775	10,542	11,514	11,546	12,444	11,772
Moist-cured	9,228	9,277	11,522	11,236	12,376	12,448	13,852	13,776	14,660	13,951
Compressive strength of cores, psi										
West face	9,407	9,312	11,118	10,743	11,598	10,964	12,970	12,400	15,080	13,775
Middle	8,660	8,959	9,674	9,724	9,833	9,656	11,635	10,720	13,404	13,003
East face	9,180	9,706	10,584	10,575	10,756	10,603	11,626	11,589	14,088	13,695
All cores	9,083	9,326	10,459	10,347	10,729	10,408	12,077	11,570	14,190	13,490
Cores/6 × 12 in moist-cured cylinders, percent										
West face	102	101	97	96	94	88	94	90	107	99
Middle	94	97	84	87	79	78	84	78	95	93
East face	99	105	92	94	87	85	87	84	100	98
All cores	98	101	91	92	87	84	87	84	101	97

* Reported strengths are average of two specimens.

on their research and other reported data.[14,15,26,38,49,83] Ahmad and Shah[8] proposed an equation to estimate the strength under very fast loading conditions. The recommended equation is

$$(f_c')_{\dot\varepsilon} = f_c'\left[0.95 + 0.27\log\frac{\dot\varepsilon}{f_c'}\right]\alpha \qquad (2.1)$$

where $\dot\varepsilon$ is the strain rate in microstrains per sec ($\mu\varepsilon$/sec).

The shape factor α accounts for the different sizes of the specimens tested by different researchers and is given by

$$\alpha = 0.85 + 0.95\,(d) - 0.02\,(h) \quad \text{for } \frac{h}{d} \leqslant 5 \qquad (2.2)$$

where d = diameter or least lateral dimension (in.), h = height (in.)

No information is available on the effect of rate of loading on the strength for concrete with strengths in excess of 10,000 psi (70 MPa).

Tensile strength

The tensile strength governs the cracking behavior and affects other properties such as stiffness, damping action, bond to embedded steel and durability of concrete. It is also of importance with regard to the behavior of concrete under shear loads. The tensile strength is determined either by direct tensile tests or by indirect tensile tests such as flexural or split cylinder tests.

Direct tensile strength

The direct tensile strength is difficult to obtain. Due to the difficulty in testing, only limited and often conflicting data is available. It is often assumed that direct tensile strength of concrete is about 10% of its compressive strength.

Two recent studies[23,31] have reported the direct tensile strength of concrete. The study at Delft University[23] utilized 4.7 in. (120 mm) diameter cylinders having a length of 11.8 in. (300 mm). The study at Northwestern[31] employed $3 \times 0.75 \times 12$ in. ($76 \times 19 \times 304$ mm) and $3 \times 1.5 \times 12$ in. ($76 \times 38 \times 304$ mm) thin plates having a notch in the central region for creating a weak section for crack initiation and propagation, and used special wedge like frictional grips. The study at Delft tested concrete of one strength which had either been sealed for four weeks or moist-cured for two weeks and air-dried for two weeks. The results indicated 18% higher tensile strength for the sealed concrete compared to the air-dried concrete. The investigation at Northwestern included different concrete strengths up to 7000 psi (48 MPa) strength, and it was concluded that the uniaxial tensile strength can be estimated by the expression $6.5\sqrt{f_c'}$.

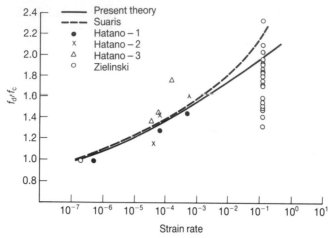

Fig. 2.20   Effect of strain rate on tensile strength of concrete[59]

Direct tensile strength data is not available for concrete with strengths in excess of 8000 psi (55 MPa).

The effect of rate of loading on the tensile strength has been the focus of some studies by Hatano,[34] Suaris and Shah,[80] and Zielinski et al.[87] The effect of fast strain rate on the tensile strength of concrete as observed by these studies is shown in Fig. 2.20. Also shown in the figure is a comparison of the predictions per a constitutive theory for concrete subjected to static uniaxial tension[59] and the experimental results.

The effect of sustained and cyclic loading on the tensile properties of concrete was investigated by Cook and Chindaprasirt.[21] Their results indicate that prior loading of any form reduces the strength of concrete on reloading. Strain at peak stress and the modulus on reloading follows the same trend as strength. This behavior can be attributed to the cumulative damage induced by repetitive loadings. Saito[69] investigated the microcracking phenomenon of concrete understatic and repeated tensile loads, and concluded that cumulative damage occurs in concrete due to reloading beyond the stage at which interfacial cracks are formed.

The effect of uniaxial impact in tension was investigated by Zielinski et al.[87] Their results indicated an increase in the tensile strength similar to the phenomenon generally observed under uniaxial impact in compression.

Indirect tensile strength

The most commonly used tests for estimating the indirect tensile strength of concrete are the splitting tension test (ASTM C496) and the third-point flexural loading test (ASTM C78).

(a)   Splitting tensile strength   As recommended by ACI Committee 363,[2]

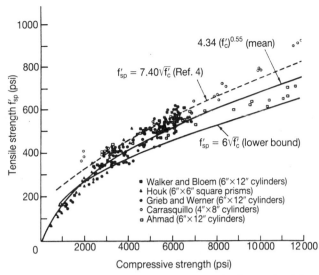

Fig. 2.21 Variation of splitting tensile strength of normal weight concrete with the compressive strength[4]

the splitting tensile strength (f_{ct}) for normal weight concrete can be estimated by

$$f_{ct} = 7.4\sqrt{f_c'} \text{ psi} \quad 3000 \leqslant f_c' \leqslant 12{,}000 \text{ psi} \quad (2.3)$$
$$(21 \leqslant f_c' \leqslant 83 \text{ MPa})$$

In 1985, based on the available experimental data of split cylinder tests on concretes of low-, medium-,[32,37,81] and high strengths,[7,16,25] an empirical relationship was proposed by Ahmad and Shah[4] as

$$f_{ct} = 4.34\,(f_c') \text{ psi} \quad f_c' \leqslant 12{,}000 \text{ psi} \quad (2.4)$$
$$(f_c' \leqslant 83 \text{ MPa})$$

Figure 2.21 shows the experimental data, with the predictions using the above equation and the recommendations of the ACI Committee 363. The latter appears to overestimate values of tensile strength. Recommendations of ACI Committee 363 were based on work performed at Cornell University.[16]

Figure 2.22 shows the aging effect on splitting tensile strength, which is similar to that under compressive loading. In an investigation by Ojdrovic on cracking modes,[60] it was concluded that at early ages, tensile strength of concrete is the property of the matrix which governs the cracking mode.

The effect of prior compressive loading on the split tensile strength was investigated by Liniers[46] and the results are shown in Fig. 2.23. From this figure, he concluded that limiting the compressive stresses to 60% of the strength is essential if only tolerable damage is to be accepted.

The tensile strength of condensed silica fume (CSF) concrete is related to the compressive strength in a manner similar to that of normal concrete.

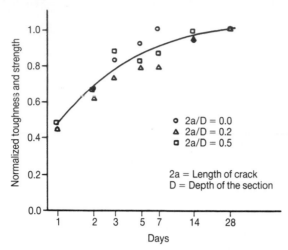

Fig. 2.22   Normalized splitting tensile strength as a function of the age at testing[60]

However if CSF concrete is exposed to drying after one day of curing in the mold, the tensile strength is reduced more than the control concrete.[28]

(b)   Flexural strength or modulus of rupture   Flexural strength or modulus of rupture is measured by a beam flexural test and is generally taken to be a more reliable indicator of the tensile strength of concrete. The modulus of rupture is also used as the flexural strength of concrete in pavement design. It is often assumed that flexural strength of concrete is about 15% of the compressive strength.

In the absence of actual test data, the modulus of rupture may be estimated by

$$f_r = k\sqrt{f_c'} \tag{2.5}$$

typically in the range of 7.5 to 12. For high strength concrete, the ACI Committee 363 *State-of-the-Art Report*[2] recommends a value of $k = 11.7$ as

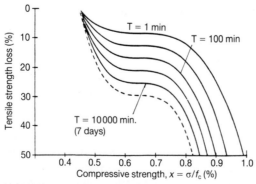

Fig. 2.23   Tensile strength loss as a function of compressive stress fraction for different duration of loading[46]

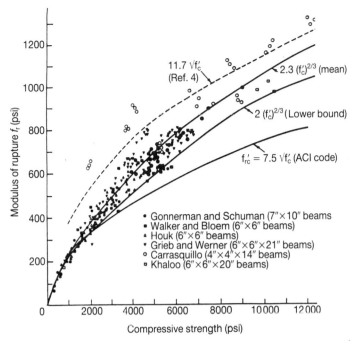

Fig. 2.24 Variation of modulus of rupture with the compressive strength[4]

appropriate for concrete with compressive strength in the range of 3000 psi to 12,000 psi (21 MPa to 83 MPa).

Based on the available data of beam flexural tests on concretes of low, medium[32,37,81] and high strengths,[7,16,25] an empirical equation to predict the flexural strength (modulus of rupture) was proposed[4] as

$$f_r = 2.30\,(f_c')^{2/3} \tag{2.6}$$

where f_c' is the compressive strength in psi.

The above equation is of the same form as proposed by Jerome,[39] which was developed on the basis of data for concretes of strengths up to 8000 psi (56 MPa). Figure 2.24 shows the plot of the experimental data and the proposed equation[4] for predicting the modulus of rupture of concretes with strengths up to 12,000 psi (83 MPa). Also shown in the figure is the expression recommended by Carrasquillo and Nilson.[16]

The results of uniaxial and biaxial flexural tests[86] indicated that the tensile strength was 38% higher in the uniaxial stress state than in the biaxial stress state.

Flexural strength is higher for moist-cured as compared to field cured specimens.[19] However, wet-cured specimens containing condensed silica fume (CSF) exhibit a lower ratio of tensile to compressive strength than dry-stored concrete specimens with silica fume.[47] For all concretes, allowing a moist cured beam to dry during testing will result in lowered measured strength, due to the addition of applied load and drying shrinkage stresses on the tensile face. The flexural strength of condensed silica fume (CSF) concrete is related to the compressive strength in a manner similar to that of concrete without silica fume; however, if CSF concrete is exposed to drying after only one day of curing in the mold, the flexural strength reduces more than the control concrete.[28]

2.3   Deformation

The deformation of concrete depends on short-term properties such as the static and dynamic modulus, as well as strain capacity. It is also affected by time dependent properties such as shrinkage and creep.

Static and dynamic elastic modulus

The modulus of elasticity is generally related to the compressive strength of concrete. This relationship depends on the aggregate type, the mix proportions, curing conditions, rate of loading and method of measurement. More information is available on the static modulus than on the dynamic modulus since the measurement of elastic modulus can be routinely performed whereas the measurement of dynamic modulus is relatively more complex.

Static modulus

The static modulus of elasticity can be expressed as secant, chord or tangent modulus. According to the ACI Building Code (ACI-318-89),[1] E_c, the static, secant modulus of elasticity, is defined as the ratio of the stress at 45% of the strength to the corresponding strain. Static, chord modulus of elasticity, as determined by ASTM C469, is defined as the ratio of the difference of the stress at 40% of the ultimate strength and the stress at 50 millionths strain to the difference in strain corresponding to the stress at 40% of ultimate strength and 50 millionths strain.

At present there are two empirical relationships that can be used for design when the static modulus of elasticity has not been determined by tests. They are the ACI Code formula[1]

$$E_c = 33\omega^{1.5}\sqrt{f_c'} \text{ psi} \tag{2.7}$$

where ω = unit weight in pounds per cubic foot (pcf) and the formula recommended by the ACI Committee 363 on High Strength Concrete[2] for concrete with unit weight of 145 pcf.

$$E_c = 1.0 \times 10^6 + 40,000\sqrt{f_c'} \text{ psi} \tag{2.8}$$

This formula is based on work performed at Cornell University.[16]

Figure 2.25 shows the range of scatter of data with the predictions of the ACI equation and the ACI Committee 363 equation. A third equation was

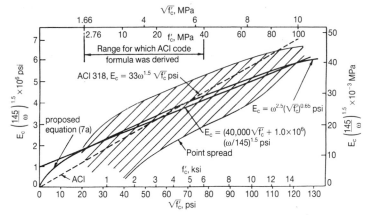

Fig. 2.25   Secant modulus of elasticity versus concrete strength[4]

recommended by Ahmad and Shah[4] which seems to be more representative of the trend of the data. The equation is

$$E_c = \omega^{2.5}(\sqrt{f_c'})^{0.65} \text{ psi} \tag{2.9}$$

where ω = unit weight of concrete in pcf.

Figure 2.26 gives a comparison of experimental values of elastic modulus collected by Cook[22] with the predictions by the ACI 318-89 Code and the ACI Committee 363 equations. The concrete contained aggregates from South Carolina, Tennessee, Texas and Arizona. Aggregate sizes varied from 3/8 in. to 1 in. (10 to 25 mm) and consisted primarily of crushed limestones, granites and native gravels. Cook recommended the following equation which gives a better fit for the particular set of experimental data.

$$E_c = \omega^{2.5}(\sqrt{f_c'})^{0.315} \text{ psi} \tag{2.10}$$

where ω = 151 pcf.

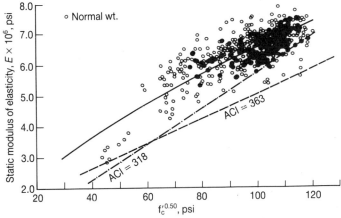

Fig. 2.26   Secant modulus of elasticity versus concrete strength for normal weight concrete[22]

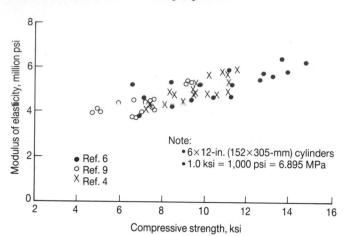

Fig. 2.27   Secant modulus of elasticity as a function of strength[29]

Figure 2.27 summarizes test results of modulus of elasticity as a function of compressive strength. These results confirm the increased stiffness at higher strengths. Modulus of elasticity of very high strength concrete up to 17,000 psi (117 MPa) is shown in Fig. 2.28. According to Moreno,[54] the results are generally closer to the predictions of the ACI Code (ACI 318-89) equation. However, at strength higher than 15,000 psi (105 MPa), the ACI Code equation overestimates the test results. Moreno also contends that ACI Committee 363 equation[2] always predicts results lower than the test data even for 17,000 psi (117 MPa) concrete, and hence it was concluded that the equation recommended by the ACI Committee 363 is more appropriate for higher-strength concrete.

In a recent study at NCSU[45] based on the results of 16 specimens with strengths varying between 8000 psi (55 MPa) at 28 days and 18,000 psi (124 MPa) at one year, it was concluded that ACI Committee 363[2] formula gave closer predictions of experimental results obtained from 6×12 in. $(150 \times 300$ mm$)$ cylinders.

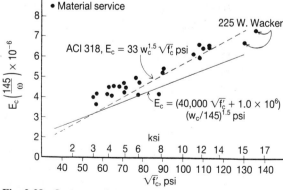

Fig. 2.28   Secant modulus of elasticity variation with square root of the compressive strength[54]

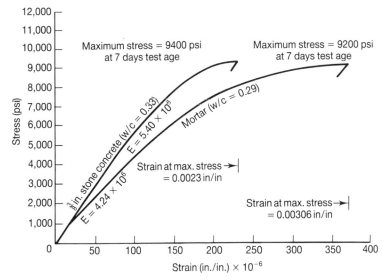

Fig. 2.29   Effect of coarse aggregate and mix proportions on the modulus of elasticity[22]

The modulus of elasticity of concrete is affected by the properties of the coarse aggregate. The higher the modulus of elasticity of the aggregate the higher the modulus of the resulting concrete. The shape of the coarse aggregate particles and their surface characteristics may also influence the value of the modulus of elasticity of concrete. Figure 2.29 shows the effect of the coarse aggregate type and the mix proportions on the modulus of elasticity. From this figure it can be concluded that, in general, the larger the amount of coarse aggregate with a high elastic modulus, the higher would be the modulus of elasticity of concrete. The use of four different types of coarse aggregates in a very high strength concrete mixture (w/c = 0.27) showed that elastic modulus was significantly influenced by the mineralogical characteristics of the aggregates.[12] Limestone and crushed aggregates from fine-grained diabase gave higher modulus than a smooth river gravel and crushed granite that contained inclusions of a soft mineral.

It is generally accepted that regardless of the mix proportions or curing age, concrete specimens tested in wet conditions show about 15% higher elastic modulus than the corresponding specimens tested in dry conditions.[53] This is attributed to the effect of drying on the transition zone. Because of drying, there is microcracking in the transition zone due to shrinkage, which reduces the modulus of elasticity.

As strain rate is increased, the measured modulus of elasticity increases. Based on the available experimental data for concrete with strength up to 7000 psi (48 MPa),[8,14,15,26,38,49,83] the following empirical equation was proposed by Ahmad and Shah[4] for estimating the modulus of elasticity under very high strain rates.

$$(E_c)_{\dot{\varepsilon}} = E_c\left[0.96 + 0.038\frac{\log \varepsilon}{\log \varepsilon_s}\right]$$ (2.11)

where $E_c = 27.5\omega^{1.5}\sqrt{f_c'}$, $\dot{\varepsilon}$ is the strain rate in microstrains per second ($\mu\varepsilon$/sec), $\varepsilon_s = 32$ $\mu\varepsilon$/sec.

A recent paper[43] has suggested that if internal strains are measured by means of embeddable strain gauges, the measured modulus is 50% higher than that from strain measurements made on the surface. The author concluded that the reason for this observation is the non-uniform strain field across the section of the cylinders.

Dynamic modulus

The measurement of dynamic modulus corresponds to a very small instantaneous strain. Therefore the dynamic modulus is approximately equal to the initial tangent modulus. Dynamic modulus is appreciably higher than the static (secant) modulus. The difference between the two moduli is due in part to the fact that heterogeneity of concrete affects the two moduli in different ways. For low, medium and high strength concretes, the dynamic modulus is generally 40%, 30% and 20% respectively higher than the static modulus of elasticity.[53]

Popovics[66] has suggested that for both lightweight and normal weight concretes, the relation between the static and dynamic moduli is a function of density of concrete, just as is the case with relation between the static modulus and strength.[66] Popovics expressed E_c as a linear function of $E_d^{1.4}/\rho$ where ρ is the density of concrete, and E_d is the dynamic modulus.

The ratio of static to dynamic modulus is also affected by the age at testing as shown by Philleo[65] in Fig. 2.30. The figure indicates that at early ages (up to 6 months) the ratio of the two moduli increases from 0.4 to about 0.8 and becomes essentially constant thereafter.

A typical relationship between the dynamic modulus determined by the vibration of the cylinders and their compressive strength is shown in Fig. 2.31. It has been reported by Sharma and Gupta[77] that the relationship between the strength and the dynamic modulus is unaffected by air entrainment, method of curing, condition at test, or type of cement.

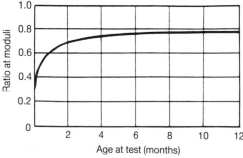

Fig. 2.30   Ratio of static and dynamic modulus of elasticity of concrete at different ages[65]

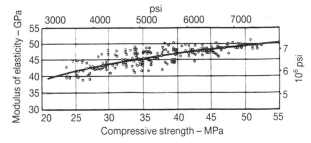

Fig. 2.31 Relation between the dynamic modulus of elasticity, determined by transverse vibration of cylinders, and their compressive strength[77]

It should also be noted that no information is available regarding the relationship between the static and dynamic modulus of elasticity for concrete with strength in excess of 8000 psi (55 MPa).

2.4 Strain capacity

The usable strain capacity of concrete can be measured either in compression or in tension. In the compression mode, it can be measured by either concentric or eccentric compression testing. In the tensile mode, the strain capacity can be either for direct tension or indirect tension. The behavior under multiaxial stress states if outside the scope of this chapter, and only the behavior under uniaxial stress condition will be discussed.

Stress-strain behavior in compression

The stress-strain behavior is dependent on a number of parameters which include material variables such as aggregate type and testing variables such as age at testing, loading rate, strain gradient and others noted above.

The effect of the aggregate type of the stress-strain curve is shown in Fig. 2.32 which indicates that higher strength and corresponding strain are achieved for crushed aggregate from fine-grained diabase and limestone, as compared to concretes made from smooth river gravel and from crushed granite that contained inclusions of a soft mineral.

A number of investigations[5,35,41,58,75,76,82,84] have been undertaken to obtain the complete stress-strain curves in compression. Axial stress-strain curves for concretes with compressive strengths up to 14,000 psi (98 MPa) concrete as obtained by different researchers are shown in Fig. 2.33.

It is generally recognized that for concrete of higher strength, the shape of the ascending part of the curve becomes more linear and steeper, the strain at maximum stress is slightly higher, and the slope of the descending part becomes steeper. The existence of the postpeak descending part of the stress-strain curve has been the focus of a recent paper.[79] It was concluded that the postpeak behavior can be quantified for inclusion in finite element

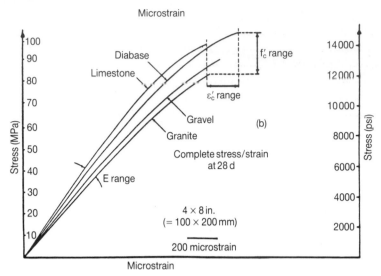

Fig. 2.32 Effect of the aggregate type on the ascending portion of the stress-strain curves of concrete at 28-days[12]

analysis and that it can have considerable influence on the predicted structural behavior and strength.[67]

To obtain the descending part of the stress-strain curve, it is necessary to avoid specimen-testing machine interaction. One approach is to use a closed-loop system with a constant rate of axial strain as a feedback signal for closed-loop operation. The difficulties of obtaining the postpeak behavior experimentally and methods of overcoming these difficulties are

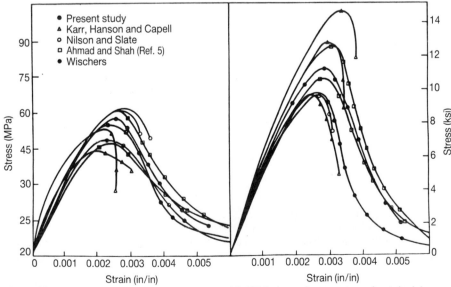

Fig. 2.33 Different stress strain curves reported for high strength concrete under uniaxial compression[4]

described in a study by Ahmad and Shah.[9] For very high strength concretes it may be necessary to use the lateral strains as a feedback signal rather than the axial strains.[74] In a paper by Kotsovos,[44] it is argued that a more realistic description of the postpeak specimen behavior may be a complete and immediate loss of load-carrying capacity as soon as the peak load is exceeded. A different point of view is reflected in another recent paper[79] which suggests that there is usable strength for concrete after peak stress. Based on the above mentioned experimental investigations, different analytical representations for the stress-strain curve have been proposed. They include use of a fractional equation,[6,70,82] or a combined power and exponential equation[75] and serpentine curve. The fractional equation is a comprehensive, yet simple way of characterizing the stress-strain response of concrete in compression.[4] The fractional equation can be written as

$$f_\varepsilon = (f_c') \frac{A(\varepsilon/\varepsilon_c') + (B-1)(\varepsilon/\varepsilon_c')^2}{1 + (A-2)(\varepsilon/\varepsilon_c') + B(\varepsilon/\varepsilon_c')} \qquad (2.12)$$

(for $f > 0.1\varepsilon$, f_c', when $\varepsilon > \varepsilon_c'$)

where f_ε is the compressive stress at strain ε, f_c' and ε_c' the maximum stress and corresponding strain,
A and B are parameters which determine the shape of the curve.

The values of the parameters A and B, which control the shape of the ascending and the descending parts, respectively, may be estimated by

$$A = E_c \frac{\varepsilon_c'}{f_c'} \qquad (2.13)$$

$$B = 0.88087 - 0.57 \times 10^{-4}(f_c') \qquad (2.14)$$

$$\varepsilon_c' = 0.001648 + 1.14 \times 10^{-7}(f_c') \qquad (2.15)$$
$$E_c = 27.55\omega^{1.5}\sqrt{f_c'} \qquad (2.16)$$

where f_c' is the compressive strength in psi and ω is the unit weight in pcf.

The parameters A, B, ε_c' and E_c are as recommended by Ahmad and Shah[4] and were determined from the statistical analysis of the experimental results on 3×6 in. (75×152 mm) concrete cylinders.[5,6] These cylinders were tested under strain controlled conditions in a closed-loop testing machine and had compressive strengths ranging from 3000 to 11,000 psi (20 to 75 MPa).

Stress-strain behavior in tension

The direct tensile stress-strain curve is difficult to obtain. Due to difficulties in testing concrete in direct tension, only limited and often conflicting data are available.

Direct tensile tests were carried out on tapered cylindrical specimens of 4.7 in. diameter and 11.8 in. length (120 mm diameter $\times$ 300 mm).[24] For

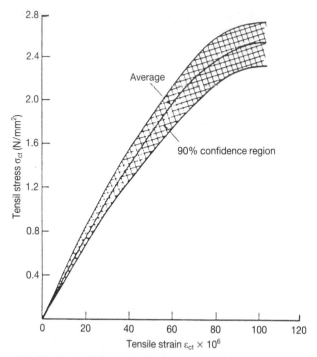

Fig. 2.34   A typical stress strain curve for concrete under uniaxial tension determined under static load controlled tests[24]

the application of the load, steel platens were glued to the top and bottom of the specimens. In order to provide plane-parallel and axial connection of these platens, a special gluing press was designed. Some 500 direct tensile tests, 300 compressive and 300 splitting tests were performed. A typical stress-strain curve with a 95% confidence region for concrete subjected to direct tension is shown in Fig. 2.34. The stress-strain curve shown in the figure is for dry specimens. The results may vary slightly for specimens tested in moist conditions.

A study at Northwestern by Gopalaratham and Shah[31] points out that due to the localized nature of the post-cracking deformations intension, no unique tensile stress-strain relationship exists. According to this study, the uniaxial tensile strength can be estimated by $\sqrt{f_c'}$, and the tangent modulus of elasticity is identical in tension and compression. The stress-strain relationship in tension before peak is less nonlinear than in compression.

Laser speckle interferometry was employed in a recent study,[13] to investigate the behavior of concrete subjected to uniaxial tension. Unique post-peak stress-strain and stress-deformation behavior were not observed. The stress-strain response of concrete was found to be sensitive to gauge-length. Strains measured within a gauge length inside the microcracking zone were two orders of magnitude higher than values previously reported.[27]

In a recent study,[20] it was shown that while the use of strain gauges would lead to non-objective constitutive stress-strain relations, interferometric measurements on notched specimens allow an indirect determination of the local stress-strain and stress-separation (deformations) relations. Guo and Zhang[33] tested 29 specimens in direct tension and obtained complete stress-deformation curves. Based on the experimental results an equation was also derived for the stress-displacement curves.

Flexural tension

While the information on the stress-strain behavior in tension is severely limited, virtually no data are available regarding the strain capacity in flexural tension. This is an area for which research is sorely needed to provide a basis for design where flexural cracking is an important consideration.

2.5   Poisson's ratio

Poisson's ratio under uniaxial loading conditions is defined as the ratio of lateral strain to strain in the direction of loading. In the inelastic range, due to volume dilation resulting from internal microcracking, the apparent Poisson's ratio is not constant but is an increasing function of the axial strain.

Experimental data on the values of Poisson's ratio for high strength concrete is very limited.[16,64] Based on the available experimental information, Poisson's ratio of higher strength concrete in the elastic range appears comparable to the expected range of values for lower-strength concrete. In the inelastic range, the relative increase in lateral strains is less for higher-strength concrete compared to concrete of lower strength.[6] That is, higher-strength concrete exhibits less volume dilation than lower-strength concrete (see Fig. 2.35). This implies less internal microcracking for concrete of higher strength.[17] The lower relative expansion during the

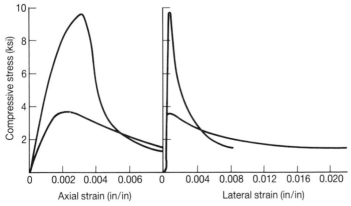

Fig. 2.35   Axial stress versus axial strain and lateral strain for normal and high-strength concrete[4]

inelastic range may mean that the effects of triaxial stresses will be proportionately different for higher-strength concrete. For example, the effectiveness of hoop confinement is reported to be less for higher-strength concrete.[6]

Information on Poisson's ratio of concrete with strength greater than 12,000 psi (83 MPa) is not available in the literature.

Acknowledgements

This work was supported by the Strategic Highway Research Program (SHRP) at North Carolina State University, Raleigh, NC.

References

1   Committee 318 (1989) *Building code requirements for reinforced concrete.* American Concrete Institute.
2   Committee 363 (1984) State-of-the-art report on high strength concrete. *ACI Journal*, **81**, 4, July–Aug, 364–411.
3   Committee 363 (1987) Research needs for high-strength concrete. *ACI Materials Journal*, November–December, 559–61.
4   Ahmad, S.H. and Shah, S.P. (1985) Structural properties of high strength concrete and its implications for precast prestressed concrete. *PCI Journal*, Nov–Dec, 91–119.
5   Ahmad, S.H. (1981) *Properties of confined concrete subjected to static and dynamic loading.* Ph.D. Thesis, University of Illinois, Chicago, March.
6   Ahmad, S.H. and Shah, S.P. (1982) Stress-strain curves of concrete confined by spiral reinforcement. *ACI Journal*, **79**, 6, Nov–Dec, 484–90.
7   Ahmad, S.H. (1982) *Optimization of mix design for high strength concrete.* Report No. CE 001-82, Department of Civil Engineering, North Carolina State University, Raleigh, NC.
8   Ahmad, S.H. and Shah, S.P. (1985) Behavior of hoop confined concrete under high strain rates. *ACI Journal*, **82**, 5, Sept–Oct, 634–47.
9   Ahmad, S.H. and Shah, S.P. (1979) Complete stress-strain curves of concrete and nonlinear design. Progress Report PFR 79-22878 to the National Science Foundation, University of Illinois, Chicago, August and also *Non-linear design of concrete structures*. University of Waterloo Press, 1980, 222–30.
10 Aitcin, P.-C. (1983) *Condensed silica fume.* University of Sherbrooke, Sherbrooke, Quebec.
11 Aitcin, P.-C., Sarkar, S.L. and Laplante, P. (1990) Long-term characteristics of a very high strength concrete. *Concrete International*, Jan, 40–4.
12 Aitcin, P.-C. and Metha, P.K. (1990) Effect of coarse-aggregate characteristics on mechanical properties of high-strength concrete. *ACI Materials Journal*, March–April, 103–107.
13 Ansari, F. (1987) Stress-strain response to microcracked concrete in direct tension. *ACI Materials Journal*, **84**, 6, Nov–Dec, 481–90.
14 Atchley, B.L. and Furr, H.L. (1967) Strength and energy absorption capabilities of plain concrete under dynamic and static loadings. *ACI Journal*, **64**, 11, Nov, 745–56.
15 Bresler, B. and Bertero, V.V. (1975) Influences of high strain rate and cyclic loading on behavior of unconfined and confined concrete in compression. *Proceedings, Second Canadian Conference on Earthquake Engineering*, MacMaster University, Hamilton, Ontario, Canada, June.

16 Carrasquillo, R.L., Nilson, A.H. and Slate, F.O. (1981) Properities of high strength concrete subject to short-term loads. *ACI Journal*, May–June, 171–8.

17 Carrasquillo, R.L., Slate, F.O. and Nilson, A.H. (1981) Microcracking and behaviour of high strength concrete subjected to short term loading. *ACI Journal*, **78**, 3, May–June, 179–86.

18 Carrasquillo, P.M. and Carrasquillo, R.L. (1988) Effect of using unbonded capping systems on the compressive strength of concrete cylinders. *ACI Materials Journal*, May–June, 141–7.

19 Carrasquillo, P.M. and Carrasquillo, R.L. (1988) Evaluation of the use of current concrete practice in the production of high-strength concrete. *ACI Materials Journal*, **85**, 1, Jan–Feb, 49–54.

20 Cedolin, L., Poli, S.D. and Iori, I. (1987) Tensile behavior of concrete. *Journal of Engineering Mechanics*, ASCE, **113**, 3, March, 431–49.

21 Cook, D.J. and Chindaprasirt, P. (1981) Influence of loading history upon the tensile properties of concrete. *Magazine of Concrete Research*, **33**, 116, Sept, 154–60.

22 Cook, J.E. (1989) Research and application of high-strength concrete: 10,000 PSI concrete. *Concrete International*, Oct, 67–75.

23 Cornelissen, H.A.W. (1984) Fatigue failure of concrete in tension. *Heron*, **29**, 4, 1–68.

24 Cornelissen, H.A.W. and Reinhardt, H.W. (1984) Uniaxial tensile fatigue failure of concrete under constant-amplitude and programme loading. *Magazine of Concrete Research*, **36**, 129, Dec, 216–26.

25 Dewar, J.D. (1964) The indirect tensile strength of concrete of high compressive strength. *Technical Report*, No. 42.377. Cement and Concrete Association, Wexham Springs, England, March

26 Dilger, W.H., Koch, R. and Andowalczyk, R. (1984) Ductility of plain and confined concrete under different strain rates. *ACI Journal*, **81**, 1, Jan–Feb, 73–81.

27 Evans, R.H. and Marathe, M.S. (1968) Microcracking and stress-strain curves for concrete in tension. *Materials and Structures, Research and Testing*, (RILEM, Paris), **1**, 1, Jan–Feb, 61–4.

28 FIP Commission on Concrete (1988) *Condensed Silica Fume in Concrete*. State of Art Report, Federation Internationale de la Precontrainte, London.

29 Fiorato, A.E. (1989) PCA research on high-strength concrete. *Concrete International*, April, 44–50.

30 Freedman, S. (1970/71) High strength concrete. *Mocern Concrete*, **34**, 6, Oct 1970, 29–36; 7, Nov 1970, 28–32; 8, Dec 1970, 21–24; 9, Jan 1971, 15–22; and 10, Feb 1971, 16–23.

31 Gopalaratham, V.S. and Shah, S.P. (1985) Softening response of concrete in direct tension. *Research Report*, Technological Institute, Northwestern University, Chicago, June 1984 also *ACI Journal*, **82**, 3, May–June 1985, 310–23.

32 Grieb, W.E. and Werner, G. (1962) Comparison of splitting tensile strength of concrete with flexural and compressive strengths. *Public roads*, **32**, 5, Dec.

33 Guo, Z-H. and Zhang, X-Q. (1987) Investigation of complete stress-deformation curves for concrete in tension. *ACI Materials Journal*, **84**, 4, July–Aug, 278–85.

34 Hatano, T. (1960) Dynamic behavior of concrete under impulsive tensile load. *Technical Report*, No. C-6002, Central Research Institute of Electric Power Industry, Tokyo, 1–15.

35 Helland, S. *et al.* (1983) Hoyfast betong. Presented at Norsk Betongdag, Trondheim, Oct. (In Norwegian).

36 (1977) *High-strength concrete in Chicago, high-rise buildings*. Task Force Report No. 5, Chicago Committee on High-Rise Buildings, Feb.

37   Houk, H. (1965) Concrete aggregates and concrete properties investigations. *Design Memorandum*, No. 16, Dworshak Dam and Reservoir, U.S. Army Engineer District, Walla, WA.

38   Hughes, B.P. and Gregory, R. (1972) Concrete subjected to high rates of loading and compression. *Magazine of concrete Research*, **24**, 78, London, March.

39   Jerome, M.R. (1984) Tensile strength of concrete. *ACI Journal*, **81**, 2, March–April, 158–65.

40   Johansen, R. (1979) Silicastov Iabrikksbetong. Langtidseffekter. *Report*, STF65 F79019, FCB/SINTEF, Norwegian Institute of Technology, Trondheim. (In Norwegian), and Johansen R, (1981) Report 6: Long-term effects. *Report*, STF65 A81031, FCB/SINTEF, Norwegian Institute of Technology, Trondheim, 1981.

41   Kaar, P.H., Hanson, N.W. and Capell, H.T. (1977) Stress-strain characteristics of high strength concrete. *Research and Development Bulletin*, RD051-01D, Portland Cement Association, Skokie, Illinois, 11 pp. also *Douglas McHenry International Symposium on Concrete and Concrete Structures*, ACI special publication, SP-55, Detroit 1978, 161–85.

42   Klieger, P. (1958) Effect of mixing and curing temperatures on concrete strength. *ACI Journal*, **54**, 12, June, 1063–81.

43   Klink, S.A. (1985) Actual elastic modulus of concrete. *ACI Journal*, Sept–Oct, 630–3.

44   Kotsovos, M.D. (1983) Effect of testing techniques on the post-ultimate behaviour of concrete in compression. *Materiaux et Constructions* (RILEM, Paris), Jan–Feb, **16**, 91, 3–12.

45   Leming, M.L. (1988) Properties of high strength concrete: an investigation of high strength concrete characteristics using materials in North Carolina. Research Report FHWA/NC/88-006, Department of Civil Engineering, North Carolina State University, Raleigh, N.C., July.

46   Liniers, A.D. (1987) Microcracking of concrete under compression and its influence on tensile strength. *Materiaux et Constructions* (RILEM, Paris), **20**, 116, Mar, 111–16.

47   Loland, K.E. and Gjørv, O.E. (1981) Silikabetong. *Nordisk Betong*, 6, 1–6 (In Norwegian).

48   Maage, M. and Hammer, T.A. (1985) Modifisert Portlandsement. Detrapport 3. Fasthetsutvikling Og E-Modul. *Report*, STF65 A85041, FCB/SINTEF, Norwegian Institute of Technology, Trondheim (In Norwegian).

49   Mainstone, R.J. (1975) Properties of materials at high rates of straining or loading. *Materieaux et Constructions* (Rilem, Paris), **8**, 44, March–April.

50   Malhotra, H.L. (1956) The effect of temperature on the compressive strength of concrete. *Magazine of Concrete Research*, V. 8, No. 23, Aug, 85–94.

51   Malhotra, V.M. (ed.) (1983) *Fly ash, silica fume, slag and other mineral by-products in concrete*, SP 79, Vol. II, American Concrete Institute.

52   Malhotra, V.M. (ed.) (1986) *Fly ash, silica fume, slag and natural pozzolans in concrete*, SP 91, Vol. II, American Concrete Institute.

53   Metha, P.K. (1986) *Concrete structures, properties and materials*. Prentice Hall, Inc, Englewood Cliffs, New Jersey.

54   Moreno, J. (1990) The state of the art of high-strength concrete in Chicago: 225 W. Wacker Drive. *Concrete International*, Jan, 35–9.

55   Nawy, E.G. *Reinforced concrete: a fundamental approach*, Second edition. Prentice Hall, New Jersey.

56   Neville, A.M. (1981) *Properties of concrete*, Third edition. Pitman Publishing Ltd., London.

57   Ngab, A.S., Nilson, A.H. and Slate, F.O. (1981) Shrinkage and creep of high strength concrete. *ACI Journal*, **78**, 4, July–Aug, 255–61.

58 Nilson, A.H. and Slate, F.O. (1979) Structural design properties of very high strength concrete. *Second Progress Report*, NSF Grant ENG 7805124, School of Civil and Environmental Engineering, Cornell University, Ithaca, New York.

59 Oh, B.H. (1987) Behavior of concrete under dynamic tensile loads. *ACI Materials Journal*, Jan–Feb, 8–13.

60 Ojorv, O.E. (1988) High strength concrete. *Nordic Betong*, **32**, 1, Stockholm, 5–9.

61 Parrot, L.J. (1969) The properties of high-strength concrete. Technical Report No. 42.417, *Cement and Concrete Association*, Wexham Springs, 12 pp.

62 Pentalla, V. (1987) Mechanical properties of high strength concretes based on different binder compositions. *Proceedings Symposium on Utilization of High Strength Concrete*, Norway, June 15–18, 123–34.

63 Perenchio, W.F. (1973) An evaluation of some of the factors involved in producing very high-strength concrete. *Research and Development Bulletin*, No. RD014.01T, Portland Cement Association, Skokie.

64 Perenchio, W.F. and Klieger, P. (1978) Some physical properties of high strength concrete. *Research and Development Bulletin*, No. RD056.01T, Portland Cement Association, Skokie, IL.

65 Philleo, R.E. (1955) Comparison of results of three methods for determining Young's Modulus of elasticity of concrete. *ACI Journal*, **51**, Jan, 461–9.

66 Popovics, S. (1975) Verification of relationships between mechanical properties of concrete-like materials. *Materials and Structures* (RILEM, Paris), **8**, 45, May–June, 183–91.

67 Pramono, E. (1988) Numerical simulation of distributed and localized failure in concrete. *Structural Research Series*, No. 88-07, C.E.A.E. Department, University of Colorado, Boulder.

68 Ronne, M. (1987) Effect of condensed silica fume and fly ash on compressive strength development of concrete. *American Concrete Institute, SP-114-8*, 175–89.

69 Saito, M. (1987) Characteristics of microcracking in concrete under static and repeated tensile loading. *Cement and Concrete Research*, **17**, 2, March, 211–18.

70 Sargin, M. (1971) *Stress-strain curves relationships for concrete and analysis of structural concrete sections*. Study No. 4, Solid Mechanics Division, University of Waterloo, Ontario, Canada.

71 Saucier, K.L. (1984) High strength concrete for peacekeeper facilities. *Final Report*, Structures Laboratory, U.S. Army Engineer Waterways Experimental Station, March.

72 Saucier, K.L., Tynes, W.O. and Smith, E.F. (1965) High compressive strength concrete-report 3, Summary Report. *Miscellaneous Paper No. 6-520, U.S. Army Engineer Waterways Experiment Station*, Vicksberg, Sept.

73 Sellevold, E.J. and Radjy, F.F. (1983) *Condensed silica fume (microsilica) in concrete: water demand and strength development*. Publication SP-79, American Concrete Institute, Vol. II, 677–94.

74 Shah, S.P., Gokos, U.N. and Ansari, F. (1981) An experimental technique for obtaining complete stress-strain curves for high strength concrete. *Cement, Concrete and Aggregates*, CCAGDP, **3**, Summer.

75 Shah, S.P., Fafitis, A. and Arnold, R. (1983) Cyclic loading of spirally reinforced concrete. *Journal of Structural Engineering*, ASCE, **109**, ST7, July, 1695–710.

76 Shah, S.P. and Sankan, R. (1987) Internal cracking and strain suffering response of concrete under uniaxial compression. *ACI Materials Journal*, **84**, May–June, 200–12.

77 Sharma, M.R. and Gupta, B.L. Sonic modulus as related to strength and static

modulus of high strength concrete. *Indian Concrete Journal*, **34**, 4, 139–41.
78   Skalny, J. and Roberts, L.R. (1987) High-strength concrete. *Ann. Rev. Mater. Sci.*, **17**, 35–56.
79   Smith, S.S., William, K.J., Gerstle, K.H. and Sture, S. (1989) Concrete over the top, or: is there life after peak? *ACI Materials Journal*, **86**, 5, Sept–Oct, 491–7.
80   Suaris, W. and Shah, S.P. (1983) Properties of concrete subjected to impact. *Journal of Structural Engineering*, ASCE, **109**, 7, July, 1727–41.
81   Walker, S. and Bloem, D.L. (1960) Effects of aggregate size on properties of concrete. *ACI Journal*, **32**, 3, September, 283–98.
82   Wang, P.T., Shah, S.P. and Naaman, A.E. (1978) Stress-strain of normal and lightweight concrete in compression. *ACI Journal*, **75**, 11, November, 603–11.
83   Watstein, D. Effect of straining rate on the compressive strength and elastic properties of concrete. *ACI Journal*, **49**, 8, April, 729–44.
84   Wischers, G. (1979) Application and effects on compressive loads on concrete. Betontechnische Berichte, 1978, Betonverlag Gmbh, Dusseldorf, 31–56.
85   Yogenendran, V., Langan, B.W. and Ward, M.A. (1987) Utilization of silica fume in high strength concrete. *ACI Materials Journal*, **84**, 2, March–April, 85–97, also *Proceedings Symposium on Utilization of High Strength Concrete*, Norway, June 15–18, 85–97.
86   Zielinski, Z.A. and Spiropoulos, I. (1983) An experimental study on the uniaxial and biaxial flexural tensile strength of concrete. *Canadian Journal of Civil Engineering*, **10**, 104–15.
87   Zielinski, A.J., Reinhardt, H.W. and Kormeling, H.A. (1981) Experiments on concrete under uniaxial impact tensile loading. *Materiaux et Constructions* (RILEM, Paris), **14**, 80, Mar–Apr, 103–12.
88   Zielinski, A.J., Reinhardt, H.W. and Kormeling, H.A. (1981) Experiments on concrete under repeated uniaxial impact tensile loading. *Materiaux et Constructions* (RILEM, Paris), **14**, 81, May–June, 163–9.

3 Shrinkage creep and thermal properties

F de Larrard, P Acker and R Le Roy

3.1 Introduction

Shrinkage and creep are time-dependent deformations that, along with cracking, provide the greatest concerns for the designers because of uncertainty associated with their prediction. Concrete exhibits elastic deformations only under loads of short duration and, due to additional deformation with time, the effective behavior is that of an inelastic and time-dependent material. A quantitative knowledge of mechanical behavior, including delayed deformations and thermal effects, is necessary for a number of structures: bridges, buildings etc. In other cases, control of short- and long-term cracking requires an accurate modelling of strains and stresses at all ages of the structure.

High performance concretes (HPC) are concretes with attributes (including creep, shrinkage and thermal effects) that are superior to conventional concretes. It is generally recognized that the higher the compressive strength of concrete, the better the other attributes. For the purpose of this chapter, high strength concrete (HSC) is defined as concrete with 28 day compressive strength in excess of 6000 psi (42 MPa). With the increasing use of concretes of higher strengths, the knowledge of time dependent behavior and the thermal properties is becoming important to ascertain the long term behavior of structures utilizing high strength concrete.

In this chapter, the mechanisms of shrinkage and creep are presented, along with experimental data which enunciate the effects of different parameters on the time dependent properties of concrete. The thermal properties are summarized. Also some field case histories are presented in which concretes to meet the higher performance requirements with special attention to shrinkage, creep and strength characteristics were designed and used. In these field cases, thermal effects and long-term strains have been monitored, and compared with predictions based on laboratory data.

3.2   Shrinkage

Shrinkage is the decrease of concrete volume with time. This decrease is due to changes in the moisture content of the concrete and physio chemical changes, which occur without stress attributable to actions external to the concrete. Swelling is the increase of concrete volume with time. Shrinkage and swelling are usually expressed as a dimensionless strain (in./in. or mm/mm) under given conditions of relative humidity and temperature. Shrinkage is primarily a function of the paste, but is significantly influenced by the stiffness of the coarse aggregate. The interdependence of many factors creates difficulty in isolating causes and effectively predicting shrinkage without extensive testing. The principal variables that affect shrinkage are summarized in Table 3.1.[1] The key factors affecting the magnitude of shrinkage are:

Aggregate

The aggregate acts to restrain the shrinkage of cement paste; hence concrete with higher aggregate content exhibits smaller shrinkage. In addition, concrete with aggregates of higher modulus of elasticity or of rougher surfaces is more resistant to the shrinkage process.

Water-cementitious material ratio

The higher the w/c ratio is, the higher the shrinkage. This occurs due to two interrelated effects. As w/c increases, paste strength and stiffness decrease; and as water content increases, shrinkage potential increases.

Member size

Both the rate and the total magnitude of shrinkage decrease with an increase in the volume of the concrete member. However, the duration of shrinkage is longer for large members since more time is needed for shrinkage effects to reach the interior regions.

Medium ambient conditions

The relative humidity greatly affects the magnitude of shrinkage; the rate of shrinkage is lower at higher values of relative humidity. Shrinkage becomes stabilized at low temperatures.

Table 3.1 Factors affecting concrete creep and shrinkage and variables considered in the recommended prediction method[1]

		Factors	Variables Considered	Standard Conditions
Concrete (Creep & Shrinkage)	Concrete Composition	Cement Paste Content	Type of cement	Type I and III
		Water-Cement Ratio	Slump	2.7 in. (70 mm)
		Mix Proportions	Air Content	≤ 6 percent
		Aggregate Characteristics	Fine Aggregate Percentage	50 percent
		Degree of Compaction	Cement Content	470 to 752 lb/cu.yd (279 to 446 kg/m³)
	Initial Curing	Length of Initial Curing	Moist Cured	7 days
			Steam Cured	1–3 days
		Curing Temperature	Moist Cured	73.4 ± 4°F (23 ± 2°C)
			Steam Cured	≤ 212°F, (≤ 100°C)
		Curing Humidity	Relative Humidity	≥ 95 percent
Member Geometry & Environment (Creep & Shrinkage)	Environment	Concrete Temperature	Concrete Temperature	73.4 ÷ 4 °F, (23 ÷ 2 °C)
		Concrete Water Content	Ambient Relative Humidity	40%
	Geometry Size and Shape	Volume-Surface Ratio, (V/s) or Minimum Thickness		v/s = 1.5 in (v/s = 38 mm) 6in, (150 mm)
Loading (Only Creep)	Loading History	Concrete Age at Load	Moist Cured	7 days
		Application	Steam Cured	1–3 days
		Duration of Loading Period	Sustained Load	Sustained Load
		Duration of Unloading Period Number of Load Cycles	—	—
	Stress Conditions	Type of Stress and Distribution Across the Section	Compressive Stress	Axial Compression
		Stress/Strength Ratio	Stress/Strength Ratio	≥ 0.50

Admixtures

Admixture effect varies from admixture to admixture. Any material which substantially changes the pore structure of the paste will affect the shrinkage characteristics of the concrete. In general, as pore refinement is enhanced shrinkage is increased.

Pozzolans typically increase the drying shrinkage, due to several factors. With adequate curing, pozzolans generally increase pore refinement. Use of a pozzolan results in an increase in the relative paste volume due to two mechanisms; pozzolans have a lower specific gravity than portland cement and, in practice, more slowly reacting pozzolans (such as Class F fly ash) are frequently added at better than one-to-one replacement factor, in order to attain specified strength at 28 days. Additionally, since pozzolans such as fly ash and slag do not contribute significantly to early strength, pastes containing pozzolans generally have a lower stiffness at earlier ages as well, making them more susceptible to increased shrinkage under standard testing conditions. Silica fume will contribute to strength at an earlier age than other pozzolans but may still increase shrinkage due to pore refinement.

Chemical admixtures will tend to increase shrinkage unless they are used in such a way as to reduce the evaporate water content of the mix, in which case the shrinkage will be reduced. Calcium chloride, used to accelerate the hardening and setting of concrete, increases the shrinkage. Air-entraining agents, however, seem to have little effect.

Cement type

The effects of cement type are generally negligible except as rate-of-strength-gain changes. Even here the interdependence of several factors make it difficult to isolate causes. Rapid hardening cement gains strength more rapidly than ordinary cement but shrinks somewhat more than other types, primarily due to an increase in the water demand with increasing fineness. Shrinkage compensating cements can be used to minimize or eliminate shrinkage cracking if they are used with restraining reinforcement.

Carbonation

Carbonation shrinkage is caused by the reaction between carbon dioxide (CO_2) present in the atmosphere and calcium hydroxide ($CaOH_2$) present in the cement paste. The amount of combined shrinkage varies according to the sequence of occurrence of carbonation and drying process. If both phenomena take place simultaneously, less shrinkage develops. The process of carbonation, however, is dramatically reduced at relative humidities below 50%.

The effect of the aggregate content and the w/c ratio on the shrinkage

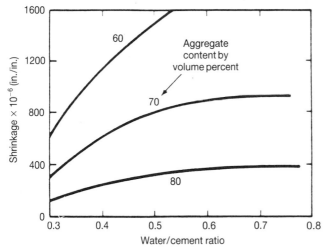

Fig. 3.1   Effect of w/c materials ratio and aggregate cement on shrinkage[23]

deformations is shown in Fig. 3.1. The figure reinforces the generally recognized fact that shrinkage deformations decrease with a higher aggregate content and a lower w/c ratio.

The shrinkage properties of concretes with higher compressive strengths are summarized in an ACI State-of-the-Art Report.[2] The basic conclusions were: (1) Shrinkage is unaffected by the w/c ratio[3] but is approximately proportional to the percentage of water by volume in concrete, (2) Laboratory[4] and field studies[5,6] have shown that shrinkage of higher strength concrete is similar to that of lower-strength concrete, (3) Shrinkage of high strength concrete containing high range water reducers is less than for lower strength concrete, (4) Higher strength concrete exhibits relatively higher initial rate of shrinkage,[7,8] but after drying for 180 days, there is little difference between the shrinkage of higher strength concrete and lower strength concrete made with dolomite or limestone. Shrinkage of high performance concrete may be expected to differ from conventional concrete in three broad areas: plastic shrinkage, autogenous shrinkage, and drying shrinkage.

Plastic shrinkage occurs during the first few days after fresh concrete is placed. During this period, moisture may evaporate faster from the concrete surface than it is replaced by bleed water from lower layers of the concrete mass. Paste-rich mixes, such as high performance concretes, will be more susceptible to plastic shrinkage than conventional concretes. Drying shrinkage occurs after the concrete has already attained its final set and a good portion of the chemical hydration process in the cement gel has been accomplished. Drying shrinkage of high strength concretes, although perhaps potentially larger due to higher paste volumes, do not, in fact, appear to be appreciably larger than conventional concretes. This is probably due to the increase in stiffness of the stronger mixes. Data for early strength high performance concretes is limited. Autogenous shrinkage due to self-desiccation is perhaps more likely with very low w/c ratio concretes, although there is little data outside indirect evidence with certain high strength concrete research.[9]

Mechanisms of setting and hardening

The main physical mechanisms that occur during the setting and the hardening process of concrete include:

Sedimentation

Before setting, concrete constituents are in suspension and, in certain cases nonoptimized packing of granular skeleton. A vertical displacement of the constituents occurs by gravity:[10] downward for the larger grains, upward for the water which entraps the elements having a lower sedimentation velocity; a film of very clean water appears at the top surface (bleeding).

Le Chatelier's contraction

The volume of hyrdrates formed in the hydration reaction is substantially smaller than the sum of the volumes of the two components (anhydrous cement and water) entering into the reaction. The range is generally between 8 to 12%, depending on the properties of the cement. The potential lineic shrinkage of the cement paste is assumed to be about 3 to 4% (these values cannot be produced experimentally, due to the mechanical stiffness of the hardened paste itself).

Heat of hydration

The hydration reaction is highly exothermic (150 to 350 joules per gram of cement) which elevates the temperature under adiabatic conditions to values between 25 and 55 K in the concrete.

Self-desiccation

Only a fraction of the hydration is completed during the setting process. The hydration continues inside a rigid porous skeleton, resulting in a reduction in the water content in the pore space, a reduction that has the same mechanical effect as drying.[11]

Desiccation

This occurs in ordinary concrete, when about twice as much water as is strictly necessary for the hydration of the cement is used for reasons of workability, whereas this is not the case for high performance concretes

(HPC). After demolding, a drying process begins from the surfaces in contact with the ambient atmosphere.[12]

Setting and early hardening kinetics

At the start of setting process, there are isolated solid grains in a connected liquid phase. Hydration starts from the surface of the cement grains; they are covered with a layer of hydrates, a crust that grows thicker and increasingly retards the very hydration reaction that causes it to form.

The formation of the hydrates around the grains then leads to the establishment of contacts, and the crystals coalesce; in less than one hour, the concrete changes from a suspension to a continuous solid.[13]

In ordinary or conventional concretes hydration never ends; the layer of hydrates that forms around the cement grains becomes thicker and thicker and more and more impermeable, slowing down the reaction, but never actually cuasing it to stop. Anhydrous cores which are residues of cement grains can be found in very old concrete, explaining that physical and chemical properties continue to evolve over a long time.[14] From the structural standpoint, this post-setting hydration process has two consequences. First an internal growth of the skeleton (hardening) and second a simultaneous reduction of the evaporate water content in the pore space (self-desiccation). This occurs due to the negative volumetric balance of the chemical reaction. From a mechanical point of view the reduction of the evaporated water has the same effect as the drying, which is essentially the shrinkage of the skeleton of the hydrating materials.

For low water/cement ratios, the values of this autogenous shrinkage are of the same order as those of a drying shrinkage resulting from a reduction in water content equal to that produced by the hydration itself. Therefore, the observation of a macroscopic shrinkage of the hardened cement paste indicates that the mineral matrix as a whole is under compression.[15] This compression can be attributed to the force exerted by the pore water, through the surface tensions developed at the liquid–vapor interface.

Just after setting, the porous network is completely filled with water, and the self-desiccation process begins. During a long time, *the liquid water is connected* (there exists a continuous path from each point to each other in this phase). This does not mean that the gaseous phase is not also connected. Several connected networks can coexist in three dimensions, and are therefore relatively free to move under a pressure gradient (Darcy's process). The mobility of the liquid phase however decreases as fast as the water content decreases.[16]

As self-desiccation and drying proceed, the mobility of the liquid phase decreases rapidly and the dominant mechanism of transfer is no longer in the liquid phase, but in the gaseous phase. Then the phenomenon becomes completely different. It is no longer the total pressure that sets the water molecules in motion, as in the liquid phase, but their concentration. This is because of the trapped air, and any disturbance of total pressure equilibrium in the gaseous phase sets the air in motion with no distinction among molecules. If water moves in the gaseous phase, it is because there is a disturbance of equilibrium of the concentrations (Fick's process). This results in much slower movements. The tendency toward the equilibrium of concentrations results only from the fact that the probability of the presence of a molecule in the space accessible to it tends toward a uniform function, and its kinetics mainly depend on thermal random agitation.

High strength concrete (HSC) in general appears to be more sensitive to early drying. During the recent construction of a HSC nuclear containment, cracks were seen to form before setting.[17] These cracks completely disappeared when water curing was applied, showing that here, unlike the previous case, desiccation alone was involved (in addition to the effect of curing, it should be pointed out that the cracks were not very deep). The effect was very similar to the desiccation cracking of a soil, with the presence of ultrafines. Before setting, this gives the concrete material a structure close to that of the soils sensitive to this type of cracking. It should be emphasized that, for each type of HSC, an effective curing is essential.

High strength concrete exhibits several specific aspects such as high sensitivity to early drying (and the absolute necessity of an efficient curing) and faster autogenous shrinkage.

Heat development

The setting of cement is a highly exothermic chemical process which generates about 150 to 350 Joules per gram of cement. This generates to a temperature rise of between 25 and 55 K in the concrete under adiabatic conditions. Setting may then occur (in the core of massive parts, for instance) at a higher temperature and the differential contractions that occur during the subsequent cooling may lead to very significant cracking.[18] Hydration heat in concrete is roughly proportional to the amount, fineness and chemical composition (C_3A content) of the cement used. It increases with the amount, the fineness of the cement and the C_3A content in the cement. For the same cement content, less water leads to a reduction in the degree of hydration and an increase in the tensile strength of the hardened concrete.

Heat of hydration is affected not only by the cement content, but also by the water/cementitious ratio $w/(c+s)$ and the silica fume content.[19] Smeplass and Maage investigated the heat of hydration for a number of high strength concrete mixes by the means of a so-called semi-adiabatic calorimeter test. The results of Smeplass and Maage[19] indicate that the heat of hydration can be affected within a relatively wide range by the utilization of traditional mix-design parameters (Figs. 3.2 and 3.3). The heat of hydration per cement unit decreases approximately by 9% when the $w/(c+s)$ ratio is reduced from 0.36 to 0.27. At a constant paste/aggregate ratio, the resulting temperature rise increases by 6% (3 °C).

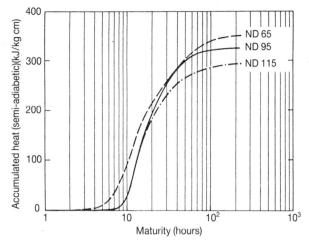

Fig. 3.2 Semi-isothermal heat development of three high strength concretes with different values of w/(c + s) ratio: 0.50 (ND65), 0.36 (ND95) and 0.27 (ND115)

Consequently, the effect of an increase in cement content normally is stronger than the effect of a reduced degree of hydration.

Smeplass and Maage[19] also found that at w/(c + s) ratio of 0.50, the replacement of cement by silica fume on a 1:1 basis induces a signifcant increase in the heat evolved per cement unit. The use of silica fume reduces the retarding effect of the dispersing agent, but also the magnitude of the secondary accelerating effect. The effect is observed for w/(c + s) ratios ranging from 0.50 to 0.27. At w/(c + s) ratios about 0.50 the effect of the silica fume on the total heat evolvement is just as strong as the effect of the cement. The effect weakens with decreasing w/(c + s) ratio. At w/(c + s) ratio 0.27, silica fume does not affect the heat evolvement significantly. Below w/(c + s) ratio 0.40, cement replacements by silica fume result in a reduced 'final' semi-adiabatic temperature.

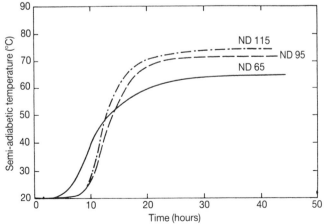

Fig. 3.3 Semi-adiabatic temperature development for three high strength concretes with different values of w/(c + s) ratio: 0.50 (ND65), 0.36 (ND95) and 0.27 (ND115)

From these results, it appears that HSC is not necessarily more sensitive to thermal cracking under all conditions. Furthermore, it should be noticed that thermal effects are not significant for concrete thickness less than 1.6 ft (40 cm). This is because of a drastic size effect due to the fact that the rate of heat diffusion increases like the square of the thickness (and this effect is accentuated by thermo-activation).[20] For specific industrial applications more sensitive to thermal cracking, like reservoirs, appropriate HSC mix-designs may be utilized with relatively low heat of hydration.[17]

Figure 3.4 shows that for a given temperature rise, the setting occurs sooner in high strength concretes than in normal strength concretes (NSC).

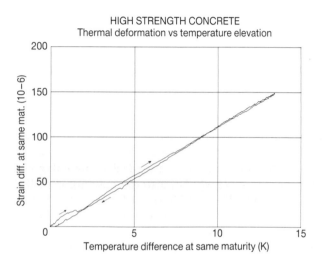

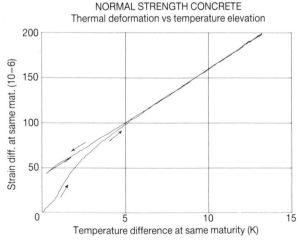

Fig. 3.4 Strain difference between two samples (at the same maturity but with different curing contractions) versus temperature difference, for a normal strength concrete (4a) and a high strength concrete (4b)

Figure 3.4a and 3.4b shows the thermal strain difference versus the temperature difference for normal as well as high strength concretes. The slope of curves shown in these figures is the coefficient of thermal expansion. The time over which the coefficient of thermal expansion becomes constant is essentially the time required to complete the transition from the suspension state to the solid state. In an experiment where a concrete specimen is hardening while being completely restrained, the final stress will be controlled by the difference between the temperature at the time of setting and the temperature at equilibrium with the ambient medium. Therefore, if the setting is lower, the cracking hazard during the setting and the early age is reduced.

Hardening kinetics

In order to obtain a parametric description of the mechanical behavior such as strength, modulus of elasticity, autogenous shrinkage and basic creep of high strength concrete, a total of ten mixes were examined by de Larrard and Le Roy.[21] The variables were the water/cement ratio (w/c = 0.28, 0.33, 0.38, 0.42), the paste volume ratio (V_p = 0.269, 0.286, 0.313, 0.325), and the silica fume ratio (s/c = 0.00, 0.05, 0.10, 0.15). The mix proportions and the results are given in Table 3.2.[21]

The hardening kinetics during early age and up to 28 days were investigated by studying the effect of aging (between 1 and 28 days) on the compressive strength and the modulus of elasticity for the ten HSC mixes and control reference mix [Figs. 3.5 and 3.6]. From these figures, it clearly appears that the hardening kinetics of high strength concretes significantly differs from that of ordinary concrete during the early age (up to 1 day), because of the reduction of water/cement ratio, and after one day, due to the hydration process of silica.

Autogenous shrinkage

The high autogenous (or self-desiccation) shrinkage of silica-fume HSC was first pointed out by Paillere, Buil and Serrano.[22] A very early shrinkage, even for non-drying specimens, was monitored in free specimens, leading to early cracking in restrained ones. More recently, de Larrard and Le Roy[21] have carried out measurements of autogenous shrinkage on ten HSC mixes (Table 3.2), and they proposed a relationship between mix-design and autogenous shrinkage, for HSC in the range of 60 to 100 MPa.[23] The final form of the equation can be written as

$$\varepsilon = \frac{(1 + E_d/E_g)(1 - g/g^*) + [4(1 - g^*)g/g^*]/[(1 - 2g^*) + g^* E_g/E_d]}{1 + g/g^* + (1 - g/g^*) E_d/E_g}$$

$$\frac{(1 - 0.65\exp[-11s/c])(1 - 1.43w/c) K_c}{E_c} \tag{3.1}$$

Table 3.2 Compositions and results of tests on 10 HSC mixes. Concrete C_0 is a control concrete, without admixture nor silica fume

Formulae lb/yd³ (kg/m³)	C_0	C_1	C_2	C_3	C_4	C_5	C_6	C_7	C_8	C_9	C_{10}
Coarse aggregate	2022 (1200)	2049 (1216)	2088 (1239)	1975 (1172)	1914 (1136)	2029 (1204)	2022 (1200)	2025 (1202)	2022 (1200)	2022 (1200)	2022 (1200)
Sand	1129 (670)	1127 (669)	1144 (679)	1083 (643)	1050 (623)	1112 (660)	1109 (658)	1110 (659)	1109 (658)	1109 (658)	1109 (658)
Cement	576 (342)	671 (398)	617 (366)	711 (422)	770 (457)	598 (355)	723 (429)	629 (373)	718 (426)	694 (412)	650 (386)
Silica Fume	0 (0)	67.0 (39,8)	61.7 (36,6)	71.1 (42,2)	77.0 (45,7)	59.8 (35,5)	72.3 (42,9)	62.9 (37,3)	0 (0)	34.7 (20,6)	97.6 (57,9)
Superplast.*	0 (0)	32.5 (19,3)	30.0 (17,8)	34.5 (20,5)	37.4 (22,2)	29.0 (17,2)	33.9 (20,1)	30.5 (18,1)	34.9 (20,7)	33.7 (20,0)	31.7 (18,8)
Added water	288 (171)	199 (118)	182 (108)	211 (125)	229 (136)	231 (137)	179 (106)	217 (129)	212 (126)	206 (122)	192 (114)
Density	2,36	2,43	2,45	2,42	2,42	2,41	2,43	2,41	2,43	2,43	2,43
Entrap. air %	1,9	0,6	1,2	0,7	0,4	0,7	0,9	0,5	1,2	0,8	0,6
Slump in (mm)	2.4 (60)	7.9 (200)	7.1 (180)	8.7 (220)	9.8 (250)	8.7 (220)		8.7 (220)	7.9 (200)		
Aggregate proportions	0.705	0.714	0.731	0.687	0.675	0.712	0.711	0.715	0.708	0.712	0.714
w/c**	0.50	0.33	0.33	0.33	0.33	0.42	0.28	0.38	0.33	0.33	0.33
s/c	0.0	0.1	0.1	0.1	0.1	0.1	0.1	0.1	0.0	0.05	0.15
Compressive strengths in ksi (MPa)											
fc_1	1.65 (11,4)	3.68 (25,4)	3.88 (26,8)	3.31 (22,8)	4.54 (31,3)	2.99 (20,6)	4.96 (34,2)	2.07 (14,3)	3.68 (25,4)	3.70 (25,5)	4.18 (28,8)
fc_3	3.68 (25,4)	7.47 (51,5)	7.21 (49,7)	6.97 (48,1)	7.12 (49,1)	5.16 (35,6)	7.79 (53,7)	5.48 (37,8)	5.54 (38,2)	6.21 (42,8)	6.18 (42,6)
fc_7	4.6 (32,0)	10.3 (70,7)	10.0 (69,1)	10.1 (69,5)	10.2 (70,3)	8.21 (56,6)	10.9 (75,6)	8.38 (57,8)	8.29 (57,2)	9.35 (64,5)	9.71 (67,0)
fc_{28}	6.31 (43,5)	13.4 (92,1)	13.7 (94,3)	13.5 (93,3)	14.4 (99,4)	10.8 (74,6)	14.1 (97,3)	11.5 (79,5)	9.74 (67,2)	10.8 (74,6)	13.7 (94,3)

Young's modulus in ksi $\times 10^3$ (GPa)

Ei_1	3.93 (27,1)	4.47 (30,8)	4.60 (31,7)	4.16 (28,7)	5.15 (35,5)	4.23 (29,2)	5.60 (38,6)	3.35 (23,1)	4.38 (30,2)	4.52 (31,2)	4.92 (33,9)
Ei_3	4.95 (34,1)	6.25 (43,1)	6.25 (43,2)	5.97 (41,2)	6.13 (42,3)	5.96 (41,1)	6.82 (47,0)	5.90 (40,7)	6.05 (41,7)	6.15 (42,4)	6.38 (44,0)
Ei_7	5.28 (36,4)	6.73 (46,4)	7.09 (48,9)	6.77 (46,7)	6.77 (46,7)	6.47 (44,6)	7.40 (51,0)	6.34 (43,7)	6.67 (46,0)	6.93 (47,8)	6.93 (47,8)
Ei_{28}	5.99 (41,3)	7.31 (50,4)	7.51 (51,8)	7.70 (53,1)	7.05 (48,6)	6.80 (46,9)	7.74 (53,4)	6.57 (45,3)	6.67 (46,0)	7.22 (49,8)	7.50 (51,7)
Autogenous shrinkage between 72 and 5000 hours (10^{-6})	41	89	76	111	108	67	82	91	44	64	94

* Melamine resin at 30.9% dry extract

** w/c = total water/cement alone

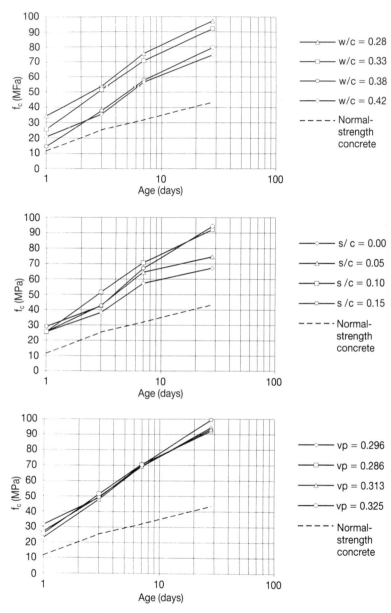

Fig. 3.5   Compressive strength development for the ten concrete mixes given in Table 3.2. Dash lines represent average values of the control mix, a 5000 psi (35 MPa) normal strength concrete made with the same components

where E_d = delayed modulus of the paste = $E_p/6$, where E_p is the paste
E_g = modulus
g = modulus of the aggregate
g^* = actual aggregate volume

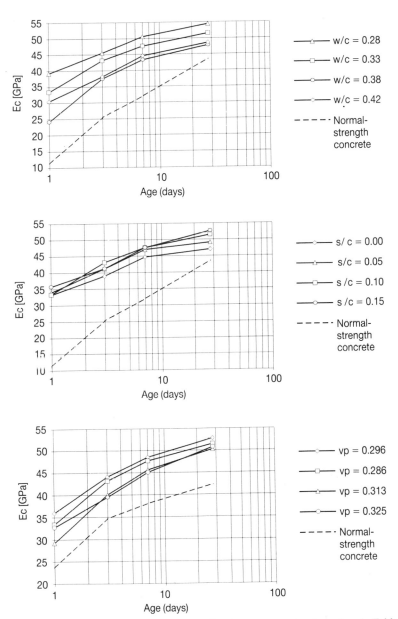

Fig. 3.6 Young's modulus development for the ten concrete mixes given in Table 3.2. Dash lines represent average values of the control mix, a 5000 psi (35 MPa) normal strength concrete made with the same components

s/c = maximum aggregate volume
w/c = silica-cement ratio
K_c = water-cement ratio
       Calibrating coefficient.

The value of K_c depends on the cement and the time over which the shrinkage is evaluated. The value of K_c recommended was 1.23 ksi (8.5 MPa) which was computed for the shrinkage occurring between 72 and 5000 hours for the test data of de Larrard and Le Roy.[21]

It can be concluded that a high autogenous shrinkage leading to early age cracking occurs in concretes with both low w/c and presence of silica fume. Therefore, in the cases where this cracking is likely and harmful, it is better not to use high content of using silica fume, or concretes with very low water/cement ratios.[17]

3.3   Creep

When a viscoelastic material is subjected to a stress, from a time t_0, its strain, measured parallel to the axis of the stress, changes over time. This time dependent increase in strain of hardened concrete subjected to sustained stress is termed creep. It is usually determined by subtracting, from the total measured strain in a loaded specimen, the sum of the initial instantaneous strain (usually considered elastic) due to sustained stress, the shrinkage, and any thermal strain in an identical load-free specimen, subjected to the same history of relative humidity and temperature conditions. The principal variables that affect creep are summarized in Table 3.1.

Creep is closely related to shrinkage and both phenomenon are related to the hydrated cement paste. As a rule, a concrete that is resistant to shrinkage also has a low creep potential. The principal parameter influencing creep is the load intensity as a function of time; however, creep is also influenced by the composition of the concrete, the environmental conditions and the size of the specimen.

A distinction must first be made between *basic* creep, which occurs in the absence of hygrometric exchanges with the ambient medium, and *drying* creep, that appears when the material dries. The former corresponds to the creep of a very thick actual structural element, whereas the latter, generally measured in the laboratory on small specimens – transverse dimensions less than 0.5 ft (15 cm) – reflects the maximum creep exhibited by thin structures. The basic creep and the total creep are the limits, between which lies the creep of a part having any shape, drying at between 50 and 100% relative humidity.

There is an abundant literature on creep of normal strength concretes and Neville[23] has given an excellent summary. Information on creep of concretes of higher strengths is limited.[2,24]

Relation between the basic creep and microstructure

Concrete consists of aggregate and cement paste, which itself consists of hydrates (CSH), anhydrous grains, free water and air bubbles. The basic

Table 3.3 Degrees of hydration and remaining anhydrous minerals in various cement pastes after 150 days (data taken from Sellevold and Justnes[25]). The compressive strengths have been evaluated thanks to Feret's modified formula[17,45,71,84] with Kg = 4.91 and R_c = 55 MPa

Pastes	W/C+S	W/C	S/C	W	C	SF	α_c	α_{sf}	Anhyd.	f_c ksi (MPa)
A0	0.2	0.200	0.00	0.387	0.613	0.000	0.46	–	0.331	14.8 (102)
A8	0.2	0.216	0.8	0.379	0.557	0.064	0.46	0.87	0.308	16.2 (112)
A16	0.2	0.232	0.16	0.373	0.510	0.117	0.42	0.80	0.319	16.2 (112)
B0	0.3	0.300	0.00	0.486	0.514	0.000	0.61	–	0.200	10.3 (71)
B8	0.3	0.324	0.8	0.478	0.468	0.054	0.57	0.95	0.204	11.7 (81)
B16	0.3	0.348	0.16	0.472	0.430	0.099	0.53	0.90	0.212	11.7 (81)
C0	0.4	0.400	0.00	0.558	0.442	0.000	0.82	–	0.080	7.7 (53)
C8	0.4	0.432	0.08	0.550	0.404	0.046	0.72	1.00	0.113	8.8 (61)
C16	0.4	0.464	0.16	0.544	0.372	0.085	0.70	1.00	0.111	8.8 (61)

Headers: Initial prop. (W, C, SF); D° of hydr. (α_c, α_{sf}, Anhyd.)

creep is intrinsic to the hydrates (the creep of the aggregates being zero or negligible with respect to that of the matrix; the same for the anhydrous cement). Its physical origin is poorly understood, and in any case no consensus has been reached concerning it. But it must be controlled to a great extent by the total volume of hydrates in the concrete microstructure. The amplitude of the basic creep is also directly influenced by the presence of free water in the microstructure of the material: a dry concrete exhibits no creep.[25]

Therefore, two features of HSC microstructure are expected to modify the creep behavior (compared with the one of NSC): the volume of hydrates and the free water content. In Table 3.3, one can see the volume of hydrates for cement pastes with water/binder ratio ranging between 0.2 to 0.4, and silica/cement ratio ranging from 0 to 16% (data taken from Sellevold and Justnes[26]). Obtaining a higher strength concrete entails imposing a low water/binder ratio; this gives a reduced volume of the hydrate and the free water content. Moreover, self-desiccation appears. These are factors which reduce creep.

Basic creep versus mix design and age at loading

In a recent study,[21] a series of ten concretes were tested in order to quantify the influence of mix-design on basic creep of HSC. The mix proportions are presented in Table 3.2. Data on the materials are given elsewhere.[21] For each concrete, 6.3 in. × 39.5 in. (160 × 1000 mm) cylindrical specimens were loaded at 1, 3, 7 and 28 days, and the creep strains were measured from the time of loading up to 18 months. Based on the results the following equation was proposed:

$$\varepsilon_{cr}(t, t_0) = \frac{K_{cr} \cdot \sigma \cdot [(t - t_0)^\alpha / \beta + (t - t_0)^\alpha]}{E_{i28}} \tag{3.2}$$

where t is the time (in hours), t_0 is the age of concrete at loading, σ is the applied stress, E_{i28} is the elastic modulus, ε_{cr} is the creep strain, K_{cr} is the creep coefficient, α and β are material parameters depending on t_0.

The results of the creep tests, with specific creep values at the end of the tests are given in Table 3.4. Creep curves for concrete C1 are shown in Fig. 3.7. For each mix, it can be noted that the amplitude of specific creep decreases with the age of the material. The ratio of specific creep of concrete loaded at 1 day to the one of concrete loaded at 28 days is approximately equal to 2. Another general feature is the lowering of the kinetics: the concrete creeps faster when it is loaded at early age. As a consequence, for some mixes, the specimens loaded at 28 days continued to creep quite quickly at the end of the tests, so that the extrapolation (K_{cr} value) is not very reliable. The results also indicate that creep is sensitive to the paste content of concrete (Fig. 3.8) and it decreases when the water/cement ratio decreases (Fig. 3.9). A low dosage of silica fume (5% by weight of cement) leads to a decrease of the specific creep, but higher dosage (up to 15%) increases the deformation (Fig. 3.10).

In summary, any change of the mix-design parameters involving an increase of strength also leads to a decrease of creep, except when the silica fume amount is more than 10%. On the other hand, the kinetics of deformations do not show any obvious tendency when the values of mix-design parameters are changed.

Creep under high stress

The creep of concrete exhibits a quasi-linear domain in which the delayed strain (not counting shrinkage) is proportional to the applied stress. The limit of this domain, of the order of 40% of the strength at failure for conventional concretes, seems to be a little higher for HSC,[27,28] as, moreover, is its instantaneous behavior. Above this threshold, creep increases faster than applied stress. Above 75%, a delayed failure may occur.[29]

Models for basic creep

A number of analytical models are available for estimating the creep behavior and are summarized in the ACI committee report.[1] In addition,

Table 3.4 Results of creep tests for high performance concretes (the mix-compositions for which appear in Table 3.2)

Concrete	Age of loading (days)	Béta (ħalpha)	Alpha	Kcr	Specific creep at 24 months [10–6/mPa]
C1	1	7.3	0.569	1.09	20.49
	3	37.9	0.698	0.49	9.14
	7	30.7	0.609	0.70	12.53
	28	26.2	0.474	0.64	9.96
C2	1	1.7	0.425	0.93	17.08
	3	51.1	0.822	0.60	11.26
	7	42.1	0.686	0.55	9.95
	28	21.7	0.345	0.85	9.21
C3	1	3.7	0.677	1.14	22.54
	3	24	0.662	1.16	22.15
	7	49.6	0.781	0.60	11.65
	28	18.2	0.385	0.65	9.09
C_4	1	5.1	0.539	1.18	22.56
	3	–	–	–	–
	7	16.4	0.567	0.81	14.92
	28	7.1	0.342	0.33	5.15
C5	1	6	0.625	1.45	29.41
	3	5.1	0.521	0.67	13.30
	7	10.3	0.474	0.93	17.36
	28	21.9	0.306	1.47	14.42
C6	1	5.2	0.579	1.28	23.17
	3	97.1	1.052	0.70	12.85
	7	10.1	0.494	0.73	12.42
	28	32.5	0.39	0.67	7.14
C7	1	15.7	0.788	0.88	18.24
	3	46	0.754	0.64	13.01
	7	44.8	0.837	0.56	11.54
	28	35.7	0.371	1.04	11.16
C8	1	17.1	0.846	1.04	22.02
	3	23.7	0.534	0.58	10.88
	7	18	0.459	0.76	13.44
	28	22.8	0.306	1.46	14.43
C9	1	2.6	0.596	1.06	21.38
	3	14.8	0.541	1.16	22.02
	7	20.5	0.585	0.64	12.21
	28	42	0.433	0.74	9.39
C10	1	3.9	0.616	1.09	20.51
	3	55.1	0.79	0.74	13.76
	7	10.5	0.453	0.91	15.30
	28	31.1	0.328	1.25	10.52

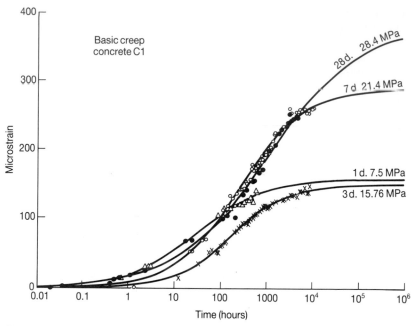

Fig. 3.7   Basic creep of concrete C1 (mix composition, see Table 3.2)

there are other available models.[29,30,31,32] The model of Bazant and Chern based on log double-power law appears to be the quite effective analytically.

At LCPC (Paris), Auperin, de Larrard *et al.*[33] have conducted in the past a preliminary program of experiments on a silica fume concrete having a strength of 11.6 ksi (80 MPa) at 28 days, with a view to characterizing its creep vs. the age of loading. Based on the trend of the experimental data. The following expression was proposed:

$$\varepsilon_{cr}(t, t_0) = K_{cr}[\varepsilon/E_i(28)]f(t - t_0) \tag{3.3}$$

where K_{cr} is the creep coefficient, E_i is the modulus and $f(t - t_0)$ a kinetic

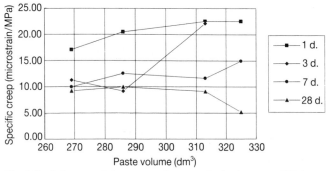

Fig. 3.8   Influence of the paste volume on the basic creep of high strength concrete (specific creep after 2 years)

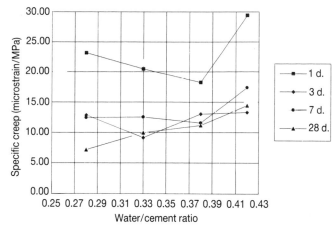

Fig. 3.9 Influence of the w/c ratio on the basic creep of high strength concrete (specific creep after two years)

function of time t, which is expressed in days. For the modulus, the creep coefficient and the kinetic function, the following expressions were proposed.

$$E_i(t_0) = 1.132 \exp[-0.42/(t_0 - 0.4)^{1/2}] E_{i28} \qquad (3.4)$$

$$K_c(t_0) = 0.363 \exp[9.3/(t_0 - 0.34)]^{1/2} \qquad (3.5)$$

$$f(t - t_0) = \frac{\exp[-1.7/(t - t_0 + 0.027)]^{1/3} - \exp(-4)}{1 - \exp(-4)} \qquad (3.6)$$

Investigation of different loading paths (for HSC concrete 760 lbs/cyd, 484.4 kg/m³, of cement, 8% silica fume and w/c of 0.38) showed that superposition was also valid for higher strength concretes (Fig. 3.11). Comparison of data of this high strength concrete with other data from the literature revealed that its creep kinetics was exceptionally fast. The creep amplitude for early loading (at one day) was also unusually high.

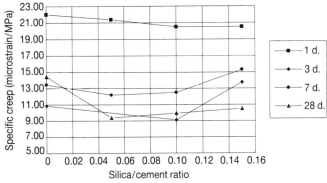

Fig. 3.10 Influence of the silica/cement on the basic creep of high strength concrete (specific creep after two years)

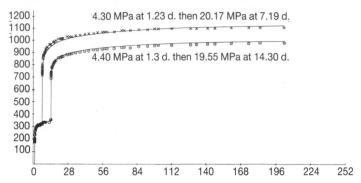

Fig. 3.11 Creep of high strength concrete subjected to two stress steps, and application of the superposition method. fC 28 = 12 ksi (83áMPa), cement: 760 lb/yd (450 kg/m), silica fume: 8%, w/c = 0.38; deformation in microstrains, time in days (linear scale)

A more comprehensive program has been recently carried out at LCPC.[21] As already pointed that the compressive strength appears to be a convenient parameter in order to evaluate the basic creep.[29] After a smoothing of the test results of 40 specimens,[21] the following equations were proposed in the LCPC model:

$$\varepsilon_{cr}(t, t_0) = (\sigma/E_{i28}) K_{cr}(t - t_0) \alpha/[(t - t)^\alpha + \beta] \tag{3.7}$$

$$K_{cr} = 1.2 - 3.93 \times 10^{-2} f_c(t_0) \tag{3.8}$$

$$\beta = 1.5 \exp 3 f_c(t_0)/f_{c28}] \tag{3.9}$$

where $\alpha = 0.5$ and $f_c(t_0)$ is the compressive strength (mean cylinder value in ksi) at the age at loading t_0. To utilize the model, one must evaluate the E-modulus at 28 days. A sophisticated formula, based upon the homogenization theory, has been recently proposed by the authors[21] and can be used if no experimental data is available. More simple models are available in the building codes. The basic creep of high strength concretes seems to strongly decrease for loading after 28 days, while the evolution of strength is poor. This trend is similar to normal strength concretes.

3.4 Drying shrinkage and drying creep

Drying process

There have so far been few investigations on the drying of high strength concretes (HSCs). As pointed out earlier, HSC is subjected to self-desiccation. Thus, the water consumed by hydration cannot go out of the material, and the water loss of hardened HSC is generally smaller than that of NSC, but not zero. For continuation of the drying process, external humidity must be lower than the internal one. In the case of very-high strength silica fume concrete, the internal humidity falls below 80%.[34] Therefore, such a concrete will not show any drying shrinkage when exposed to an external humidity above 80% relative humidity.

For most high strength concretes, the pore structure is very fine, due to the low porosity of the paste, and to the presence of cementitious admixtures. This leads to a very low gas-permeability.[35] Thus, the drying kinetics is expected to be very slow. In order to check these assumptions, the drying process of two concretes – a NSC and a silica fume very high strength concrete (VHSC) – has been monitored by means of gammadensimetry.[36] Two different curing regimes were employed on 6.3 in. (160 mm) diameter-cylinders: drying at 50% relative humidity and 68 °F (20 °C) after demolding (at 24 hours), and sealed during 28 days, then dried in the same conditions as the previous regime. The distributions of water losses are shown in Figs. 3.12 and 3.13.

With the latter curing regime, the hydration of the covercrete is more complete, so that the drying kinetics is slower than in the former case. After four years, only three centimeters of the very high strength concrete (VHSC) had begun to dry, while the NSC had reached its hygral equilibrium with the ambient air. For long-term extrapolation, say 50 or 100 years, it is not easy to evaluate the future water field in the VHSC. A conservative hypothesis is that the whole specimen will dry at the same level than the one of the covercrete (in this example, the uniform water loss would be about 2%, the value of 3.5% corresponding to a region modified by the wall effect). However, as the material continues to hydrate, and becomes tighter after 28 days, this hypothesis could be pessimistic.

Drying shrinkage

Very little information is available concerning drying shrinkage of HSC, as most tests in the literature are performed on drying specimens, without sealed companions (which should allow to separate the part due to self-desiccation). Moreover, it is difficult to propose mathematical models giving sound extrapolations, because of the great slowness of the drying process. At the moment, it is only possible to indicate some tendencies, related to strains measured during a limited time on small specimens.

Field measurements of surface shrinkage strains on a mock column, fabricated with high strength concrete, after two and four years and the comparison with measurements on specimens under laboratory conditions[9] showed that the surface shrinkage strains under field conditions are considerably lower than those measured under laboratory conditions(Fig. 3.14).

There is conflicting information on the drying shrinkage for concrete with high range water reducers.[37,38,39] Flowing concrete, for a given strength, is likely to require a slightly higher cement content and therefore will exhibit somewhat higher shrinkage.[40] Use of a HRWR to reduce water content can be expected to reduce shrinkage in most cases. Tests over a period of one year showed that the effect of naphthalene-based HRWR, with 840 to 1000 pcy (500 to 600 kg/m^3-cement) was to reduce shrinkage.[37]

However, there was an increase in swelling after a year of storage in water. The swelling of concrete with superplasticizer was approximately 50% greater than that of control concrete. Since swelling increases with an increase in cement content, it can be postulated that the higher swelling of the concrete with superplasticizer is due to a higher hydrated-cement paste content because of a rapid early-age development of strength. An alternative explanation is that the admixture modifies the paste structure so that its swelling capacity is increased.

The drying shrinkage in high strength concrete with silica fume is either equal to or somewhat less than that of concrete without silica fume but containing fly ash. This was concluded by Luther and Hansen,[41] based on their results on five high strength concrete mixtures which were monitored for 400 days. The use of high fly ash content (50% cement replacement by weight with low calcium, Class F fly ash) for 5800 psi and 8700 psi (40 MPa and 60 MPa) concretes,[42] resulted in ultimate shrinkage values from 400 microstrains to 500 microstrains, with swelling amounting to 40% to 55% of the shrinkage value. The study also indicated the importance of continued water curing for full pozzolanic reaction of fly ash. The data[42] shows that shrinkage of concrete properly proportioned with high fly ash content compares favorably with that of portland cement concrete.

The influence of drying and of sustained compressive stresses, at and in excess of normal working stress levels, on shrinkage properties of high, medium and low strength concretes were investigated by Smadi, Slate and Nilson.[27] The 28 day compressive strength ranged from 3000 psi to 10,000 psi (21 to 69 MPa). The long-term shrinkage was found to be greater for low strength concrete and smaller for medium and high strength concretes. The study also indicated that the effect of aggregate content on the shrinkage of low strength and medium strength concrete is less significant than the effect of w/c ratio.

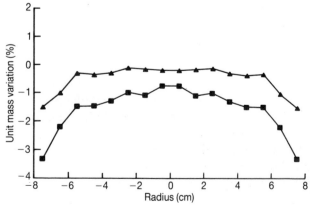

Water content distribution in control concrete
demoulded at 1 day. ▲ 28 days, ■ 80 days

(a)

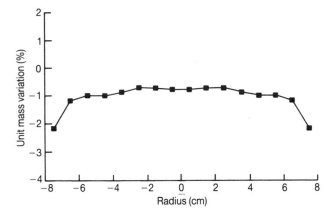

Water content distribution in control concrete
at 90 days (demoulded at 28 days)

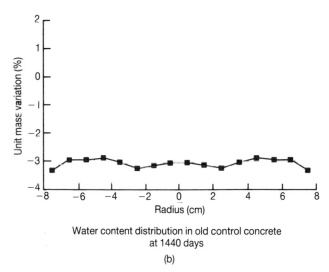

Water content distribution in old control concrete
at 1440 days

(b)

Fig. 3.12 Water content distribution of NSC with two curing regimes. (a) Drying from 1 day at 20 ÁC and 50% RH; (b) sealed during 28 days then drying

In a recent study,[43] it was shown that for five mixes with 28-day design strengths ranging from 8700 psi to 9300 psi (60 to 64 MPa), the shrinkage deformation was inversely proportional to the moist-curing time (the longer the curing time, the lower the shrinkage). It was also concluded that shrinkage was somewhat less for concrete mixtures with lower cement paste and larger (1.5 in. or 38 mm) aggregate size. In addition, the use of a high range water reducing admixture did not have a significant effect on the shrinkage deformation.

Observed shrinkage deformations of higher strength concrete (12,000 psi to 19,700 psi or 83 MPa to 136 MPa) were compared by Pentalla[44] to the predictions based on CEB Code recommendation.[30] It was concluded that shrinkage deformation of higher strength concrete took place much faster than predicted by the CEB formula.

The drying shrinkage characteristics for seven HSC mixes were recently investigated.[45] The data of seven HSC mixes are presented in Table 3.5. Mix No. 1 is a control NSC. Mix No. 2 is a HSC without silica fume (same composition as the one used for Joigny bridge). Mixes No. 3 to 5 are silica fume HSCs with constant binder dosages, and variable water/content ratios. Mix No. 6 is a VHSC, and mix No. 7 is a particular silica fume HSC, where a part of the cement has been replaced by a limestone filler (in order to minimize the thermal cracking). Shrinkage of mixes No. 1 and 6 are plotted in Fig. 3.15. The drying shrinkage obtained for mix No. 2 is quite low, but was still continuing at the end of the test. With mixes No. 3 to 5, the balance between the two kinds of shrinkage is emphasized. When the water/cement ratio increases, there is less autogenous shrinkage and more drying shrinkage, the sum remaining roughly constant. The optimal mix-design of mix No. 6 leads to a moderate autogenous shrinkage (in spite of the very low water/cement ratio), and a very low drying shrinkage. As for the last mix (No. 7), the negligible autogenous shrinkage entails a quite large drying one (the highest of the seven concretes). However, gammadensimetry measurements have shown, in this particular case, that the drying process was very rapid, with practically no hygral gradients.[36] Therefore, this high value of drying shrinkage is propbably near the asymptotic one, unlike the other mixes.

Alfes[47] recently proposed a quantitative model for total shrinkage of HSC, taking the aggregate restraining effect into account. This model is similar to the model of de Larrard and Le Roy,[21] except that the latter predicts the autogenous shrinkage from the whole mix composition, with

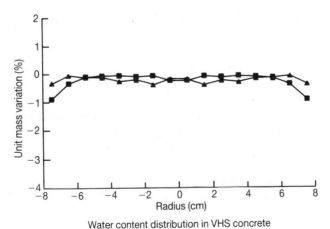

Water content distribution in VHS concrete
demoulded at 1 day. ▲28 days, ■90 days

(a)

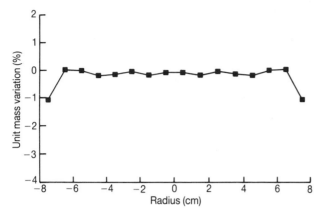

Water content distribution in VHS concrete

at 90 days (demoulded at 28 days)

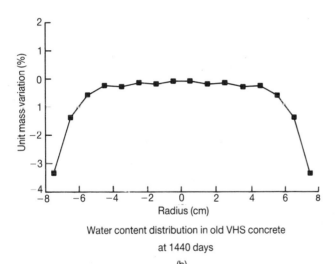

Water content distribution in old VHS concrete

at 1440 days

(b)

Fig. 3.13   Water content distribution of VHSC with two curing regimes. (a) Drying from 1 day at 20 ÁC and 50% RH; (b) sealed during 28 days then drying

only one parameter to fit, while Alfes' model bases the computations on the paste shrinkage. The equation for Alfes' model is:

$$\varepsilon_s = \varepsilon^{sm} . V_m{}^n \tag{3.10}$$

$$n = \frac{1}{[0.23 + 1.1/[1 + V_m(E_g/E_m - +)]} \tag{3.11}$$

where ε_s is the concrete shrinkage, ε_{sm} the paste shrinkage, V_m is the paste volume, E_g and E_m are the moduli of aggregate and paste respectively.

In recent papers[48,49] a large amount of data on carefully controlled

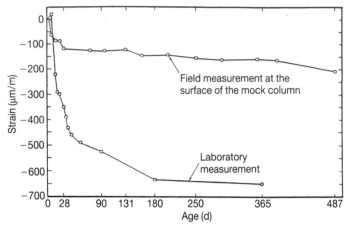

Fig. 3.14   Comparison of field and laboratory drying shrinkage[9]

shrinkage tests of concrete involving a large number of identical specimens were used to compare the predictive equations of ACI, CEB-FIP, and Bazant and Panula.[31,50–52] Bazant and Panula claimed that their equation (which has a large number of empirical constants) gave the best agreement with the experimental data. The validity of the Bazant and Panula equation for high strength concrete was explored by Bazant *et al.*,[52] and it was reported that the equation could be made applicable to higher strength concrete with minor adjustments. Similar conclusions by Almudaiheem and Hansen[53] also showed good correlations with the experimental data.

Drying creep

There is no consensus on the physico-chemical origin of the phenomenon of drying creep. According to Bazant and Chern,[32] the increase in the delayed *intrinsic* strain of concrete caused by drying is greater under load than in the absence of loading. A reduction of drying creep should, therefore, be expected. According to other researchers,[15,20,54,55] drying creep is related primarily to a *structural effect*: differential drying induces a state of skin cracking that relaxes part of the self-stresses and so decreases the deformation caused by self-desiccation) (i.e. drying shrinkage).

The composition of concrete can essentially be defined by the w/c ratio, aggregate and cement types and quantities. Therefore, as with shrinkage, an increase in w/c ratio and in cement content generally results in an increase in creep. Also, as with shrinkage, the aggregate induces a restraining effect so that an increase in aggregate content reduces creep. Numerous tests have indicated that creep deformations are proportional to the applied stress at low stress levels. The valid upper limit of the relationship can vary between 0.2 and 0.5 of the compressive strength. This range of the proportionality is expected due to the large extent of microcracks in concrete at about 40 to 45% of the strength.

Table 3.5   Shrinkage of various HSC mixes[45]

Mixes no.	1 NSC	2 HSC	3 HSC	4 HSC	5 HSC	6 VHSC	7 HSC
Cement type	BPC*	OPC	OPC	OPC	OPC	OPC	BPC*
Strength in ksi (MPa)	7.97 (55)	9.43 (65)	9.43 (65)	9.43 (65)	9.43 (65)	7.97 (55)	7.97 (55)
Dosage lb/yd³ (kg/m³)	590 (350)	758 (450)	768 (456)	763 (453)	763 (453)	709 (421)	448 (266)
Limestone filler lb/yd³ (kg/m³)	– –	– –	– –	– –	– –	– –	111 (66)
Silica fume lb/yd³ (kg/m³)	– –	– –	60.7 (36)	60.7 (36)	60.7 (36)	70.8 (42)	67.4 (40)
Superplasticizer** lb/yd³ (kg/m³)	– –	7.58 (4,5)	11.8 (7,0)	11.1 (6,6)	6.06 (3,6)	13.3 (7,9)	6.07 (3,6)
Retarder** lb/yd³ (kg/m³)	– –	1.52 (0,9)	0.84 (0,5)	0.84 (0,5)	0.84 (0,5)	– –	– –
Water lb/yd³ (kg/m³)	329 (195)	283 (168)	254 (151)	295 (175)	316 (188)	189 (112)	280 (166)
W/C	0,56	0,37	0,33	0,39	0,42	0,27	0,62
Slump, in (mm)	2.7 (70)	>7.9 (>200)	>7.1 (>180)	>7.1 (>180)	>7.1 (>180)	>7.9 (>200)	>7.1 (180)
Mean cylinder strength at 28 days in ksi (MPa)	5.8 (40)	11.3 (78)	13.6 (94)	12.0 (83)	10.7 (74)	14.6 (101)	9.7 (67)
Autogenous shrinkage at 1 year (10^{-6})	90	90	290	200	140	150	30
Drying shrinkage at 1 year (10^{-6})	290	90	120	190	260	110	310
Total shrinkage for drying specimens (10^{-6})	380	180	410	390	400	260	340

*Blended Portland cement containing 9% of limestone
**Equivalent dry extract

Very few data are reported on creep of concrete containing condensed silica fume. In one study,[56] creep tests were performed on a concrete in which 25% of the cement was replaced by silica fume and a naphthalene-based high-range water reducer was added. The results showed that total deformation was reduced under drying conditions, with no significant reduction in basic creep.

Results of studies on creep of concrete containing fly ash[57-59] indicate that creep of sealed specimens can be reduced in the same proportions as the ratio of replacement of portland cement by fly ash, ranging between 0 and 30%, if water content is reduced substantially. However, fly ash mixes

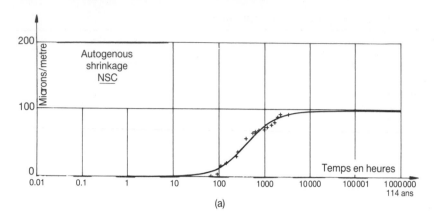

(a)

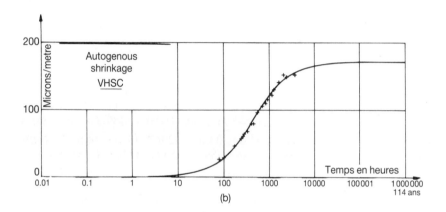

(b)

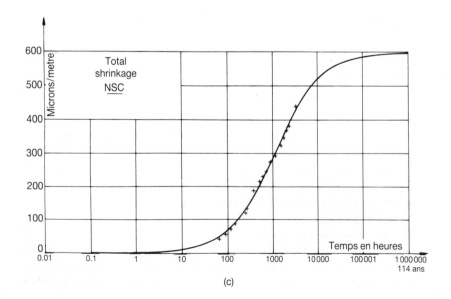

(c)

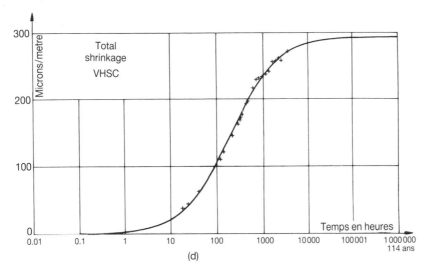

Fig. 3.15   (a) Autogenous shrinkage of NSC after demolding at 24 hours[46]; (b) autogenous shrinkage of VHSC after demolding at 24 hours; (c) total shrinkage of drying specimens of NSC at 50%, 68 ÁF (20 ÁC); (d) total shrinkage of drying specimens of VHSC at 50% RH, 68 ÁF (20 ÁC)

frequently show slightly higher creep under drying conditions than control mixes at the same 28-day strength, due to somewhat slower initial strength gain combined with early drying associated with standard test procedures.

Information on creep of concrete containing a high-range water reducer[39,60] is restricted in scope. The creep of concrete with melamine based high-range water reducer was reported to be 10% higher than control concrete.[61] Tests over a period of one year[37] showed that mixes with 840 lbs/cu yd (500 kg/m³) cement containing a naphthalene-based high-range water reducer exhibited creep characteristics similar to control concrete; mixes with 1000 lbs/cu yd (600 kg/m³) cement and the same high-range water reducer exhibited greater creep than control mixes. Flowing concrete for a given strength is likely to require a slightly higher cement content and it exhibits lower creep, while its creep recovery and post creep recovery elastic deformation are generally comparable to control concrete.[40]

In a recent study by Luther and Hansen,[41] the specific creep of five high strength concrete mixtures ($7350 < f_c' < 69$ MPa) was monitored for 400 days. It was concluded that creep of the silica fume (SF) concrete was not significantly different from that of fly ash concrete. The relationship between the specific creep and compressive strength is shown in Fig. 3.16, which also includes other data.[4,62–65] The data in the figure show a hyperbolic relationship between specific creep and compressive strength. Also, these data[41] nestle between the Neville data (dashed curve on the left) and the Perenchio and Klieger data (dashed curve on the right), and they agree with the applicable Ngab, Nilson and Slate results. Furth-

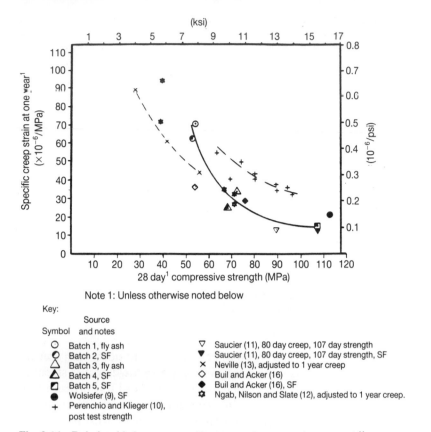

Fig. 3.16   Relationship between specific creep and compressive strength[41]

ermore, these data near high-strength levels are very close to Wolsiefer's[62] and Saucier's[64] results. Thus all of the concretes show similar specific creep to compressive strength relationships. Therefore, there is no apparent difference between the specific creep of silica fume (SF) concrete, portland cement concrete, or fly ash concrete.

It has been reported by Pentalla[44,66] that the creep deformation of higher-strength concrete with strengths from 12,000 psi to 19,700 psi (83 to 136 MPa) takes place much faster than the prediction by the CEB Code recommendations.[30]

The influence of drying and of sustained compressive stresses, at and in excess of normal working stress levels, on creep properties of high, medium and low strength concretes was investigated by Smadi, Slate and Nilson.[27] The 28 day compressive strength ranged from 3000 psi to 10,000 psi (21 MPa to 69 MPa). Creep strains, creep coefficient and specific creep were all smaller for high strength concrete than for concretes of medium and low strengths at different stress levels, and at any time after loading. The creep-to-stress proportionality limit was higher for the high strength concrete than for the others by about 20%.[27]

The effect of mix proportions on creep characteristics was investigated in a study,[43] in which five mixes with 28-day strengths ranging from 8700 psi to 9300 psi (60 MPa to 64 MPa) were used. The results indicated that creep is somewhat less for concrete mixtures with lower cement paste and large aggregate size. It was also shown that the use of high-range water-reducing admixture did not show a significant effect on the creep deformations.

A very limited amount of work is reported on tensile creep tests.[67] Tensile creep tests at 35% of the ultimate short-term strength show that specific creep increases with an increasing w/c ratio and decreasing aggregate-cement ratio. These trends are similar to those of compressive creep and the levels of specific creep are similar. Sealed concrete creeps less than immersed concrete.[67] Also, tensile creep generally increases with concurrent shrinkage and swelling. These effects are similar to those that occur in compression.[67] In another study relations between the relative stress level and the time of failure were derived, from uniaxial tensile creep tests.[68] These relations were not found to be affected by temperature, cement type or concrete quality.

For high strength concrete with silica fume, drying creep is poor in cylindrical specimens with 4.33 in. (110 mm) diameter[29,44] and practically non-existent for 6.3 in. (160 mm) diameter cylindrical specimens.[33,69,70] In HSC without silica fume, the drying effect remains, but is smaller than in NSC. In the same way as for drying shrinkage, the data are too scarce and the phenomenon too slow for allowing the development of reliable and comprehensive mathematical models, which could be applicable for different mixture proportions, different sizes of specimens and different abient conditions.

Figure 3.17 shows the basic and the total creep of several high strength concrete mixes reported in the literature along with the prediction of the LCPC model.[21] As expected, it can be seen that the contribution to overall shrinkage due to drying decreases when the strength of concrete increases (the drying process becoming slower, as the concrete is tighter). It can be noted that the compressive strength is far from being the only variable controlling the total creep of HSC. It appears that information regarding this relationship between the mix-design and creep phenomenon for HSC is lacking and there is a need for further investigations.

3.5   Thermal properties

The thermal properties of concrete are of special concern in structures where thermal differentials may occur from environmental effects, including solar heating of pavements and bridge decks. The thermal properties of concrete are more complex than for most other materials, because not only is concrete a composite material the components of which have different thermal properties, but its properties also depend on moisture content and porosity. Data on thermal properties of high performance concrete are limited, although the thermal properties of high strength concrete fall

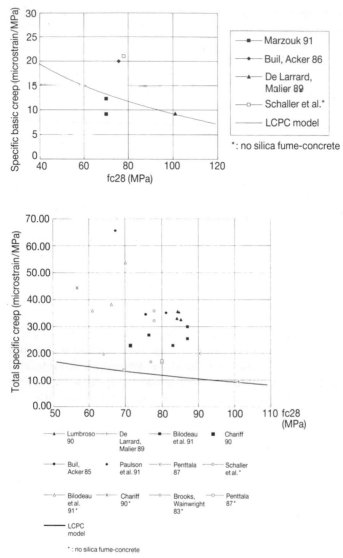

Fig. 3.17   Comparison of creep strains at 1 year reported in the literature with the prediction of LCPC model; (a) for specimens loaded near 28 days, and sealed; (b) specimens loaded at 28 days and kept at 50% RH

approximately within the same range as those of lower strength concrete,[7,71] for characteristics such as specific heat, diffusivity, thermal conductivity and coefficient of thermal expansion.

Three types of test are commonly used to study the effect of transient high temperature on the stress-strain properties of concrete under axial compression: (1) unstressed tests where specimens are heated under no initial stress and loaded to failure at the desired elevated temperature; (2)

stressed tests, where a fraction of the compressive strength capacity at room temperature is applied and sustained during heating and, when the target temperature is reached, the specimens are loaded to failure; and (3) residual unstressed tests, where the specimens are heated without any load, cooled down to room temperature, and then loaded to failure.

For normal strength concretes, exposed to temperatures above 450 °C, the residual unstressed strength has been observed to drop sharply due to loss of bond between the aggregate and cement paste. If concrete is stressed while being heated, the presence of compressive stresses retards the growth of cracks, resulting in a smaller loss of strength.[72]

The moisture content at the time of testing has a significant effect on the strength of concrete at elevated temperatures. Tests on sealed and unsealed specimens have shown that higher strength is obtained if the moisture is allowed to escape.[73,74]

For a given temperature, as the preload level is increased, the ultimate strength and the stiffness of normal strength concrete has been observed to decrease while the ultimate strain also decreases.[75]

In a recent study, Castillo and Durrani[76] tested concrete with strengths from 4500 psi to 12,900 psi (31 to 89 MPa) under temperatures ranging from 23 °C to 800 °C. The presence of loads in real structures were simulated by preloading the specimens before exposure to elevated temperatures. The strength of stressed and unstressed specimens of both normal and high strength concretes at different temperatures is shown in Fig. 3.18. Each point represents an average of at least three specimens normalized with respect to maximum compressive strength at room temperature. From this figure it can be seen that exposure to temperatures in the range of 100 °C to 300 °C decreased the compressive strength of high strength concrete by 15% to 20%. As the strength increased, the loss of strength from exposure to high temperature also increased. At tempera-

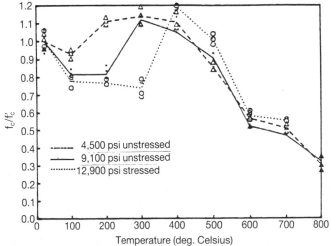

Fig. 3.18   Variation of compressive strength with increase in temperature[76]

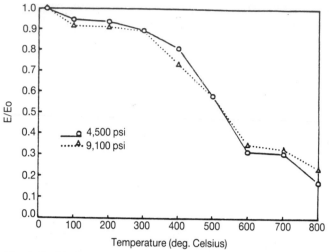

Fig. 3.19   Variation in modulus of elasticity of normal and high-strength concrete with increase in temperature[76]

tures above 400 °C, the high strength concrete progressively lost its compressive strength which at 800 °C dropped to about 30% of the room-temperature strength. The study also observed that none of the preloaded specimens were able to sustain the load beyond 700 °C. About one third of these specimens failed in explosive manner in the temperature range of 320 °C to 360 °C while being heated under a constant preload. The variation of the modulus of elasticity of normal and high strength concretes with increasing temperature is shown in Fig. 3.19. The modulus decreases between 5% to 15% when exposed to temperatures in the range of 100 °C to 300 °C. This trend is similar for normal and high strength concretes. At 800 °C, for both the normal and high strength concretes, the modulus of elasticity was only 20% to 25% of the value at room temperature.

In a recent study at Helsinki University,[77] high temperature behavior of three high strength concretes made with different binder combinations was investigated. Ordinary cements, silica fume and class F fly ash with superplasticizers were used as binders. The aggregates were granite-based sand and crushed diabase. Compressive strengths of 4 in. (100 mm) cubes at 28 days were 12,300 psi to 16,000 psi (85 MPa to 111 MPa). The study consisted of investigations of mechanical properties at elevated temperatures, and attention was also given to the chemical and physical background of their alteration due to heating. The thermal stability and alterations were also investigated. The mechanical properties at high temperatures were studied by determining the stress-strain relationship. The risk of spalling was also studied primarily with small specimens. The three high strength concretes showed very similar temperature behavior. They all show, in the temperature region from 100 °C to 350 °C, more loss of strength than normal strength concrete. This is caused by temperature-dependent destruction of cement paste. Its influence on the strength of high strength concrete is more decisive than on the normal strength concrete, because the cement paste matrix of high strength concrete must carry higher loads than in normal strength concrete (more homogeneous stress distribution between the aggregate and cement paste). The denser cement paste and overall microstructure of the higher strength concrete result in slower drying. So the higher risk of destructive spalling must be taken into account in structures exposed to fire.

It is generally recognized that concrete cast and cured at low temperatures develops strengths at a significantly slower rate than similar concrete placed at room temperature. Lee[78] conducted a study on mechanical properties of high strength concrete in the temperature range between +20 °C and −70 °C (68 °F and −94 °F) without considering the effect of freezing cycles. Test results showed that the values of compressive and tensile strength, modulus of elasticity and Poisson's ratio increased as the temperature decreased. The ratio of bond strength at low temperature is generally larger under reversed cyclic loading than under monotonic or repeated cyclic loads. The rate of increase in compressive, splitting tensile strength, Young's modulus, local bond strength and Poisson's ratio for high strength concrete at corresponding low temperature (−10 °C, −30 °C, −50 °C, −70 °C) is generally lower than for normal strength concrete. The effect of decreasing temperature on the compressive strength and the tensile strength is shown in Figs. 3.20 and 3.21, respectively. From these figures it can be seen that higher strength concrete shows less susceptibility to decreasing temperatures compared to normal strength concrete.

Price,[79] and Klieger[80] determined that concrete mixed and placed at 4 °C had a 28-day compressive strength which was 22% lower than concrete cast and continuously cured at 21 °C. However, recent work[81,82] indicated that expected slow strength development at low temperatures was not realized for cold cast and cured concretes.

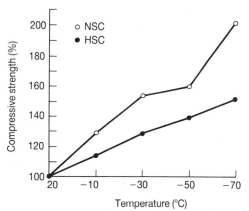

Fig. 3.20 Effect of low temperature variation on the percentage of compressive strength increase for normal and high-strength concrete[78]

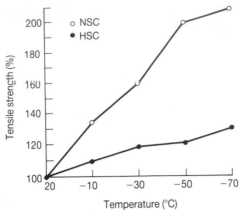

Fig. 3.21 Effect of low temperature on the percentage of tensile stress increase for normal and high-strength concrete[78]

3.6   Structural effects: case studies

In this section, two examples of field applications where the concrete had to be designed to meet the performance specifications are presented. The examples are of a nuclear containment structure in Civaux (France) and a prestressed concrete bridge at Joigny (France). For the nuclear containment structure, there was a great need to improve the air-tightness and reduction of thermal cracking. For the prestressed concrete bridge, there was a need for higher strength concrete with reduced heat of hydration, shrinkage and creep properties.

Thermal cracking – Civaux nuclear containment structure (France)

The reactors of the nuclear power plants currently being built in France are enclosed by double concrete containments. The purpose of these prestressed concrete structures are to protect the nuclear reactor from possible outside dangers and most important of all, to protect the environment (and its population) from any radioactive release should an event or accident occur during the operation of the plant.

In practice, the critical requirement is of limiting the leaks to 1.5% of the total mass of the gas contained in the vessel, for 24 hours under an internal absolute pressure of 72.5 psi (0.5 MPa) and a temperature close to 300 °F (150 °C) (design accident conditions). The axisymmetric structure is therefore massively prestressed (along the meridians and the parallels), so much so that the concrete remains in compression in all design conditions and particularly when the vessel is pressurized to the critical pressure. In the absence of internal pressure, the concrete must be able to withstand this prestress and this results in wall thickness of 4 ft to 5 ft (120 to 15 cm). As for the raft, earthquake considerations result in thickness of about 10 ft (3 m).

Nevertheless, this massive amount of concrete does not guarantee air-tightness, primarily because of systematic thermal cracking. The numerical analysis of computed thermal stresses clearly showed that this damage is not a skin cracking phenomenon (due to the local gradient) but crossing cracks phenomenon, which is due to the restraint of the mean thermal contraction of the last layer of fresh concrete by the previous one. For this reason, cracks are vertical, widely spaced and visible. After complete cooling of the structure, the opening of these cracks is reduced, but only to a small extent, for several reasons. These include the setting (and initialization of strains) which does not occur at the ambient temperature, the changing value of the modulus of concrete (it being quite a lot higher during cooling than during temperature rise, the duration of cooling of concrete during which thermal stresses are generated due to visco-elastic behavior of concrete.

The specifications required for a concrete to meet the performance criteria for this application were: (a) high stability and workability, to avoid formation of porous zones in concrete; (b) limited shrinkage, including autogenous shrinkage and thermal contraction; (c) improved resistance to tensile stresses resulting from restrained deformations; (d) lower material permeability to air, which would cut losses through the plain concrete (i.e. outside of the cracked zones; (e) improved anchorage on passive reinforcements, which should reduce crack opening; and (f) higher compressive strength, making it possible to alter the design of the structure, starting with the elimination of such extra thickness as the wall/raft and the wall/dome gussets. Cost was also a consideration and the aim was to keep the extra cost per cubic meter to a minimum, given the large quantities involved in the structure, about 16,000 yd^3 (12,000 m^3).

A comprehensive approach (described by de Larrard[17,45,71,83]) led to a high performance concrete (HPC) for the Civaux containment project with silica fume and a large reduction in the cement content, compensated, for its 'grading' role, by a calcareous filler (HSC1). Finally, two HPC mixes and two conventional concrete mixes (OC1 and OC2) were designed for this project (Table 3.6). The contractor preferred a concrete without any retarder, i.e. mix HSC2. The heat of hydration of OC2 and HSC2 were investigated, by the means of a semi-adiabatic calorimeter, on the basis of the same spproach similar to Smeplass.[19] The results (Fig. 3.22) show a significant reduction (20 to 25% on the final heat values per cubic meter).

Comparative testing of concretes

The principal mechanical and physical properties of these four concretes are shown in Table 3.7. The compressive strength values of the HPC mixes for this project increased rather slowly with time because of the high proportion of the binder constituents having a slow rate of hydration and

Table 3.6   Composition of the four concretes investigated: masses given in lb/yd^3 (kg/m^3)

	OC1	OC2	HSC1	HSC2
Aggregates 12.5/25[1]	1499	1296	1591	1499
	(890)	(769)	(944)	(890)
Aggregates 4/12.5[1]	350	511	372	352
	(208)	(303)	(221)	(209)
Sand 0/5[2]	1222	1178	1296	1353
	(725)	(699)	(769)	(803)
Cement[3]	632	590	448	448
	(375)	(350)	(266)	(266)
Filler[4]		102	112	118
		(60.8)	(66.4)	(69.8)
Silica fume			67.9	67.9
			(40.3)	(40.3)
Superplasticizer[5]	6.32[6]	2.06[5]	58.3[7]	15.3
	(3.75)	(1.22)	(34.6)	(9.08)
Retarder[8]			3.59	
			(2.13)	
Mixing water	320	329	214	271
	(190)	(195)	(127)	(161)
Slump in in (mm)	3.1	2.8	3.5	7.1
	(80)	(70)	(90)	(180)

[1] 'Arlaut' crushed calcareous aggregates
[2] Crushed calcareous sand containing 6% filler
[3] 'Airvault CPJ 55' Portland cement with 9% calcareous filler
[4] 'Cical' calcareous filler
[5] Melamine type 'Melment' 20% dry content in weight
[6] Lignosulfonate type plasticizer
[7] 'Rhéobuild 1000' naphtalene type superplasticizer
[8] 'Melretard'

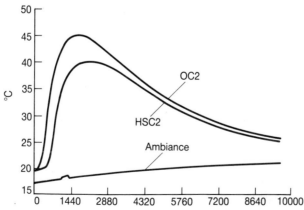

Fig. 3.22   Semi-adiabatic temperature of normal strength concrete, OC2, and high strength concrete, HSC2. (The mix designs are shown in Table 3.6)

Table 3.7   Comparative tests on the four concretes given in Table 3.6

	OC1	OC2	HSC1	HSC2
Compressive strength in ksi (MPa)				
f_{c1}	–	–	1.74	–
	–	–	(12)	–
f_{c3}	–	–	5.22	–
	–	–	(36)	–
f_{c7}	–	–	7.25	–
	–	–	(50)	–
f_{c28}	5.8	6.67	10.2	9.72
	(40)	(46)	(70)	(67)
Splitting strength in ksi (MPa)				
f_{t1}	–	–	0.17	–
	–	–	(1.2)	–
f_{t3}	–	–	0.45	–
	–	–	(3.1)	–
f_{t7}	–	–	0.57	–
	–	–	(3.9)	–
f_{t28}	0.54	0.58	0.65	0.59
	(3.7)	(4.0)	(4.5)	(4.1)
Modulus in ksi $\times 10^3$ (GPa)				
E_{i3}	–	–	4.93	
	–	–	(34)	–
E_{i8}	–	–	5.37	–
	–	–	(37)	–
E_{i28}	4.93	4.93	5.51	5.22
	(34)	(34)	(38)	(36)
Flexural strength in ksi (MPa)				
F_{f18h}	0.36	0.32	0.09	0.32
	(2.5)	(2.2)	(0.6)	(2.2)
f_{f1}	0.44	0.45	0.39	0.41
	(3.0)	(3.1)	(2.7)	(2.8)
f_{f3}	0.61	0.58	0.61	0.59
	(4.2)	(4.0)	(4.2)	(4.1)
f_{f8}	0.65	0.59	0.78	0.68
	(4.5)	(4.1)	(5.4)	(4.7)
f_{f28}	–	0.67	–	0.80
	–	(4.6)	–	(5.5)
Tensile strain at failure (10^{-6})				
ϵ_{18h}	120	120	105	100
ϵ_1	130	125	130	120
ϵ_3	155	160	155	150
ϵ_8	150	140	165	160
ϵ_{28}	–	155	–	160
Shrinkage between 28 and 90 days (10^{-6})				
– of protected specimen		20		10
– of a drying specimen		150		80

Table 3.7 *cont.*

	OC1	OC2	HSC1	HSC2
Creep + shrinkage (10^{-6})				
– of protected specimen		250		280
– of a drying specimen		700		360
under a stress of (MPa):		12		20
Basic creep (10^{-6})				
– per ksi (MPa):		230		270
		132		93.1
		(19.2)		(13.5)
Additional drying dreep (10^{-6})				
– per ksi (MPa):		320		10
		184		3.45
		(26.7)		(0.5)
Creep/elastic strain ratio K_C				
– without drying		0.65		0.49
– with exchange of water		1.56		0.50
Air permeability ($10^{-18} ft^2$ (m^2)				
– under 4.3 psi	69.9		8.82	
(0.03 MPa)	(6.5)		(0.82)	
– under 29 psi	32.3		1.61	
(0.20 MPa)	(3.0)		(0.15)	
– under 58 psi	153		1.29	
(0.40 MPa)	(14.3)		(0.12)	

because of the relatively low cement content. The increase in the splitting tensile strength values with time was proportional to the compressive strength gain with time. The modulus of elasticity for the high strength mixes was relatively smaller than expected, in spite of the small volume of binder paste in these concretes.

The amplitude of the autogenous shrinkage is governed by the ability of the binder to continue to hydrate and consume water after setting. In the HPC mixes designed for this project, part of the cement was replaced by a chemically inert filler, and by silica fume, which reacts more slowly, and probably without consuming water. Therefore for these concrete mixes, the self-desiccation was low, unlike most silica-fume high strength concrete mixes. This was confirmed experimentally by using embedded gauges to measure early age shrinkage on 6.3×39.5 in. (160×1000 mm) cylinders, at $68 \pm 1.8\,°C$ ($20 \pm 1\,°F$) and $50 \pm 10\%$ relative humidity. For each concrete, one of two specimens was protected against drying by two coats of resin separated by aluminium foil; the other specimen was dried freely. The autogenous shrinkage for HPC mixes for this project was lower than that for ordinary concrete and practically zero (Table 3.7). In the first month,

HPC mixes for this project exhibited more drying shrinkage, but the opposite is observed after 28 days.

In order to obtain basic and drying creep of OC2 and HSC2 concretes, the companion specimens were loaded at 28 days under 30% of their instantaneous failure load. Creep results (where autogenous and drying shrinkage were respectively deduced) are shown in Table 3.7. Like all silica-fume concrete results reported in the literature, the HSC exhibits negligible drying creep. It should be pointed out that the reduction of total creep resulting from the use of HSC is therefore between 30 and 70%, depending on the thickness of the structure.

Temperature and restrained deformations fields were calculated using the CESAR-LCPC finite-element program[83]. Temperatures were calculated with constant diffusivity coefficient and a second term representing the heat of hydration, the rate of which depends on the temperature according to Arrhenius's law. An on-site model representing two 66 ft (20 m)-sections of a containment shell 4 ft (1.20 m)-thick with its reinforcements and prestressing sheets were built, one with the conventional concrete, the other with the HPC. The maximum temperature rise in the first core was 72 °F (40 °C), and the temperature rise in the high strength concrete core was 54 °F (30 °C). The conventional concrete wall showed vertical cracks, at half of the vertical prestressing sheets, three of these cracks exceeding 0.008 in. (200 μm). By contrast, only one microcrack, with an opening of 0.004 in. (100 μm), occurred in the shell with HPC. These results confirm the hypothesis regarding thermal cracking, i.e. if young concrete is able to support the tensile stresses which pass through a maximal value at a very early age during its maturing, thermal cracking is avoided.

HPC without silica fume – Joigny bridge (France)

In order to obtain a better knowledge of the serviceability of HSC and its interest for prestressed structures, the construction of the Joigny bridge (1988–1989) was accompanied by laboratory tests and on site monitoring, particularly regarding the effects of hydration heat, shrinkage and creep.

The composition and strength values of the concrete are given in Table 3.8. The high cement content was used to obtain the desired high strength

Table 3.8   Composition of the Joigny Bridge concrete: in lb/yd^3 (kg/m^3)

Yonne coarse aggregates 5/20	1730	(1027)
Yonne sand 0/4	1092	(648)
Fine sand 0/1	177	(105)
'CPA HP' Portland cement from Cormeilles	758	(450)
'Melment' superplasticizer (40% dry content)	18.96	(11.25)
'Melretard' retarder	7.58	(4.5)
Added water	266	(158)
Slump	9 in	(230 mm)

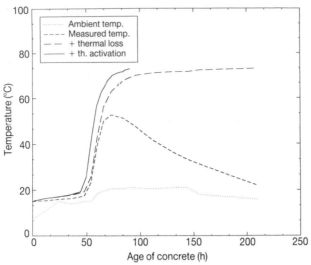

Fig. 3.23   Semi-adiabatic temperature of the high strength concrete used in the construction of Joigny Bridge, France

values, which also made it necessary to use a large proportion of superplasticizer. A retarding agent was also used to prevent the poured concrete from setting before the casting of the deck was completed (about 24 h). The aim was in fact to allow the concrete to adapt to the deformations of the formwork during the casting of the three spans of the structure. Four 6.3×39.5 in. (150×1000 mm) concrete cylinders were cast on site and raw materials were stored for additional laboratory tests.

The heat of hydration was measured by the means of a semi-adiabatic calorimeter. The final values are very high, due to the high cement content (Fig. 3.23). Autogenous and drying shrinkage, measured on protected (with self-adhesive aluminium sheets) and non-protected 6.3×39.5 in. (160×1000 cm) concrete cylinders at $68\,°F$ ($20\,°C$) and 50% relative humidity are shown in Fig. 3.24. The final values are close to current values of normal strength concrete. As pointed out previously, high autogenous shrinkage of certain HSC is mainly related to the presence of silica fume, together with low water-cement ratio.

Experimental results of creep and shrinkage results allow us to predict the delayed behavior of the structure, and can be compared to the values measured on site. Figure 3.25 shows the mean deformation of the mid-span section versus time along with the predicted results. The difference between the calculated and the measured deformation can be attributed in part to the climatic effect (annual variations).

3.7   Summary and conclusions

The physical phenomenon controlling the shrinkage and creep deformations in concrete properties can be broadly categorized in the following

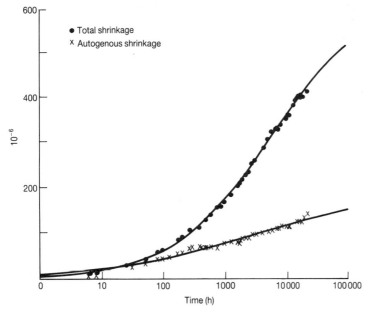

Fig. 3.24 Autogenous and drying shrinkage of the high strength concrete used in the construction of Joigny Bridge, France

stages. These phenomena allow the main features of delayed strains in higher strength concretes to be understood.

Stage 1 – Just after casting, the concrete remains in a suspension state, where little or no segregation occurs due to the very compact state of the cement paste. This dormant period is lengthened by the retarding effect of superplasticizers.

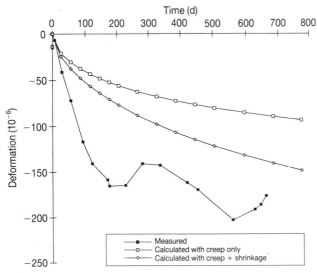

Fig. 3.25 Comparison of the observed and predicted mean deformation of the mid-span section of Joigny Bridge, France

Stage 2 – When the cement hydration appears, the setting starts sooner due to the closeness of cement grains; self-desiccation of the internal porosity, and a rapid heat development occurs at the same time.

Stage 3 – Hydraton continues until the internal humidity falls below a certain level (say, 70–80%). During the hardening phase, concrete becomes rapidly watertight. If the concrete is submitted to drying, the effect will be harmful on fresh or plastic concrete (because of the lack of bleeding), but the effect is very limited after the hydration has begun. The endogenous phenomena of higher strength concretes indicate that autogenous shrinkage is higher in HSC than normal strength concretes, especially when silica fume is incorporated. The heat development is higher in HSC because of the higher cement dosage, but limited by incomplete hydration. The basic creep is reduced due to the decreased volume of hydrates, and to the self-desiccation process.

The deformations for higher strength concretes related to drying are generally reduced, particularly for short and medium term. Even, the drying creep of silica-fume HSC could be practically insignificant. However, the very long-term drying shrinkage of HSC needs more investigations.

The laboratory experimental observations have been validated by monitoring of some full-scale structures (during several years). Parametric models have been proposed for autogenous shrinkage and basic creep. More accurate models, and models for drying-related deformations, are still to be developed. These will provide tools for engineers to accurately predict long term time dependent performance for designing and maintaining reinforced and prestressed concrete structures.

References

1   ACI Committee 209 (1990) Prediction of creep, shrinkage and temperature effects in concrete structures, *Manual of concrete practice*, Part 1. American Concrete Institute, 209R 1–92.
2   ACI Committee 363 (1984) State-of-the-art report on high strength concrete. *ACI Journal*, **81**, 4, 364–411.
3   Freedman, S. (1970/71) High strength concrete. *Modern Concrete*, **34**, 6, 1970, 29–36; 7, 1970, 28–32; 8 1970, 21–4; 9, 1971, 15–22; 10, 1971, 16–23.
4   Ngab, A.S., Nilson, A.H. and Slate, F.O. (1981) Shrinkage and creep of high strength concrete. *ACI Journal*, **78**, 4, 255–61.
5   Chicago Committee on High-Rise Buildings (1977) *High strength concrete in Chicago, high-rise buildings*.
6   Pfeifer, D.W., Nagura, D.D., Russell, H.G. and Corley, W.G. (1971) Time dependent deformations in a 70 story structure, in *Designing for effects of creep, shrinkage, temperature in concrete structures, SP-37*. American Concrete Institute, Detroit, 159–85.
7   Parrot, L.J. (1969) *The properties of high-strength concrete*, Technical Report No. 42.417. Cement and Concrete Association, Wexham Springs.
8   Swamy, R.N. and Anand, K.L. (1973) Shrinkage and creep in high strength concrete. *Civil Engineering and Public Works Review* (London), **68**, 807, 859–65; 867–8.
9   Aitcin, P.-C., Sarkar, S.L. and Laplante, P. (1990) Long-term characteristics of a very high strength concrete. *Concrete International*, 40–44.

10 Powers, T.C. (1968) *The properties of fresh concrete.* Wiley & Sons, New York.

11 Buil, M. (1979) *Contribution à l'étude du retrait de la pâte de ciment durcissante.* Rapport de Recherche No. 92, LCPC, Paris.

12 L'Hermite, R.G., Mamillan, M. (1973) Répartition de la teneur en eau dans le béton durci. *Annales de l'ITBTP*, 309–10, 30–34.

13 Scrivener, K.L. (1989) The microstructure of concrete, in *Materials science of concrete, I,* Skalny, J.P. (ed.), The American Ceramic Society Inc.

14 Byfors, J. (1980) *Plain concrete at early ages,* Report of the Swedish Cement and Concrete Research Institute, Stockholm.

15 Acker, P. (1992) Physicochemical mechanisms of concrete cracking, in *Materials science of concrete, II,* Skalny, J.P. (ed.), The American Ceramics Society Inc.

16 Daian, J.F. and Saliba, J. (1990) Using a pore-network model to simulate the experimental drying of cement mortar. *Proceedings of the 7th International Drying Symposium,* IDS'90/CHISA'90, Prague.

17 de Larrard, F., Ithurralde, G., Acker, P. and Chauvel, D. (1990) High performance concrete for a nuclear containment. *Second International Symposium on Utilization of High-Strength Concrete,* Berkeley, ACI SP 121–27.

18 Bamforth, P.B. (1982) *Early age thermal cracking in concrete.* Slough Institute of Concrete Technology, Tech. No. TN/2.

19 Smeplass, S. and Maage, M. (1990) Heat of hydration of high-strength concretes, in *Proceedings IABSE Colloquium on High-Strength Concrete,* Berkeley, 433–56.

20 Acker, P., Foucrier, C. and Malier, Y. (1986) Temperature-related mechanical effects in concrete elements and optimization of the manufacturing process, in *Properties of concrete at early ages,* Young, J.F. (ed.), ACI Publ, SP-95, Detroit, 33–47.

21 de Larrard, F. and Le Roy, R. (1992) *The influence of mix-composition on the mechanical properties of silica-fume high-performance concrete.* Fourth International ACI CANMET Conference on Fly Ash, Silica Fume, Slag and Natural Pozzolans in Concrete, Istanbul.

22 Paillere, A.M., Buil, M. and Serrano, J.J. (1987) Durabilité du béton à très hautes performances: incidence du retrait d'hydratation sur la fissuration au jeune âge [Durability of VHSC: effect of hydration shrinkage on early-age cracking], First International RILEM Conference, *From Materials Science to Construction Materials Engineering,* Versailles, Chapman and Hall, Vol. 3, 990–97 [in French].

23 Neville, A.M., Dilger, W.H. and Books, J.J. (1983) *Creep of plain and structural concrete.* Construction Press.

24 ACI Committee 363 (1987) Research needs for high-strength concrete. *ACI Materials Journal,* 559–61.

25 Bazant, Z.P., Asghari, A.A. and Schmidt, J. (1976) Experimental study of creep of Portland cement paste at variable water content. *Materials and Structures, RILEM,* **9,** 52, 279–90.

26 Sellevold, E. and Justnes, H. (1992) *High-strength concrete binders: nonevaporable water, self-desiccation and porosity of cement pastes with and without condensed silica fume.* Fourth International ACI-CANMET Conference on Fly Ash, Silica Fume, Slag and Natural Pozzolans in Concrete, Istanbul.

27 Smadi, M.M., Slate, F.O. and Nilson, A.H. (1987) Shrinkage and creep of high-, medium- and low-strength concretes, including overloads. *ACI Materials Journal,* 224–34.

28 Bjerkeli, L., Tomaszewicz, A. and Jensen, J.J. (1990) Deformation properties and ductility of high-strength concrete. *Second International Conference on Utilization of High Strength Concrete,* ACI Special Publication, Berkeley.

29   Lumbroso, V. (1990) 'Réponse différée du béton à hautes performances soumis à un chargement stationnaire – influence des conditions d'environnement et de la composition' [Delayed behavior of high-strength concrete submitted to stationary loading]. Ph.D. doctoral thesis, Institut National des Sciences Appliquées, Toulouse.

30   Comite European du Béton Federation Internationale de la Precontrainte (1978) *CEB-FIP Model Code for Concrete Structures*, 3rd edn, Paris.

31   Bazant, Z.P. and Panula, L. (1980) Creep and shrinkage characterization for analyzing prestressed concrete structures. *PCI Journal*, **25**, 3, 86–122.

32   Bazant, Z.P. and Chern, J.C. (1985) Log double power law for concrete creep. *ACI Journal*, **82**, 5, 665–75.

33   Auperin, M., De Larrard, F., Richard, P. and Acker, P. (1989) Retrait et fluage de bétons à hautes performances – influence de l'âge au chargement [Shrinkage and creep of high-strength concretes – influence of the age at loading]. *Annales de l'Institut Technique du Bâtiment et des Travaux Publics*, 474, 50–75.

34   Buil, M. (1990) Le comportement physicochimique du système ciment-fumée de silice [Physico-chemical behavior of the cement/silica fume system]. *Annales de l'Institut Technique du Bâtiment et des Travaux Publics*, 483 [in French].

35   Perraton, D. and Aitcin, P.-C. (1990) La perméabilité au gaz des BHP [Gas-permeability of HSC] in *Les Bétons à hautes Performances*, Malier, Y. (ed.), Presses de l'ENPC, Paris [in French].

36   De Larrard, F. and Bostvironnois, J.L. (1991) On the long-term strength losses of silica-fume high-strength concretes. *Magazine of Concrete Research*, **43**, 155, 109–19.

37   Books, J.J. and Wainwright, P.J. (1983) Properties of ulta-high-strength concrete containing a superplasticizer. *Magazine of Concrete Research*, **35**, 125.

38   Johansson, A. and Petersons, A. (1979) *Flowing concrete, advances in concrete slab technology*. Pergamon Press, Oxford, 58–65.

39   Omojola, A. (1974) *The effect of cormix SPI super workability aid on creep and shrinkage of concrete*. Research and Development Report No. 014J/74/1683, Taylor Woodrow, London.

40   Dhir, R.K. and Yap, A.W.F. (1984) Superplasticized flowing concrete: strength and deformation properties. *Magazine of Concrete Research*, **36**, 129, 203–15.

41   Luther, M.D. and Hansen, W. (1989) Comparison of creep and shrinkage of high-strength silica fume concretes with fly ash concretes of similar strengths, in *Fly ash, silica fume, slag and natural pozzolans in concrete, SP 114, Vol. 1*. American Concrete Institute, 573–91.

42   Swamy, R.N. and Mahmud, H.B. (1989) Shrinkage and creep behaviour of high fly ash content concrete, in *Fly ash, silica fume, slag and natural pozzolans in concrete, SP 114, Vol. 1*. American Concrete Institute, 453–75.

43   Collins, T.M. (1989) Proportioning high-strength concrete to control creep and shrinkage. *ACI Materials Journal*.

44   Pentalla, V. (1987) Mechanical properties of high-strength concretes based on different binder combinatons. *Stavanger Conference Utilization of High-Strength Concrete*, Tapir Ed., Trondheim, Norway.

45   De Larrard, F. and Acker, P. (1990) *Déformations Libres des Bétons à Hautes Performances* [Free deformations of HSC], Séminaire sur la Durabilité des BHP, Cachan [in French].

46   De Larrard, F. and Malier, Y. (1989) *Propriétés constructives des bétons à très hautes performances – de la micro- à la macrostructure* [Engineering properties of very-high-strength concretes, from the micro- to the macrostructure],

Annales de l'Institut Techniques des Travaux Publics, No. 479.

47 Alfes, C. (1989) High-strength, silica-fume concretes of low deformability. *Betonwerk + Fertigteil-Technik*, No. 11.

48 Bazant, Z.P., Kim, J.K., Wittmann, F.H. and Alou, F. (1987) Statistical extrapolation of shrinkage data – Part II: Bayesian updating. *ACI Materials Journal*, **84**, 2, 83–91.

49 Bazant, Z.P., Wittmann, F.H., Kim, J.K. and Alou, F. (1987) Statistical extrapolation of shrinkage data – Part I: Regression. *ACI Materials Journal*, 20–34.

50 Bazant, Z.P. and Panula, L. (1978/1979) Practical prediction of time-dependent deformations of concrete. *Materiaux et Constructions*, (RILEM) Paris, **11**, 65, 307–28 and 66, 415–34, 1978; **12**, 69, 169–83, 1979.

51 Bazant, Z.P. and Panula, L. (1982) New model for practical 'Prediction of creep and shrinkage', in *Designing for creep and shrinkage in concrete structures*. SP 76, American Concrete Institute, 7–23.

52 Bazant, Z.P. and Panula, L. (1984) Practical preediction of creep and shrinkage of high strength concrete. *Materiaux et Constructions*, (RILEM) Paris, **17**, 101, 375–78.

53 Almudaiheem, J.A. and Hansen, W. (1989) Prediction of concrete drying shrinkage from short-term measurements. *ACI Materials Journal*, 401–8.

54 Acker, P. (1980) The drying of concrete – consequences on creep tests interpretation, *Re-evaluation of the time-dependent behavior of concrete*. CEB Task Group, Lausanne.

55 Wittmann, F.H. and Roelfstra, P.E. (1980) Total deformation of loaded drying concrete. *Cement and Concrete Research*, **10**, 5.

56 Buil, M. and Acker, P. (1985) Creep of a silica fume concrete. *Cement and Concrete Research*, **15**, 463–6.

57 Bamforth, P.B. (1980) In situ measurements of the effect of partial Portland cement replacement using either fly ash or granulated blast furnace slag on the performance of mass concrete. *Proceedings of Institution of Civil Engineers, Part 2*, **69**, 777–800.

58 FIP Commission on Concrete (1988) *Condensed silica fume in concrete*, State of art report. Federation Internationale de la Precontrainte, London.

59 Yamato, T. and Sugita, H. (1983) Fly ash, silica fume, slag and other mineral by-products in concrete, in *Fly ash, silica fume, slag and other mineral by-products in concrete, Vol. 1*, SP 79. American Concrete Institute, 87–102.

60 Brooks, J.J., Wainwright, P.J. and Neville, A.M. (1979) Time-dependent properties of concrete containing a superplasticizing admixture, *Superplasticizers in concrete*, SP 62. American Concrete Institute, 293–314.

61 Alexander, R.M., Bruere, G.M. and Ivanusec, I. (1980) The creep and related properties of very high-strength syper-plasticized concrete. *Cement and Concrete Research*, **10**, 2, 131–7.

62 Wolsiefer, J. (1984) Ultra high-strength field placeable concrete with silica fume admixture. *Concrete International*, **6**, 4, 25–31.

63 Perenchio, W.F. and Klieger, P. (1978) *Some physical properties of high strength concrete*. Research and Development Bulletin No. RD 056.01T, Portland Cement Association, Skokie, IL.

64 Saucier, K.L. (1984) *High strength concrete for peacekeeper facilities*, Final report. Structures Laboratory, US Army Engineer Waterways Experimental Station.

65 Neville, A.M. (1981) *Properties of concrete*, 3rd edn, Pitman Publishing Ltd, London.

66 Pentalla, V. and Rautanen, T. (1990) Microporosity, creep and shrinkage of high-strength concretes. *Second International Conference on Utilization of High-Strength Concrete*, Berkeley, ACI Special Publication.

67   Domone, P.L. (1974) Uniaxial tensile creep and failure of concrete. *Magazine of Concrete Research*, **20**, 88, 144–52.
68   Cornelissen, H.A.W. and Siemes, A.J.M. (1985) Plain concrete under sustained tensile or tensile and compressive fatigue loadings, in *Behaviour of offshore structures*. Elsevier Science Publishers, Amsterdam, 487–98.
69   De Larrard, F., Acker, P., Malier, Y. and Attolou, A. (1988) *Creep of very-high-strength concretes*, 13th IABSE Conference, Helsinki.
70   De Larrard, F. (1990) *Creep and shrinkage of high-strength field concretes*, Second International Conference on Utilization of High-Strength Concrete, Berkeley, ACI SP 121–28.
71   Saucier, K.L., Tynes, W.O. and Smith, E.F. (1965) *High compressive strength concrete – report 3, summary report*, Miscellaneous Paper No. 6-520, US Army Engineer Waterways Experiment Station, Vicksburg.
72   Abrams, M.S. (1971) Compressive strength of concrete at temperatures to 1600 °F, in *Temperature and concrete*, SP 25. American Concrete Institute, 33–58.
73   Hannant, D.J. Effects of heat on concrete strength. *Engineering*, London, **203**, 21, 302.
74   Lankard, D.R., Birkimer, D.L., Fondfriest, F.F. and Snyder, M.J. (1971) Effects of moisture content on the structural properties of Portland cement concrete exposed to temperatures up to 500 °F, in *Temperature and concrete*, S 25. American Concrete Institute, 59–102.
75   Schneider, U. (1976) Behavior of concrete under thermal steady state and non-steady state conditions. *Fire and Materials*, **1**, 103–15.
76   Castillo, C. and Duranni, A.J. (1990) Effect of transient high temperature on high-strength concrete. *ACI Materials Journal*, **87**, 1, 47–53.
77   Diederichs, U., Jumppanen, U.-M. and Pentalla, V. (1989) *Behaviour of high strength concrete at high temperatures*, Report 92. Department of Structural Engineering, Helsinki University, Espoo.
78   Lee, G.C., Shih, T.S. and Chang, K.C. (1988) Mechanical properties of high-strength concrete at low temperature. *Journal of Cold Regions Engineering*, **2**, 4, 169–79.
79   Price, W.H. (1951) Factors influencing concrete strength. *ACI Journal*, **47**, 6, 417–31.
80   Klieger, P. (1958) Effect of mixing and curing temperatures on concrete strength. *ACI Journal*, **54**, 12, 1063–81.
81   Aitcin, P.-C., Cheung, M.S. and Sha, V.S. (1985) Strength development of concrete cured under Arctic sea conditions, in *Temperature effects on buildings*, STP 858, ASTM, Philadelphia, 3–20.
82   Gardner, N.J., Sau, P.L. and Cheung, M.S. (1988) Strength development and durability of concretes cast and cured at 0 °C. *ACI Materials Journal*, **85**, 6, 529–36.
83   De Larrard, F. (1990) Creep and shrinkage of high-strength field concretes. *Second International Conference in Utilization of High-Strength Concrete*, Berkeley, ACI SP 121–28.

4 Fatigue and bond properties

A S Ezeldin and P N Balaguru

4.1 Introduction

Many concrete structural members are subjected to repeated fluctuating loads the magnitude of which is well below the maximum load under monotonic loading. This type of loading is typically known as fatigue loading. Contrary to static loading where sustained loads remain constant with time, fatigue loading varies with time in an arbitrary manner. Fatigue is a special case of dynamic loading in which inertia forces do not influence the stresses. Examples of structures that are subjected to fatigue loading include bridges, offshore structures and machine foundations.

Fatigue is one property of concrete that is not well understood, specially in terms of the mechanism of failure, because of the difficult and tedious experiments required for conducting research investigations. Detailed presentations of research findings concerning the fatigue behavior of concrete are included in the references.[2,3,21,38,40] Use of high performance concrete in structures subjected to fatigue loading requires knowledge about its behavior under such loading. As mentioned previously (Chapter 2) the performance characteristics of concrete generally improve with the strength attribute. Unfortunately, the available data on the fatigue behavior of high-strength concrete is very limited.[4,5] This chapter presents an overview of the results available in the published literature on concrete fatigue in general, with emphasis on the behavior of high-strength concrete (compressive strength >6000 psi (42 MPa)).

Fatigue is the process of cumulative damage that is caused by repeated fluctuating loads. Fatigue loading types are generally distinctly divided between high-cycle low amplitude and low-cycle high amplitude. Hsu[20] has summarized the range of cyclic loading into a spectrum of cycles. He classified the fatigue loads into three ranges, Table 4.1. The low-cycle fatigue loading occurs with less than 1000 cycles. The high-cycle fatigue loading is defined in the range of 10^3 to 10^7 cycles. This range of fatigue

Table 4.1   Fatigue load spectrum[20]

Low-cycle fatigue	High-cycle fatigue		Super-high-cycle fatigue
Structures subjected to earthquake	Airport pavements and bridges	Highway and railway bridges, highway pavements, concrete railroad ties	Mass rapid transmit structures Sea structures

0	10^1	10^2	10^3	10^4	10^5	10^6	10^7	10^8

Number of cycles 5×10^7 5×10^8

loading occurs in bridges, highways, airport runways and machine foundations. The super-high-cycle fatigue loading is characterized by even higher cycles of fatigue loads. This category was established in recent years for the newly developed sophisticated modern structures such as elevated sections on expressways and offshore structures.

Fatigue damage occurs at non-linear deformation regions under the applied fluctuating load. However, fatigue damage for members that are subjected to elastic fluctuating stresses can occur at regions of stress concentrations where localized stresses exceed the linear limit of the material. After a certain number of load fluctuation, the accumulated damage causes the initiation and/or propagation of cracks in the concrete matrix. This results in an increase in deflection and crack-width and in many cases can cause the fracture of the structural member.

The total fatigue life N, is the number of cycles required to cause failure of a structural member or a certain structure. Many parameters affect the fatigue strength of concrete members. These parameters can be related to stress (state of stress, stress range, stress ratio, frequency, and maximum strength), geometry of the element (stress concentration location), concrete properties (linear and non-linear behavior), and external environment (temperature and aggressive elements).

Structural members are usually subjected to a variety of stress histories. The simplest form of these stress histories is the constant-amplitude cyclic-stress fluctuation shown in Fig. 4.1. This type of loading, which usually occurs in heavy machinery foundations, can be represented by a constant stress range, Δf; a mean stress f_{mean}; a stress amplitude, f_{amp}; and a stress ratio R. These values can be obtained using the following equations, Fig. 4.1.

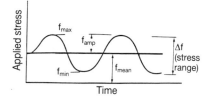

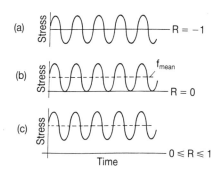

Fig. 4.1 Constant amplitude fatigue loading. (a) $R = -1$; (b) $R = 0$; (c) $0 < R < 1$

$$\Delta f = f_{\max} - f_{\min} \tag{4.1}$$

$$f_{\text{mean}} = \frac{f_{\max} - f_{\min}}{2} \tag{4.2}$$

$$f_{\text{amp}} = \frac{f_{\max} - f_{\min}}{2} \tag{4.3}$$

$$R = \frac{f_{\min}}{f_{\max}} \tag{4.4}$$

Thus, a complete reversal of load from a minimum stress to an equal maximum stress corresponds to an $R = -1$ and a mean stress of zero, as shown in Fig. 4.1(a). A cyclic stress from zero to a peak value corresponds to an $R = 0$ and a mean stress equal to half the peak stress value, Fig. 4.1(b). $R = 1$ represents a case of constant applied stress with no intensity fluctuation. Generally, range of fatigue load for concrete structures is between $R = 0$ and $R = 1$, Fig. 4.1(c). However, special cases such as lateral wave loading on offshore structures can produce a reversed loading condition with a mean stress close to zero ($R = -1$).

Variable-amplitude random-sequence stress histories are very complex, (Fig. 4.2). This type of stress history is experienced by offshore structures. During this stress history, the probability of the same sequence and magnitude of stress ranges to occur during a particular time interval is very small. To predict the fatigue behavior under such loading a method known as the Palmgren–Miner hypothesis has been proposed. This hypothesis, first suggested by Palmgren[35] and then used by Miner[32] to test notched

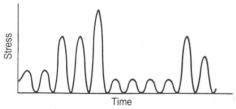

Fig. 4.2   Variable-amplitude random sequence fatigue loading

aluminium specimens, is quite simple because it assumes that damage accumulates linearly with the number of cycles applied at a particular load level. The failure equation is represented as:

$$\sum_{i=1}^{k} \frac{n_i}{N_i} = 1.0 \tag{4.5}$$

where n_i is the number of constant amplitude cycles at stress level i; N_i is the number of cycles that will cause failure at that stress level i, and k is the number of stress levels. Several researchers have checked the validity of the P–M hypothesis for concrete.[18,22,28] Their results indicated that the equality to the unity is not always true. Holmen[19] proposed a modified P–M hypothesis where an interaction factor w that depends on the loading parameters was introduced

$$\sum_{i=1}^{k} \frac{n_i}{N_i} = w \tag{4.6}$$

The factor w has been expressed as a function of the ratio (S_{min}/S_c), Fig. 4.3.

Concern with fatigue damage of concrete was recognized early in this century. Van Ornum[50,51] observed that cementitious composites possessed the properties of progressive failure, which become total under the repetition of load well below the ultimate strength of the material. He also noticed that the stress-strain curve of concrete varies with the number of repetitions, changing from concave towards the strain axis (with a hysteresis loop on unloading) to a straight line, which shifts at a decreasing rate (plastic permanent deformation) and finally to concave toward the stress axis, Fig. 4.4. The degree of this latter concavity is an indication of how near the concrete is to failure.

The fatigue strength can be represented by means of S-N curves (known also as the f–N curve or the Wohler's curve of the fatigue curve). In these curves (Fig. 4.5.) S is a characteristic stress of the loading cycle, usually indicating a stress range or a function of the maximum and minimum stress and N is the number of cycles to failure. S is expressed with a linear scale while N is presented using log scale. Using this format, usually the data can be approximated to a straight regression line. At any point on the curve, the stress value is the 'fatigue strength' (i.e., the value of stress range that

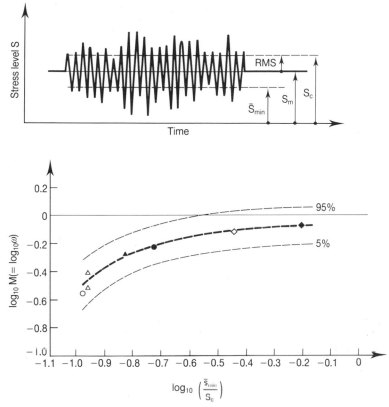

Fig. 4.3 Empirical relationship between fatigue loading parameters and the factor w (= Miner sum at failure)[19]

will cause failure at a given number of stress cycles and at a given stress ratio) and the number of cycles is the 'fatigue life' (i.e., the number of stress cycles that will cause failure at a given stress range and stress ratio). To include the effect of the minimum stress f_{min}, and the stress range $f_{max} - f_{min}$, the fatigue strength can also be represented by means of a

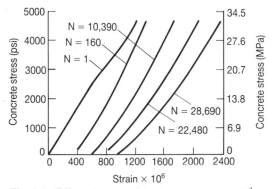

Fig. 4.4 Effect of repeated load on concrete strain[3]

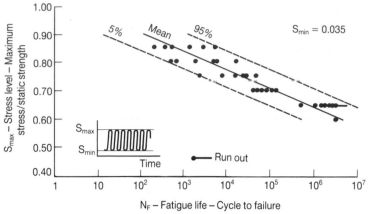

Fig. 4.5   Typical *S-N* relationship for concrete in compression[28]

$f_{max} - f_{min}$ modified Goodman diagram, shown in Fig. 4.6. The maximum and minimum stress levels in this diagram are expressed in terms of the percentage of the static strength. The diagram is for a given number of cycles to failure (e.g., 2 million cycles). From such a diagram, for a specified f_{min}/f_c' value, the allowable ratio f_{max}/f_c' can be determined.

4.2   Mechanism of fatigue

Considerable research has been done to study the nature of fatigue failure. However, the mechanism of fatigue failure is still not clearly understood. Researchers have measured surface strains, change in pulse velocity, internal and surface cracks in an attempt to understand the phenomenon of fatigue fracture. Large increase in the longitudinal and transverse strains and decrease in pulse velocity have been reported prior to fatigue failure. No satisfactory theory of the mechanics of fatigue failure has yet been

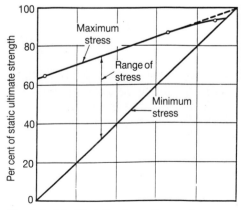

Fig. 4.6   Modified Goodman diagram for concrete subjected to repeated axial loading[16]

proposed for either normal or high strength concrete. However, with the present stage of knowledge the following two observations can be stated.[42]

1   The fatigue of concrete is associated with initiation and propagation of internal microcracks at the cement paste-aggregate interface and/or within the cement paste itself.
2   Cracks due to fatigue failure are more extensive than cracks initiated by static compressive failure.

4.3   Cyclic compression

Graf and Brenner[16] studied the effect of the minimum stress and stress range on the fatigue strength of concrete. Based on a fatigue criteria of 2 million cycles of loading, they developed a modified Goodman diagram for the repeated compressive loading, Fig. 4.6. Both maximum stress and minimum stress level were expressed in terms of the percentage of the static compressive strength.

Aas-Jakobson[1] studied the effect of the minimum stress. He observed that the relationship between f_{max}/f_c' and f_{min}/f_c' was linear for fatigue failure at 2 million cycles of loads. Based on statistical analysis of data, he proposed a general model between $\log N$ and the cycles of stresses using a factor β. He gave a value of $\beta = 0.064$.

$$\log N = \frac{1}{\beta}\left[\frac{1-f_{max}/f_c'}{1-f_{min}/f_{max}}\right] \tag{4.7}$$

Tepfers and Kutti[48] compared their experimental fatigue strength data of plain normal and lightweight concrete with the equation proposed by Aas-Jakobson.[1] Their results indicated the use of $\beta = 0.0679$ for normal weight concrete and $\beta = 0.0694$ for lightweight concrete when $R < 0.80$. They recommended to use a mean value of $\beta = 0.0685$ for both normal and lightweight concrete.

Kakuta et al.,[23] based mainly on the Japanese tests, proposed the following expression for the fatigue of concrete

$$\log N = 17\left[1 - \frac{(f_{max}-f_{min})/f_c'}{1-f_{min}/f_c'}\right] \tag{4.8}$$

Gray et al.[17] conducted an experimental investigation on the fatigue properties of high strength lightweight aggregate concrete under cyclic compression. They tested 150 3×6 in. $(75 \times 150\,mm)$ cylinders using maximum stress levels of 40, 50, 60, 70, and 80% of the static ultimate strength and minimum stress levels of 70 and 170 psi (0.5 to 1.2 MPa). Their test results indicated that there was no difference in the fatigue properties between normal and high strength lighweight concrete. The variation of rate of loading between 500–1000 cycles/minute had no effect on the fatigue properties. They also observed no fatigue limit up to 10 million cycles of loading.

Bennett and Muir[7] studied the fatigue strength in axial compression of high strength concrete using 4-in. (100 mm) cubes. The compressive strength was as high as 11,155 psi (78 MPa). They found that after one million cycles, the strength of specimens subjected to repeated load varied between 66% and 71% of the static strength for a minimum stress level of 1250 psi (8.75 MPa). The lower values were found for the higher-strength concretes and for concrete made with smaller-size coarse aggregate. However, the actual magnitude of the difference was small.

Equations (4.7) and (4.8) represent a significant contribution towards summarizing the available test data on normal strength concrete. However, as pointed out by Hsu,[20] the equation has two main limitation, namely: the static strength f_c' which is used to normalize the maximum stress f_{max} is time-dependent, and the rate of loading is not included. Hsu[20] introduced the effect of time (T) into the (stress-number of cycles) relationship where T is the period of the repetitive loads expressed in sec/cycle. With this approach, a three-dimensional space is created consisting of non-dimensional stress ratio, f, as the vertical axis with $\log N$ and $\log T$ as the two orthogonal horizontal axes. A graphical representation of such a space is shown in Fig. 4.7. Based on the available experimental data in literature, Hsu proposed two equations, one for high-cycle fatigue and the other for low-cycle fatigue.

For high-cycle fatigue

$$\frac{f_{max}}{f_c'} = 1 - 0.0662\,(1 - 0.566R)\log N - 0.0294\log T \qquad (4.9a)$$

For low-cycle fatigue

$$\frac{f_{max}}{f_c'} = 1.2 - 0.2R - 0.133(1 - 0.779R)\log N - 0.05301\,(1 - 0.445R)\log T$$
$$(4.9b)$$

These equations were found to be applicable for the following conditions:

(a)   normal weight concrete with f_c' up to 8000 psi (56 MPa)
(b)   for stress range $0 < R < 1$
(c)   for load frequency range between 0 to 150 cycles/sec
(d)   for number of cycles from 1 to 20 million cycles
(e)   for compression and flexure fatigue.

Chimamphant[10] conducted series of experiments on uniaxial cyclic compression of high-strength concrete. Concrete strength varied from 7500 psi (52 MPa) to 12,000 psi (84 MPa). Maximum stress level varied from $0.4 f_c'$ to $0.9 f_c'$ while the minimum stress level was kept constant at $0.1 f_c'$. Two different rates of loading were used, namely: 6 cycles/sec and 12 cycles/sec. He observed no significant difference in the fatigue behavior of high strength concrete when compared to normal strength concrete. He reported that up to 1 million cycles the S-N curve of high-strength concrete was linear. He also observed no measurable effect on the fatigue strength

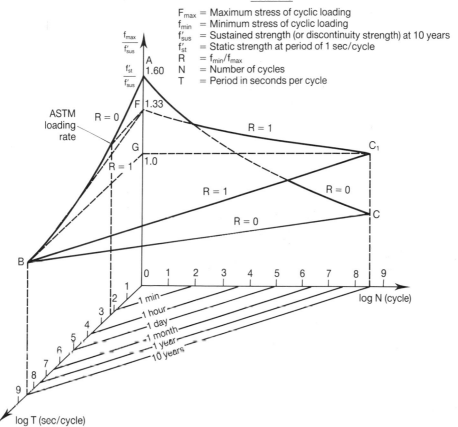

Fig. 4.7   Graphical representation of *f-N-T-R* relationship[20]

of high-strength concrete when the loading rate was changed from 6 to 12 cycles/sec.

Petkovic *et al.*[36] performed studies on the fatigue properties of high-strength concrete in compression. Two types of normal-weight concrete with compressive strengths of 8000 psi (56 MPa) and 10,900 psi (76 MPa) and one type of lightweight aggregate concrete with a compressive strength of 11,600 psi (81 MPa) were tested. Their experimental results gave an indication of the existence of a fatigue limit, which cannot be defined as one level of loading, since its correlation to different loading parameters must be taken into account. They presented design rules for fatigue in compression, which could be applied to the three types of concrete tested (Fig. 4.8). The following three ranges can be distinguished based on the number of cycles. Retion 1: from the beginning of loading to $\log N = 6$, the *S-N* curve lines follow the expression

$$\log N = (1 - S_{max}) \times (12 + 16 S_{min} + 8 S_{min}^2) \qquad (4.10a)$$

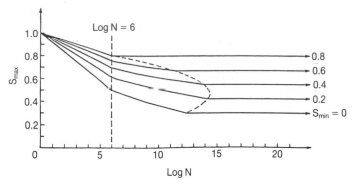

Fig. 4.8   The *S-N* diagram for failure of high strength concrete in compression[36]

Region 2: beyond the value of $\log N = 6$, the inclination of these lines is changed and the $\log N$ value is found by multiplying the basic expression in Equation (4.10a) by the coefficient from

$$C = 1 + 0.2(\log N - 6) \tag{4.10b}$$

Region 3: the fatigue limit for different values of S_{min}. The position of S_{min} the transfer to the fatigue limit is given by the stippled curve. Its shape was only a consequence of the chosen expressions and hence, has no physical significance.

4.4   Cyclic tension

Because of difficulties encountered in applying direct tensile cyclic loads on concrete specimens and avoiding eccentricity of loading, most tension fatigue studies were conducted using indirect tension tests such as splitting tests or beams in flexure. Tepfers[44] performed splitting fatigue tests on 6 in. (150 mm) concrete cubes. Two types of concrete with strengths of 5900 psi (41 MPa) and 8200 psi (57 MPa) were used. The selected stress ratios were 0.20, 0.30 and 0.40. He observed that the fatigue strength equation similar to Equations (4.7) and (4.8) could be used for tension fatigue. He also found that the concrete strength had no effect on fatigue strength when the non-dimensional form (f_{max}/f_c') was used.

Clemmer[12] studied the flexure fatigue of plain concrete. He reported that the fatigue limit for concrete was 55% of the static ultimate flexural strength. Kesler[25] found no fatigue limit but established fatigue strength at 10 million cycles of stress ranging from a small tension value to a maximum value. He reported a fatigue strength of 62% of the static ultimate flexural strength. Williams[52] found no fatigue limit for lightweight aggregate beams. A comprehensive experimental study was conducted by Murdock and Kesler[33] on 175 concrete $6 \times 6 \times 60$ in. ($152 \times 152 \times 1520$ mm) prisms having a compressive strength of 4500 psi (31 MPa). Specimens were loaded at the third points in order to avoid shear stresses at the middle span. Three stress ratios were used, namely: 0.25, 0.50 and 0.75. They observed no fatigue limit for plain normal weight concrete subjected to repeated flexure loading of at least 10 million cycles. They found also that stress range has a significant influence on fatigue strength. They proposed the following fatigue strength equation in terms of stress range at 10 million cycles

$$F_{10} = 0.56 + 0.44M \quad \text{or} \quad F_{10} = \frac{1.3}{(2.3 - R)} \qquad (4.11)$$

where $M = \dfrac{f_{r_{min}}}{f_r}$ and $R = \dfrac{f_{r_{max}}}{f_{r_{min}}}$

for values of M and R between 0 and 1.
for stress reversal, they proposed

$$F_{10} = 0.56$$

where $-0.56 < M < 0$ and $-1 < R < 0$

In order to evaluate the probability of fatigue failure, McCall[29] performed studies on air-entrained $3 \times 3 \times 14.5$ in. $(75 \times 75 \times 368$ mm) concrete prisms. The mean modulus of rupture was 0.68 ksi. Using a loading rate of 1800 cycles/minute, the beams were tested to failure or to 20 million cycles whichever occurred first. The maximum flexure stress varied from 45% to 70% of the concrete modulus of rupture. He observed no fatigue limit in a range up to 20 million cycles. Based on his results, he proposed the following mathematical model for flexure fatigue strength

$$L = 10^{-0.0957} R^{3.32} (\log n)^{3.17} \qquad (4.12)$$

where L = probability of survival = $1 - P$
             P = probability of failure
             $R = S/S_{rp}$
             S = stress range used in the test
         S_{rp} = mean static strength

They found the probability of failure at 20 million cycles to be slightly less than 0.5 for concrete tested at a stress level of 50% of modulus of rupture.

Direct tensile fatigue tests were first conducted by Kolias and Williams.[27] Using constant-amplitude stress range and a constant minimum stress level, they conducted tests on concrete mixes with different coarse aggregates. Their results showed that finely graded concrete has a shorter fatigue life. They suggested that this behavior is due to the more brittle nature of concrete with small aggregate size. Cornelissen[13] conducted an extensive experimental study on 250 necked cylindrical specimens $(4.8 \times 12$ in., 122×305 mm). Using a constant amplitude loading at a frequency of 6 cycles/sec., he studied the effect of maximum and minimum stress levels. The direct uniform tensile stress was applied by bonding (using epoxy) the loading plates on to the top and bottom ends of the cylinders. He derived two stress-number of cycles equations; one for dried specimens (4.13a) and one for sealed specimens (4.13b)

$$\log N = 14.81 - 14.42 \left| \frac{\sigma_{max}}{f_c'} \right| + 2.79 \left| \frac{\sigma_{min}}{f_c'} \right| \qquad (4.13a)$$

$$\log N = 13.92 - 14.42 \left| \frac{\sigma_{max}}{f_c'} \right| + 2.79 \left| \frac{\sigma_{min}}{f_c'} \right| \qquad (4.13b)$$

Using these equations, the tensile fatigue strength would be 60% and 54%, respectively, of the static strength for 2 million cycles of loading.

Saito and Imai[39] used friction grips to conduct direct tension fatigue tests on $2.8 \times 2.8 \times 29$ in. $(71 \times 71 \times 736 \text{ mm})$ concrete prisms with enlarged ends having a compressive strength of 5600 psi (39 MPa). Sinusoidal pulsating loads were applied at a constant rate of 240 cycles/minute. Maximum stress levels varied from 75% to 87.5% of the static strength while minimum stress level was maintained at 8%. The ratio of minimum to maximum stress, R, was in the range of 0.09 to 0.11. The surfaces of all specimens were coated with paraffin wax to prevent drying during fatigue test. Based on their results, they proposed the following *S-N* relationship for a 50% probability of failure

$$S = 98.73 - 4.12 \log N \qquad (4.14)$$

where S = maximum applied stress range (as percentage of f_c')
           N = number of cycles to failure.

Using their equation, they estimated the fatigue strength for 2 million cycles under direct tensile loading to be 72.8% of the static strength. They observed that this fatigue strength was considerably higher than fatigue strength under indirect tension tests.

4.5   Reversed loading

Tepfers[45] studied the fatigue of plain concrete with static compressive strength in the range of 3000 to 10,000 psi (21 to 70 MPa) subjected to cyclic compression-tension stresses. He used two different testing configurations. One was transversely compressed concrete cubes subjected a pulsating splitting load, and the other was concrete prisms with axial pulsating compressive loads and central splitting line loads. He observed a slight reduction in the fatigue strength of concrete subjected to reversed cyclic loading when compared to the fatigue strength of concrete in compression. He suggested that this reduction could be due to the difficulties in loading the specimens precisely on the tensile side of the pulse. He concluded that the fatigue strength equation proposed by other investigators[1,23,47] could be used to predict the fatigue strength due to reversed stresses.

4.6   Effect of loading rate

Kesler[25] evaluated the effect of loading rate on concrete fatigue strength. Using three different rate of loading, namely: 70, 230 and 440 cycles/minute, and two different concrete strength ($f_c' = 3600$ and 4600 psi (25 to 32 MPa)), he concluded that the rate of loading within the range used in the investigation had little or no effect on the fatigue strength.

Sparks and Menzies[41] used a triangular wave form with constant loading and unloading to study the effect of loading rate on the fatigue compression strength of plain concrete made of three different types of coarse aggregate. The rate of loading were 70 and 7000 psi (0.5 to 49 MPa)/s. The minimum stress was constant ($0.33f_c'$) while the maximum stress varied between $0.70f_c'$ and $0.90f_c'$. They observed an increase in the fatigue strength with an increase in rate of loading of the fatigue load.

4.7   Effect of stress gradient

Ople and Hulsbos[34] studied the effect of stress gradient on the fatigue strength of plain concrete. They tested $4 \times 6 \times 12$ in. ($100 \times 152 \times 304$ mm) prisms under repeated compression at arate of 500 cycles/minute with three different eccentricities (0, 1/3 and 1 in. (0, 8.4, 25.4 mm)). The tests were performed until failure or up to 2 million cycles. Keeping the minimum stress constant at 10% of the static strength, they varied the maximum stress from 65% to 95% the static strength. They found that the mean S-N curves of both concentrically and eccentrically loaded samples were parallel. They concluded that the fatigue strength of eccentrically stresses specimens was higher than that of concentrically stressed speimens by about 17% of the static strength. They also reported that the fatigue life of both type of specimens was highly sensitive to small variations in maximum stress levels.

4.8   Effect of rest periods

Hilsdorf and Kesler[18] investigated the fatigue strength of 185 $6 \times 6 \times 60$ in. ($150 \times 150 \times 1500$ mm) plain concrete prisms subjected to varying flexure stresses. Using a rate of 450 cycle/minute and a ratio of minimum/maximum stress of 0.17, they loaded the specimens until failure or 1 million cycles with five different rest periods of 1, 5, 10 and 27 minutes. Their results indicated that the increase in the rest period increased the fatigue strength for a specified fatigue life. This was clear when the length of rest period increased from 1 to 5 minutes. From 5 to 27 minutes of rest periods, the fatigue strength did not show any variation.

4.9   Effect of loading waveform

Tepfers *et al.*[46] studied the effect of loading waveforms on the fatigue strength. They used three different waveforms, sinusoidal, triangular and rectangular. The results indicated that the triangular waveform was less damaging than the sinusoidal, while the rectangular waveform was the most damaging.

4.10   Effect of minimum stress: comparison of normal and high strength concrete

Petkovic *et al.*[36] conducted tests on high strength concrete (8000 psi (64 MPa) to 11,600 psi (87.2 MPa)) subjected to constant amplitude but different levels of stress. The minimum stress level varied from 0.05 to 0.6 and the maximum stress level varied from 0.6 to 0.95 of the compressive strength. The results of the tests showed no reason to distinguish between the fatigue properties of normal and high strength concrete when the stress levels are expressed relatively to the static strength of the concrete.

4.11   Effect of concrete mixture properties and curing

Raithby and Galloway[37] studied the effects of moisture condition, and age on fatigue of plain concrete with static strength in the range of 3000 psi (21 MPa) to 6400 psi (44.8 MPa). They observed that the moisture condition significantly affected the fatigue life. Oven-dried concrete showed the longest fatigue life while partially dried concrete gave the shortest. The fully saturated concrete exhibited intermediate fatigue life. They suggested that the difference in strains generated by moisture gradient with the concrete could be the cause of such performance. They also found that the mean fatigue life increased with the age of concrete. The mean fatigue life of 2 years old concrete was 2000 times the fatigue life of 4 weeks old concrete.

Klaiber and Lee[26] studied the effect of air content, water-cement ratio, coarse-aggregate type and fine aggregate type. The concrete static strength ranged from 2000 to 7400 psi (14 to 51 MPa). After testing 350 $6 \times 6 \times 36$ in. ($152 \times 152 \times 912$ mm) beams under flexural fatigue, they found that of the variables investigated, air content and coarse aggregate type had the greatest effect on flexural fatigue strength. The fatigue strength decreased with the increase of the air content, and concrete made of gravel yielded higher fatigue strength than concrete made with limestone. Water/cement ratio also affected fatigue strength but to a lesser degree. The fatigue strength decreased for low water/cement ratio (less than 0.32) but seemed not affected for higher water/cement ratios.

In their study on the fatigue behavior of high strength concrete, Petkovic *et al.*[36] investigated the influence of different moisture conditions and size of test specimens on fatigue. Cylinder sizes of 2×6 in. (50×150 mm) and 4×12 in. (100×300 mm) for three different concrete types were studied under three moisture conditions: in air, sealed and in water. The cyclic loading was sinusoidal with the maximum and minimum load levels equal to 70% and 5%, respectively, of the static strength. They found the moisture effects on fatigue to be scale dependent. Dried specimens of small dimensions gave generally longer fatigue lives. The sealed conditions was found to give results closer to the immersed specimens than to the specimens exposed to air. They concluded that sealed cylinders are preferable for fatigue tests carried out using relatively small specimens.

4.12 Biaxial state

Takhar *et al.*[43] studied the fatigue behavior of 96 concrete cyclinders with a compressive strength of 5000 psi (35 MPa) subjected to three different confining pressures (0, 1000 and 2000 psi (14 MPa)). They used sinusoidal load at a rate of 60 cycle/minute. Keeping the minimum stress level at $0.2f_c'$ they varied the maximum stress level (0.8, 0.85 and $0.9f_c'$). They found that the increase in confining pressure prolonged the fatigue life of concrete. The effect of the lateral confining pressure was dependent on the maximum stress level of the fatigue load. For a maximum stress level of $0.90f_c'$, the difference in fatigue behavior with or without the lateral confining pressure was not significant, while for a maximum stress level of $0.80f_c'$ the difference was significant.

Traina and Jeragh[18] performed an experimental investigation to study the behavior of plain concrete with compressive strength of 4000 psi (28 MPa) subjected to slow cyclic loading in compressive biaxial states. Three inch concrete cubes were subjected to two types of biaxial stress states. The first was a proportional loading type in which two loading paths are used, namely $\sigma_2/\sigma_1 = 1.0$ and $\sigma_2/\sigma_1 = 0.5$. The second loading type consisted of a constant stress in the direction with a cyclic stress in the σ_1 direction. The cyclic stress varied from zero up to 1.2 of the unconfined ultimate compressive strength. They found that concrete tested under all biaxial states of stress exhibited higher fatigue strength than uniaxial states of stress for any given number of cycles. They also observed that the stress-strain response of concrete is dependent on the stress level and number of load repetitions for both uniaxial and biaxial states of stress. They noticed a limiting value of volume change per unit volume at which concrete may be considered either failed or near failure. This limiting value was found to be higher for all biaxial states of stress and independent of the stress level at which concrete was subjected to fatigue loading.

4.13   Bond properties

In addition to portland cement, water, aggregates, reinforcing bars and/or prestressing reinforcerment, fresh high-strength concrete usually contain chemical admixtures and pozzolanic materials. Hence, in cured high-strength concrete several types of interface exist, namely: (a) interfaces between the various chemical components that make up the hydrated cement paste (hcp), (b) interfaces between (hcp) and other unhydrated cement particles and added pozzolanic materials, (c) interfaces between (hcp) and coarse aggregates, and (d) interfaces between the concrete matrix and the steel reinforcement.

The bond at the interface at any of the preceding levels is the outcome of a combination of mechanical interlock, physical bonding involving van der Waals' forces and chemical ionic reactions between the different phases of the hydrated paste. Mindess[30,31] discussed the importance of these types of bonds with respect to the behavior of concrete. The following sections cover the first three types of interfacial bonds.[30,31]

Hydrated cement paste interfaces

Hydrated cement paste consists of individual chemical components of hydration products. The paste derives its strength from: (a) intraparticle bonds, represented by the inter-atomic forces within the individual chemical components resulting from hydration (C-S-H and $Ca(OH)_2$); chemical ionic-covalent bonds are considered to be the major 2 source of intraparticle bonds; and (b) interparticle bonds, originated due to the atoms forces which attract the individual paste particles to each other. Physical bonds of the van der Waals' type act primarily between particles.

Micro-structure studies of cement paste indicate that the C-S-H is very well bonded to the various hydration phases. In addition, a strong adhesion seems to exist between the C-S-H and the $Ca(OH)_2$. These studies suggest that the strength of the cement mortar is controlled by the total porosity, the pore size distribution and any existing macroscopic flaws, rather than by the destruction of the bond between its components.

Hydrated cement paste–pozzolanic materials interfaces

Most of the high-strength concrete currently being produced contains silica fume and/or fly ash. The micro-structure that is obtained is distinctly different from that of traditional cement pastes. The structure becomes much denser and more amorphous. Boundaries between C-S-H particles are not clearly identified. However, the strength of the resulting hydrated cement paste is not significantly improved by this interparticle bond enhancement. It was demonstrated that the cement paste with and without silica fume yields the same strength at equal water/binder ratios. This would confirm that enhancing interparticle bonds would have only a minor effect on the strength of hydrated cement paste.

Hydrated cement paste–aggregate interfaces

Many investigations have dealt with the cement–aggregate interfaces. In general, there is an agreement that for normal strength concrete this interfacial region is the weak link in the concrete matrix. This is because bleed water accumulates at the lower surface of the coarse aggregate particles creating a porous paste and planes of weakness. This interface zone is generally composed of a duplex film and a transition zone. The duplex film which is usually no more than 1 μm thick is formed of about 0.5 μm thin layer of $Ca(OH)_2$ in contact with the aggregate followed by another thin layer of C-S-H. The composition of this 'transition zone' is deeply different from the bulk cement paste.

Addition of pozzolanic materials increase the strength of concrete. This is achieved primary because they are capable of producing a great reduction in the relative thickness of the transition zone. This would yield a better overall homogeneity producing higher strength for the paste–aggregate link. The extent to which this enhancement is achieved varies according to the characteristics of the individual pozzolanic material, its addition percentage and the age of curing. The silica fume was found to be the most effective pozzolanic additive in reducing the thickness of the transition zone and in achieving better overall homogeneity of the matrix. Hence, pozzolanic admixtures in general improves the bond at the hcp–aggregate interfaces. By strengthening this weak link, higher strength concrete is obtained. This bond at the interface is, sometimes, greatly improved, resulting in strength of the coarse aggregate becoming the limiting factor of the high strength that can be achieved. There is one major disadvantage which arises from an increase in the cement–aggregate bond strength. That is the ductility of the concrete decreases. This could be attributed to the decrease in extensive microcracking at the interface before failure.

While some knowledge about the micro-structure at the cement–aggregate interfaces has been accumulated, considerable research is still required to provide a more general and global understanding of interfacial micro-structure behavior of normal and high strength concrete.

Reinforcing steel–concrete bond

Reinfoced concrete is a structural material whose effectiveness depends on the interaction between the concrete matrix and reinforcement rebars. Three mechanisms can be identified that contribute to the bond between concrete and steel reinforcement:[6]

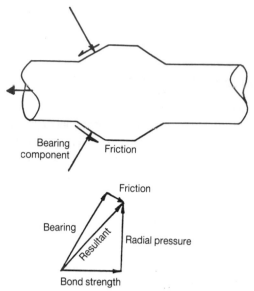

Fig. 4.9   Bond strength components for bar embedded in concrete[50]

1   Adhesive bond between steel and concrete matrix.
2   Frictional bond between the steel and the surrounding matrix.
3   Mechanical anchoring of the steel to the concrete through the bearing stresses that develop between the concrete and the deformations of the steel bars.

When a bar is pulled, the rib bears against the surrounding concrete. Friction and adhesion between the concrete and steel along the face of the rib act to prevent the rebar from sliding. This force adds vectorially to the bearing stress acting perpendicular to the rib to yield the bond strength (Fig. 4.9). The bearing stress is controlled by the radial pressure that the concrete cover and lateral reinforcement can resist before splitting and the effective shear strength capable of shearing the concrete surrounding the rebar. Pullout failure occurs when the steel bar is well confined by concrete cover or transverse reinforcement prevents a splitting failure. The pullout failure is primarily due to bearing of the ribs against the concrete causing the key between ribs to shear from the surrounding concrete.

Clark's[11] study had firmly established the effectiveness of using deformed reinforcing bars in concrete. Chapman and Shah[9] conducted an investigation to determine the bond strength between reinforcing steel and concrete at early age. They found that smooth bars did not exhibit any age effect, while the bond behavior of deformed bars was highly age dependent. They concluded that adhesion and friction contribution to the bond strength is relatively small compared to the bond strength that derives from the bearing stresses that develop between the deformations on the steel and the surrounding concrete.

Brettmann *et al.*[8] studied the effect of superplasticizers, extensively used when producing high strength concrete, on concrete–steel bond strength. They included the effect of degree of consolidation, concrete slump, concrete temperature and bar position on the bond strength of #8 deformed reinforcing bars embedded in concrete with and without super-plasticizer. The bond tests were conducted at concrete strength between 4000 and 4800 psi (28 to 33 MPa) using a modified cantilever beam.

The experimental results indicated that high-slump superplasticized concrete provided a lower bond strength than low-slump concrete of the same strength. They also observed that vibration of high-slump concrete increased the bond strength compared to high-slump concrete without vibration.

Treece and Jirsa[49] conducted experimental investigation to study the bond strength and epoxy coated reinforcing bars embedded in normal and high strength concrete, and compared it to that of uncoated bars. Twenty-one beams with splices in a constant moment region were tested to evaluate the effect of bar size, concrete strength, casting position and coating thickness. Concrete strength varied from 3860 to 10,510 psi (27 to 74 MPa). The results showed that epoxy coating significantly reduced the bond strength of reinforcing bars; for splitting failure the bond strength was about 65% of the bond strength of uncoated bars while for a pullout failure, the bond strength was about 85% of that for uncoated bars. The results also indicated that the reduction in bond strength was independent of bar size and concrete strength, and that the bond strength was not affected by the variations in the coating thickness when the average coating thickness was between 5 and 14 mil (1 mil = 0.025 mm).

Kemp[24] conducted a comprehensive experimental research plan that included a study of the influence of reinforcing bar, embedment length and spacing, stirrups, concrete cover and associatd concrete strength (up to 6000 psi (42 MPa)), and the interaction of shear and flexural bond behavior. He proposed the following design equation which he found suitable for ultimate load design.

$$(F_b)_{\text{ult}} = 232.2 + 2.716 \left(\frac{C_{bs}}{D_{ia}} \sqrt{f_c'} \left[+ 0.201 \right) \frac{A_{sst}f_{yst}}{S_p D_{ia}} \right] \tag{4.15}$$

$$+ 195.0 I_{\text{aux}} + 21.06 (F_d N)^{0.66}$$

where
A_{sst} = area of transverse reinforcement, in.^{2}
C_{bs} = the smallest concrete cover, in.
D_{ia} = diameter of reinforcing bar, in.
f_c' = concrete compressive strength, psi
F_d = dowel force psi bar, kip/bar.
f_{yst} = yield strength of transverse reinforcement, psi
I_{aux} = parameter for auxiliary reinforcement (=1 when the member has auxiliary reinforcement and 0

when the specimen is without auxiliary reinforcement)

S_p = center-to-center spacing between two adjacent
transverse reinforcement, in.

N = number of load cycles.

Gjørv et al.[15] studied the effect of silica fume on the mechanical behavior
of the steel reinforcement-concrete bond, the concrete compressive
strength from 3000 psi (42 MPa) to 12,000 psi (84 MPa) with and without
silica fume. Using a #6 bar, they found that for the same compressive
strength increased addition of silica fume up to 16% by weight of cement
showed an improving effect on the bond strength, especially in the high
compressive stress range. They justified this effect by several mechanisms:
reduced accumulation of free water at the interface during casting of the
specimens, reduced preferential orientation of CH crystals at the transition
zone, and densification of the transition zone due to pozzolanic reaction
between CH and silica fume.

Chimamphant[10] conducted pull-out tests on three different reinforcing
bar diameters (#3, #4 and #5) embedded in concrete having a compressive strength ranging from 7500 psi (52 MPa) to 12,000 psi (84 MPa). He
observed that the larger the bar diameter, the lower the average bond
strength. In this study, the average bond strength factor (average bond
strength/compressive strength) was about 0.212. Most reported values for
that factor when using nominal strength concrete range from 0.15 to 0.26.
These values get higher if lateral confinement is provided. The results of
this study indicated that normalized bond in both high strength concrete
and normal strength concrete are essentially the same.

Using 20% (by weight) of silica fume to obtain high strength concrete,
Ezeldin and Balaguru[14] conducted experimental studis on the bond
behavior of bars embedded in high tensile concrete with and without steel
fibers (compressive strength up to 11,800 psi (82 MPa)). Four bar diameters were used, namely, #3, #5, #6 and #8. They found that the
addition of silica fume increased the bond strength of concrete. However,
in presence of fibers, the proportionality constant between bond strength
and square root of the compressive strength seemed to be constant. They
concluded that the bond strength equations used for normal strength
concrete (without silica fume) could be used for high strength fiber
reinforced concrete (with silica fume).

4.14   Summary

The use of high strength concrete in modern structures and the widespread
adoption of ultimate strength design procedures renewed the interest in
studying the fatigue and bond properties of high strength concrete.

Most of the fatigue investigations have been performed on concretes
of normal strength. These investigations have established the effect of the

range of stress, load history, rate of loading, stress gradient, curing and material properties on the fatigue behavior. The limited fatigue studies conducted on high strength concrete covered the effects of these variables on the fatigue properties. Their results indicated no signifcant difference between the fatigue behavior of normal and high strength concretes when the stress levels are expressed relatively to the static strength of concrete.

The bond at the hydrated cement paste–aggregate interfaces is greatly improved in high strength concrete because of the addition of pozzolanic materials. This could result in the coarse aggregate characteristics becoming the limiting factor of the high strength that can be achieved. The average reinforcing steel–concrete matrix bond is increased for high strength concrete when compared to normal strength concrete. More research is needed on the morphology and micro-structure of the steel–cement paste transition zone in order to characterize the effect of including pozzolanic materials in high strength concrete on the bond properties.

References

1 Aas-Jakobson, K. (1970) *Fatigue of concrete beams and columns*, Bulletin No. 70-1. NTH Institute of Betonkonstruksjoner, Trondheim.
2 ACI Committee 215 (1974) *Fatigue of concrete*. American Concrete Institute, SP-41.
3 ACI Committee 215 (1974) Consideration for design of concrete structures subject to fatigue loading. *Journal of the American Concrete Institute, Proceedings*, **71**, 3, 97–121. Revised in 1986.
4 ACI Committee 363 (1984) State-of-the-art report on high-strength concrete. *ACI Journal*, Title No. 81-34, 364–411.
5 ACI Committee 363 (1987) Research needs for high-strength concrete. *ACI Journal*, Title No. 84-M49, 559–61.
6 Bartos, P. (ed.) (1982) *Bond in concrete*. Applied Science Publishers.
7 Bennett, E.W. and Muir, S.E.st.J. (1967) Some fatigue tests on high-strength concrete in axial compression. *Magazine of Concrete Research*, London, **19**, No. 59, 113–17.
8 Brettmann, B., Darwin, D. and Donahey, R. (1986) Bond of reinforcement to superplasticized concrete, Proceedings. *ACI Journal*, 98–107.
9 Chapman, R.A. and Shah, S.P. (1987) Early age bond strength in reinforced concrete. *ACI Journal*, **84**, No. 6, 501–10.
10 Chimamphant, S. (1989) Bond and Fatigue Characteristics of High-Strength Cement-Based Composites, Ph.D. Dissertation, New Jersey Institute of Technology, Newark, New Jersey.
11 Clark, P. (1949) Bond of concrete reinforcing bars. *ACI Journal, Proceedings*, **46**, 3, 161–84.
12 Clemmer, H.E. (1922) Fatigue of concrete. *Proceedings, ASTM*, **22**, Part II, 408–19.
13 Cornelissen, H.A.W. and Timmers, G. (1981) *Fatigue of plain concrete in uniaxial tension and alternating tension- compression*, Report No. 5-81-7, Stevin Laboratory, University of Technology, Delft.
14 Ezeldin, A.S. and Balaguru, P.N. (1989) Bond behavior of normal and high-strength fiber reinforced concrete. *ACI Materials Journal, Proceedings*, **86**, 5, September–October, 515–24.
15 Gjørv, O.E., Monteiro, P.J. and Mehta, P.K. (1990) Effect of condensed silica

fume on the steel–concrete bond. *ACI Materials Journal, Proceedings*, **87**, 6, Nov–Dec, 573–80.

16   Graf, O. and Brenner, E. (1934/1936) *Experiments for investigating the resistance of concrete under often repeated loads* (Versuche Zur Ermittlung der Widerstandsfahigkeit Von Beton gegen oftmals Wiederholte Druck-belastung), Bulletins No. 76 and No. 83, Deutscher Ausschuss fur Eisenbeton.

17   Gray, W.H., McLaughlin, J.F. and Antrim, J.D. (1961) Fatigue properties of lightweight aggregate concrete. *ACI Journal*, Title No. 58-6, August, 149–61.

18   Hilsdorf, H.K. and Kesler, C.E. (1966) Fatigue strength of concrete under varying flexural stresses. *Journal of American Concrete Institute, Proceedings*, **63**, 10, Oct, 1069–76.

19   Holmen, J.O. (1979) *Fatigue of concrete by constant and variable amplitude loading*, Bulletin No. 79-1, Division of Concrete Structures, Norwegian Institute of Technology, University of Trondheim.

20   Hsu, T.C. (1981) Fatigue of plain concrete. *ACI Journal, Proceedings*, Title No. 78-27, July–August, 292–305.

21   (1982) *International association for bridge and structural engineering*, Proceedings of Colloquium, Lausanne, IABSE Reports, **37**.

22   Jinawath, P. (1974) Cumulative Fatigue Damage of Plain Concrete in Compression, Ph.D. Thesis, University of Leeds.

23   Kakuta, Y. *et al.* (1982) New concepts for concrete fatigue design procedures in Japan. *Proceedings, IABSE*, Lausanne, **37**, 51–8.

24   Kemp, E. (1986) Bond on reinforced concrete: behavior and design criteria. *ACI Journal, Proceedings*, **82**, 1, Jan–Feb, 49–57.

25   Kesler, C.E. (1953) Effects of speed of testing on flexural fatigue strength of plain concrete. *Proceedings, Highway Research Board*, **32**, 251–8.

26   Klaiber, F.W. and Lee, D.Y. (1982) *The effects of air content, water-cement ratio, and aggregate type on the flexural fatigue strength of plain concrete.* American Concrete Institute, Special Publications, SP-75, 111–32.

27   Kolias, S. and Williams, R.I.T. (1978) *Cement-bound road materials: strength and elastic properties measured in the laboratory*, TRRL Report No. 344. Transport and Research Laboratory, Crowthorne, Berkshire.

28   Leeuwen, J.V. and Siemes, J.M. (1979) Miner's rule with respect to plain concrete. *Heron*, **24**, 1, 34 pp.

29   McCall, J.T. (1958) Probability of fatigue failure of plain concrete. *ACI Journal*, Aug, 233–44.

30   Mindess, S. (1989) Interfaces in concrete, in *Materials science of concrete I*, edited by Skalny, J.P. The American Ceramic Society, 163–180.

31   Mindess, S. (1988) Bonding in cementitious composites – how important is it, in *Bonding in cementitious composites*, edited by Mindess, S. and Shah, S.P. Materials Research Society, **114**, 3–10.

32   Miner, M.A. (1945) Cumulative damage in fatigue. *Transactions, American Society of Mechanical Engineers*, **67**, A159–A164.

33   Murdock, J.W. and Kesler, C.E. (1958) Effect of range of stress on fatigue strength of plain concrete beams. *ACI Journal*, August, 221–31.

34   Ople, Jr, F.S. and Hulsbos, C.L. (1966) Probable fatigue life of plain concrete with stress gradient, Research Report. *ACI Journal*, Title No. 63-2, January, 59–81.

35   Palmgren, A. (1924) 'Die Lebensdauer von Kugellagern, VDI.' *Zeitschrift Verein Deutscher Ingenieur*, **68**, 339–41.

36   Petkovic, G., Lenschow, R., Stemland, H. and Rosseland, S. (1991) *Fatigue of high-strength concrete*. American Concrete Institute, Special Publication, SP 121-25, 505–25.

37   Raithby, K.D. and Galloway, J.W. (1974) *Effects on moisture condition, age,*

and rate of loading on fatigue of plain concrete. ACI Publications, SP-41, 15–34.

38 RILEM Committee 36-RDL (1984) Long term random dynamic loading of concrete structures. *RILEMs Materials and Structures*, **17**, 97, Jan, 1–27.

39 Saito, M. and Imai, S. (1983) Direct tensile fatigue of concrete by the use of friction grips. *ACI Journal*, Title No. 80-42, Sept–Oct, 431–8.

40 Shah, S.P. (1982) *Fatigue of concrete structures.* American Concrete Institute, SP-75.

41 Sparks, P.R. and Menzies, J.B. (1973) The effect of rate of loading upon the static and fatigue strengths of plain concrete in compression. *Magazine of Concrete Research*, **25**, 83, June, 73–80.

42 Su, E.C.M. and Hsu, T.T.C. (1986) *Biaxial compression fatigue of concrete*, Research Report UHCE 86-17. University of Houston, December.

43 Takhar, S.S., Jordaan, I.J. and Gamble, B.R. (1974) *Fatigue of concrete under lateral confining pressure.* ACI Publications, SP-41, 59–69.

44 Tepfers, R. (1979) Tensile fatigue strength of plain concrete. *ACI Journal*, Title No. 76-39, August, 919–33.

45 Tepfers, R. (1986) *Fatigue of plain concrete subjected to stress reversals.* ACI Publications, SP-75, 195–215.

46 Tepfers, R., Gorlin, J. and Samuelsson, T. (1973) Concrete subjected to pulsating load and pulsating deformation of different pulse wave-form. *Nordisk Betong*, No. 4, 27–36.

47 Tepfers, R. and Kutti, T. (1979) Fatigue strength of plain, ordinary, and lightweight concrete. *ACI Journal*, Title No. 76-29, May, 635–53.

48 Traina, L.A. and Jeragh, A.A. (1982) *Fatigue of plain concrete subjected to biaxial-cyclical loading.* American Concrete Institute, Special Publications, SP-75, 217–234.

49 Treece, R.A. and Jirsa, J.O. (1989) Bond strength of epoxy-coated reinforcing bars. *Proceedings, ACI Materials Journal*, **86**, 2, March–April, 167–74.

50 Van Ornum, J.L. (1903) Fatigue of cement products. *Transactions, ASCE*, **51**, 443.

51 Van Ornum, J.L. (1907) Fatigue of concrete. *Transactions, ASCE*, **58**, 294–320.

52 Williams, H.A. (1943) Fatigue tests of lightweight aggregate concrete beams. *ACI Journal, Proceedings*, **39**, April, 441–8.

5 Durability

O E Gjørv

5.1 Introduction

Recent developments in concrete technology have made it possible to produce concrete mixtures with strength properties that are beyond the strengths that are currently used by the structural design practice. For high-quality natural mineral aggregates, compressive strengths of up to 33,000 psi (230 MPa) can now be produced under laboratory conditions.[1] If the mineral aggregate is replaced by high-quality ceramic aggregate, compressive strengths of up to 65,000 psi (460 MPa) can be achieved under controlled laboratory conditions.[2] Even with lightweight aggregate, compressive strengths of more than 14,000 psi (100 MPa), with a density of less than 3200 lb/yd^3 (1900 kg/m^3), can be obtained.[3] However, very often it is not the improved strength which is the primary objective but rather the improved durability and overall performance. Therefore, the term 'high-performance concrete', is inclusive of the term 'high-strength concrete'. In this chapter, a brief summary of the most recent developments on durability of high-strength concrete is presented.

5.2 Permeability

One of the main characteristics of high-strength concrete compared to that of the normal-strength concrete is the more uniform and homogeneous microstructure. When the portland cement is combined with the ultrafine particles of silica fume in low w/c ratios, the microstructure of such systems consists mainly of poorly crystalline hydrates forming a more dense matrix of low porosity. With increasing silica fume content, larger content of the calcium hydroxide is transformed into calcium silicate hydrates (Fig. 5.1), while the remaining calcium hydroxide tends to form smaller crystals compared to that in pure portland cement pastes. From Table 5.1 it can be seen that with the addition of higher percentage of silica fume, the

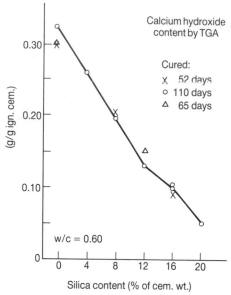

Fig. 5.1   Effect of silica fume on the calcium hydroxide content of ordinary Portland cement[4]

calcium/silicate ratio of the hydrates decreases, which allows the hydrates to incorporate ions, such as alkalis and aluminium. As a consequence, it appears that an increased resistance to aggressive ions and alkali-aggregate reaction is obtained with an increase in the silica fume content. Also, the electrical resistivity is increased.[5]

For concrete with very low w/c ratios, microcracking due to self-desiccation may affect the concrete permeability. As can be seen from Table 5.2 it is primarily the presence of silica fume in high-strength concrete, which appears to increase the autogenous shrinkage. While high-strength concrete without silica fume has an autogenous shrinkage of the same order as that of normal-strength concrete, the presence of silica fume may increase the autogenous shrinkage by up to twice that of normal-strength concrete.

One of the main effects of silica fume in high-strength concrete, however, is to improve the microstructure of the transition zone between the aggregate and the cement paste. For normal-strength concrete on a pure portland cement basis, the transition zone around the aggregate, which is 20 to 100 μm wide, has a very different microstructure compared to that of the bulk matrix.[6,7] This transition zone which is inferior in quality

Table 5.1   Effect of silica fume on the calcium/silicate-ratio of the hydrates[5]

Cement	Ca/Si-ratio
Portland cement (OPC)	1.6
OPC + 13% silica fume	1.3
OPC + 28% silica fume	0.9

Table 5.2   Effect of silica fume on the autogenous shrinkage

Type of concrete	Autogenous shrinkage μ m/m	References
NSC	90–120	6, 7
HSC	80–90	7, 8
HSC with silica fume		6, 8

and thus leads to a poorer bond between the aggregate and the cement paste, is typically characterized by the following key elements:

1   The transition zone is richer in calcium hydroxide and ettringite than the bulk phase, and the calcium hydroxide is oriented. A rim of massive calcium hydroxide can often be observed around the aggregates.
2   The porosity of the transition zone is greater than that of the bulk phase, and a gradient in porosity can be observed with a declining trend as the distance from the aggregate surface increases.

For normal-strength concrete, the special microstructure of the transition zone is apparently related to the formation of water-filled spaces around the aggregates in the fresh concrete, due to internal bleeding and to a 'wall effect', which interferes with inefficient packing of the cement particles around the aggregates. Calcium hydroxide and ettringite are known to preferably grow in large pores, which accounts for the greater contents of these phases in the transition zone.[8,9] Also, this zone has larger w/c ratio relative to the bulk[10] and is therefore characterized by higher porosity.

When silica fume is introduced into the system, and in particular in high-strength concrete, considerable changes in the microstructure of the transition zone take place. Regourd et al.[11,12] and Aitcin et al.[13,14] observed that high-strength concrete with silica fume was not as crystallized and porous as normal-strength concrete, and all of the space in the vicinity of the aggregate was occupied with amorphous and dense calcium silicate hydrates. Also, direct contact was formed between the aggregate and the calcium silicate hydrates rather than with calcium hydroxide as in normal-strength concrete. Scrivener et al.[15] quantified the interfacial microstructure and demonstrated that in high-strength concrete with silica fume, the porosity of the transition zone was practically eliminated (Fig. 5.2), and practically no gradient in porosity was observed, in contrast to normal-strength concrete.

As demonstrated in Table 5.3 the addition of silica fume may have a substantial effect on the permeability of concrete. By adding 20% of silica fume to 169 lb/yd^3 (100 kg/m^3) of cement (OPC), the same permeability as that of 421 lb/yd^3 (250 kg/m^3) cement is obtained. Addition of 10% silica fume to 421 lb/yd^3 (250 kg/m^3) cement gives a permeability as low as $1.8 . 10^{-14}$ m/s. For meeting the higher durability performance requirements for offshore concrete platforms in the North Sea, the permeability is limited to 10^{-2} m/s.[17]

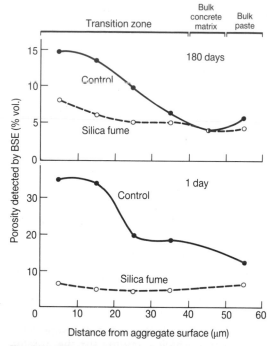

Fig. 5.2   Effect of silica fume on the porosit of the transition zone between aggregate and cement paste[15]

For cement pastes with very low w/c ratio in the range of 0.20 to 0.30, experiments[18] have shown that a 10% replacement of the cement with silica fume did only reduce the total porosity to a small extent. However, a refinement of the pore size distribution took place in such a way that the content of larger pores was reduced for decreasing w/c ratio. The effect of w/c ratio on chloride diffusivity was substantially higher at high than at low w/c ratios, but a 10% replacement with silica fume reduced the diffusivity so much that the effect of w/c ratio became less significant.

The above qualitative and quantitative observations indicate that in high-strength concrete the bulk matrix becomes very dense and typically, this dense matrix extends up to the aggregate surface, in such a way that the inhomogeneity of the transition zone is largely eliminated. It is now

Table 5.3   Effect of silica fume on permeability of concrete[16]

Cement (OPC) lb/yd^3 (kg/m^3)	Silica lb/yd^3 (kg/m^3)	Permeability m/s
168.6 (100)	0 (0)	$1{,}6 . 10^{-7}$
168.6 (100)	16,9 (10)	$4{,}0 . 10^{-10}$
168.6 (100)	33,7 (20)	$5{,}7 . 10^{-11}$
168.6 (100)	0 (0)	$4{,}8 . 10^{-11}$
421.5 (250)	42,1 (25)	$1{,}8 . 10^{-14}$

well documented that this improved microstructure is closely related to the reduced permeability and improved performance of high-strength concrete.

5.3   Corrosion resistance

Since high-strength concrete is characterized by a low porosity and a more uniform microstructure compared to that of normal-strength concrete, this indicates a high resistance to penetration of carbon dioxide and chloride ions into the concrete. However, during production of high-strength concrete, both macrocracking due to plastic shrinkage and microcracking due to self-desiccation may represent potential problems from a corrosion protection point of view. Also, somewhat depending on the amount of mineral admixture used, both the reserve basicity and the ability of the cement paste to bind chlorides may be affected.

By replacing part of the portland cement with a mineral admixture such as fly ash, only modest effects on the pore solution chemistry have been observed,[19] but this may vary with the type of fly ash. Unlike fly ash, however, the presence of silica fume has an extensive and profound effect on the pore solution chemistry.[19,20]

For an increasing replacement of the cement by silica fume, the concentrations of both K^+ and OH^- ions are substantially lowered. However, by replacing up to 20% it appears from Fig. 5.3 that the pH does not drop below that of a saturated calcium hydroxide solution which is approximately 12.5. Even at 30% cement replacement the pH does not drop below 11.5 which is considered to be a threshold value for maintaining a good passivity of embedded steel. According to Diamond[19] it appears that the removal of alkalis from the pore solution with consequent lowering

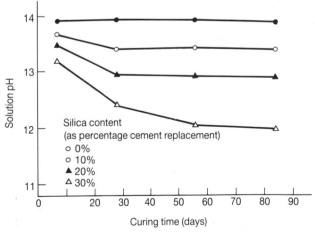

Fig. 5.3   Effect of silica on the alkalinity of cement paste with w/c = 0.50[19]

Table 5.4   Effect of slag on chloride penetration[21]

Type of mix	Mix proportion			Chloride concentration (ppm)		
Material %	W/C	C/S	HRWR %	Surface layer 0–0.27 in (0–7 mm)	Medium layer 0.27–0.5 in (7–14 mm)	Deep layer 0.55–0.79 in (14–20 mm)
Control 0	0.4	1/1.5	1.0	3970	521	42
OR slag 40	0.4	1/1.5	1.0	4570	259	62
VF slag 40	0.4	1/1.5	1.0	2910	111	34
VFG slag 40	0.4	1/1.5	1.0	3350	52	28
Silica fume 10	0.4	1/1.5	1.0	5110	312	43

of pH is less complete at low w/c ratios. Thus, it appears that it is carbonation of the concrete and penetration of chlorides which should be the controlling factors for the passivity of embedded steel. There are no data reported in the literature, however, that carbonation of high-strength concrete either with or without mineral admixtures represents any problem.

Nakamura *et al.*[21] have investigated chloride binding of slag mortars with w/c = 0.40 using alternative 24 hr. periods of immersing in seawater and oven drying at 60 °C. The specimens were then analysed for chlorides at three levels, the test results of which are given in Table 5.4. The high surface chloride content of the ordinary slag specimens is attributed to the greater binding capacity for chlorides in this blend than in the ordinary portland cement. The penetration to the middle layer suggests that there is some beneficial effect of the slag, the greatest being for the very finely ground slag with additions of anhydrite. The chloride contents of the inner layers were of the same order as of the reference specimens without exposure to chlorides. It should be noted, however, that these were immature specimens; they were only 7 days old at the beginning of the testing which lasted for only a further 14 days. A greater effect of the slag would be expected for older specimens.

When silica fume is used as a partial replacement for the cement, there is not yet a general consensus of the influence of chloride-binding. Page and Vennesland[20] determined the chloride binding capacity to be substantially reduced as shown in Fig. 5.4, whereas Byfors *et al.*[22] observed an increase in the amount of bond Cl⁻ in ordinary portland cement with 10% silica fume additions relative to the ordinary portland cement. If the concrete becomes carbonated, it is well established that the capacity for chloride binding is distinctly reduced.[23] It appears that, since carbonation is not a problem for high-strength concrete, the effect of carbonation on a possible chloride penetration should not represent any problem.

Recent investigations on offshore concrete platforms in the North Sea indicate that the rate of chloride penetration into concrete made with portland cement alone is much higher than expected earlier, but not much information on chloride penetration through high-strength concrete is

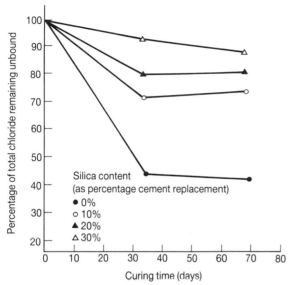

Fig. 5.4 Effect of silica fume on chloride binding capacity of hydrated cement paste of w/c = 0.50 with 0.40% chloride by weight of cement and silica fume[20]

available. Since the rate of chloride penetration is generally reduced with reduced permeability, it appears that high-strength concrete should provide a better protection against chloride penetration than normal strength concrete. Addition of mineral admixtures such as silica fume, slag or fly ash also increases the resistance to chloride penetration.[24,25] If sufficient amounts of chlorides reach the embedded steel, however, electrical resistivity and availability of oxygen are the additional factors which control the corrosion rate.

Since an electrical current passes through the concrete in the form of charged ions, it is reasonable to assume that there is close relationship between electrical resistivity, ion concentration and porosity. The moisture content of concrete is also an important factor. If the concrete is dry enough, the resistivity may be too high to allow any significant transport of ions, and significant corrosion rate with not occur.

Figure 5.5 illustrates the effect of moisture content on the electrical resistivity of normal strength concrete. By successively reducing the moisture content from 100 to 20%, it can be seen that the resistivity increases by approximately three orders of magnitude. In high-strength concrete, the w/c ratio is less than what is theoretically necessary for complete hydration, where self-desiccation of the concrete is a real possibility. Consequently, the mixing water in high-strength concrete will be used up relatively rapidly and long before the clinker hydration is complete. Excluding any interaction with the atmosphere, the relative humidity inside high-strength concrete will be lower than that in normal-strength concrete. If silica fume is also used in the concrete, both the reduced permeability and the changed pore solution chemistry will affect

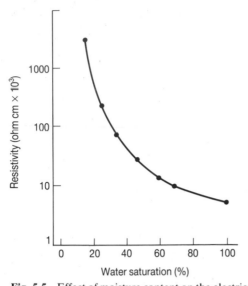

Fig. 5.5   Effect of moisture content on the electrical resistivity of concrete with w/c = 0.40[26]

the concrete in such a way that a dramatic increase in electrical resistivity may be observed, in particular at high cement contents (Fig. 5.6).

For high-strength concrete resistivities of up to 1000 ohm m even for water-saturated conditions have been observed.[27] Observations from existing concrete structures indicate that corrosion of embedded steel hardly represents any practical problem if the electrical resistivity exceeds a threshold level of 5000 to 700 ohm m.[28] Other information indicates that resistivities as low as 200 ohm m can reduce the rate of corrosion to a

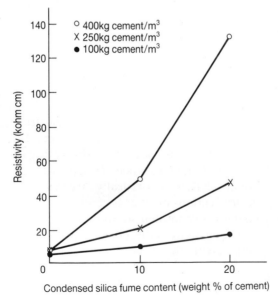

Fig. 5.6   Effect of silica fume on the electrical resistivity of concrete[27]

negligible level.[29] Therefore, it appears that high-strength concrete typically will have an electrical resistivity which is above the level where corrosion of embedded steel will represent any practical problem. If the steel should become depassivated, the rate of corrosion will also be controlled by the oxygen availability.

Even for normal strength concrete very few investigations on oxygen availability have been reported,[30] and for high-strength concrete, no particular information is available. Since the permeability of high-strength concrete is very low, the rate of oxygen transport through high-strength concrete must also be very low. For a cathodic reaction to take place, enough moisture must also be available in order to dissolve the oxygen. For a homogeneous high-strength concrete without any cracks, it is reasonable to assume, therefore, that oxygen availability will generally be very low.

Although the risk of corrosion is generally higher in cracked concrete than in uncracked concrete, the only information available on effect of cracks is based on normal concrete. Even with the comprehensive and realistic experiments carried out by Houston and Furguson[31] on normal concrete, however, it was difficult to establish a simple relationship between crack-width and risk of corrosion. When the concrete elements were investigated at an early stage of exposure, there appeared to be a distinct effect of crack width on corrosion. After longer periods of exposure, however, the effect of crack width was very small or almost non-existent. This is also demonstrated in Fig. 5.7, which shows results obtained for beams with 25 mm concrete cover after 10 years of exposure in different types of atmospheric environments. Also for submerged concrete, recent information has shown that the risk of cracks on corrosion is less than previously expected.[32,33]

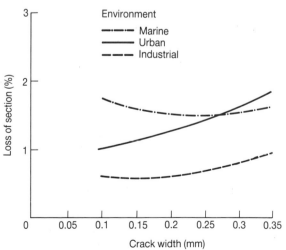

Fig. 5.7   Effect of crack width on steel corrosion in cracked beams after 10 years of exposure[32]

As far as the ability of high-strength concrete to protect embedded steel from corrosion is concerned, not much systematic research has been carried out so far. Based on the general information available, however, it appears that high-strength concrete has a high ability to protect embedded steel. This general conclusion is supported by the good performance observed for offshore concrete platforms in the North Sea, where concrete with compressive strengths of 45 to 70 MPa has been used. For these structures, no problems due to corrosion of embedded steel has been reported so far, even after 15 to 20 years of the combined exposure to heavy mechanical loading and a severe marine environment.

5.4   Frost resistance

Even for normal-strength concrete, production of concrete with a good and stable air-void system is normally a problem,[34,35] but in the presence of a high dosage of superplasticizer, the establishment of a good and stable air-void system may be an even bigger problem.[36] For production of high-strength concrete, it may also be a conflicting requirement to entrain air which will decrease the strength. Therefore, much attention has been given to finding out whether a frost resistant high-strength concrete can be produced without any air entrainment.

The general problem of assessing the frost resistance of concrete is the lack of correlation between existing laboratory test methods and field performance. Also, different test methods and different test conditions appear to give different and conflicting test results.

In 1981, Okada et al.[37] reported very good frost resistance (ASTM C666A-A) of non-air-entrained high-strength concrete with w/c ratios in the range of 0.25 to 0.35. Similar observations were later on reported by Foy et al.[38] and Gagne et al.[39] who observed good frost resistance of non-air-entrained concrete with w/c ratios of 0.25 and 0.30, respectively.

Using ASTM C666-A, Malhotra et al.[40] also tested a number of concretes with different types of cement and w/c ratios of 0.30 and 0.35. They concluded, however, that air entrainment was necessary for these concretes to be frost resistant. Hammer and Sellevold[41] tested non-air-entrained concrete with 0 and 10% silica fume and w/c ratios varying from 0.25 to 0.40. Even for the lowest w/c ratios some of the specimens were damaged during testing. These observations were in conflict with low-temperature calorimeter data, which clearly demonstrated a very low freezable water content. Based upon this, Hammer and Sellevold[41] suggested, therefore, that the observed damage could be due to thermal fatigue caused by too large differences between the thermal expansion coefficients of aggregate and binder rather than ice formation.

As far as salt scaling is concerned, there are also some conflicting results reported in the literature. In 1984 Petersson[42] reported that deterioration of high-strength concrete due to salt scaling was small for the first 50 to 100 cycles, but increased very rapidly to total destruction in the following 10 to

20 cycles. Foy *et al.*,[38] however, observed that the resistance to salt scaling of concrete with a w/c ratio of 0.25 was very good even after 150 cycles. Both Hammer and Sellevold[41] and Gagne *et al.*[39] have demonstrated that it is possible to produce high-strength concrete which is resistant to salt scaling without any air entrainment. These test results included w/c ratios of up to 0.37.

5.5 Chemical resistance

According to Biczók[43] chemical deterioration of concrete can be classified into three types of process depending on the predominant chemical reaction taking place. Leaching corrosion of concrete is the process where parts or all of the hardened cement paste are removed from the concrete. Normally, this is caused by the action of water of low carbonate hardness or carbonic acid content. The next process is corrosion by exchange reactions and by removal of readily soluble compounds from the hardened cement paste. This process occurs as a result of a base exchange reaction between the readily soluble compounds of hardened cement paste and the aggressive solution. The third process is the swelling corrosion, largely due to the formation of new and stable compounds in the hardened cement paste. This process is primarily the result of attack by certain salts. Also alkali-aggregate reactions cause expansion, where the concrete eventually is destroyed by a swelling pressure.

For all of the above deteriorating processes, the permeability of the concrete is the key factor governing the rate of deterioration (Fig. 5.8). In addition, calcium hydroxide is also an easily soluble constituent which is very vulnerable to chemical attack.

In sulfate-containing solutions the calcium hydroxide reacts with the sulfates to produce gypsum which may further react with the aluminates to form ettringite. Both gypsum and ettringite can cause disruptive expansion. Therefore, the pozzolans normally used in high-strength concrete are very effective in reducing the calcium hydroxide (Fig. 5.1), and hence also greatly enhance the resistance against sulfate attack. In addition to

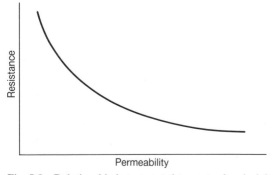

Fig. 5.8 Relationship between resistance to chemical deterioration and permeability[16]

reducing the calcium hydroxide, pozzolans, such as silica fume, also form calcium silicate hydrates which are able to incorporate aluminum ions, thus reducing the amount of aluminum available for ettringite formation.

The beneficial effect of silica fume in high sulfate-containing environments has been reported in several investigations.[44,45] In these investigations, the performance achieved has been equal or better than that obtained by use of sulfate resistant cements. Mehta[46] exposed a number of high-strength concretes to solutions of both 5% sodium sulfate and 1% concentrations of sulfuric and hydrochloric acid for a period of up to 182 days. Although the portland cement contained 7% C_3A, the results showed that w/c ratios of 0.33 to 0.35 gave too low a permeability to cause any deterioration. In more aggressive environments pure portland cements have shown some deterioration, whereas addition of silica fume has given practically unaffected performance.[45] Even for normal-strength concrete the presence of silica fume will improve the long-term performance of concrete in very aggressive environments. In 1952 a large number of concrete specimens were exposed in a field station in the underground of Oslo city, which consists of very aggressive alum shale. In spite of a sulfate content of up to 5 g/l of SO_3 and a pH of 2.8, a 15% replacement of ordinary portland cement with silica fume gave the same good performance after 26 years of exposure as that by sulfate resisting cements (ASTM Type V).[47]

The presence of pozzolans such as silica fume can also be used to control the expansion caused by alkali-aggregate reaction.[48,49] Porewater analyses of cement paste with silica fume have demonstrated the ability of silica fume to reduce the alkali concentration in the pore water rapidly, thus making it unavailable for the slower reaction with reactive silica in the aggregate.[19,20] Also, for high-strength concrete the effect of self-desiccation may reduce the moisture content to a level where no alkali-aggregate reaction can take place.

Asgeirsson and Gudmundsson[50] used silica fume with Icelandic high-alkali cement and reactive sand in mortar bars to demonstrate the ability of silica fume to reduce the expansion. Later on, field experience with Icelandic cement blended with silica fume in 200 houses constructed during the period of 1979–86 has confirmed these observations.[51]

5.6   Fire resistance

Diederichs *et al.*[52] have shown that high-strength concrete is more vulnerable to elevated temperatures compared to that of normal-strength concrete (Fig. 5.9). For high-strength concrete a distinct loss of strength (30%) was observed already at 150 °C, while normal strength concrete retained its strength up to 250 °C. The effects on compressive strength and modulus of elasticity are shown in Figs. 5.10 and 5.11, respectively.

Investigations carried out on the fire resistance of high-strength concrete

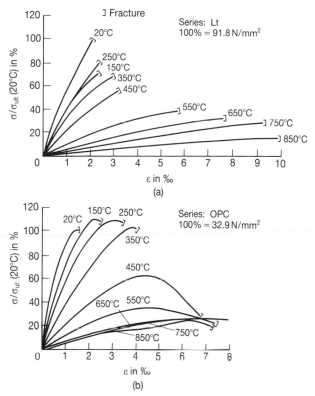

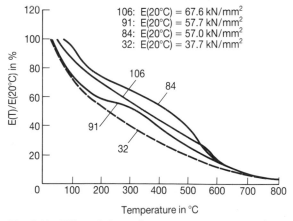

Fig. 5.9 Effect of elevated temperature on the stress-strain relationship for (a) high-strength concrete with ordinary Portlane cement and fly ash, and (b) ordinary Portlane cement[52]

are rather limited. In 1981 Pedersen[53] reported the results of an investigation where 4 in. (100 mm) cylinders of high-strength concrete with a w/c ratio of 0.16 and 20% silica fume were tested. During heating up at a rate of 33.8 °F/min (1 °C/min), several of the specimens suddenly disintegrated

Fig. 5.10 Effect of elevated temperature on compressive strength[52]

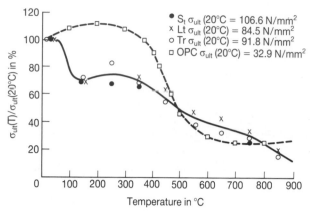

Fig. 5.11 Effect of elevated temperature on modulus of elasticity[52]

at a temperature of approximately 570 °F (300 °C). Shirley *et al.*,[54] however, tested concrete slabs according to ASTM E119,[55] where five different types of concrete with strengths varying from 7145 to 71,145 psi (50 to 120 MPa) revealed no significant difference in behavior. There was no explosive behavior, and none of the concretes showed even minor spalling on the exposed surface. Jensen *et al.*[56] reported a number of test series where different types of concrete elements were exposed to a hydrocarbon fire, where 2012 °F (1100 °C) is reached in 30 min. The strength of the elements varied from 5715 to 11,430 psi (40 to 80 MPa). The test results showed that spalling and damage occurred earlier than expected and that increasing moisture content increased the severity of the effects. Concrete with silica fume was also more sensitive to spalling. Williamson[57] tested concrete slabs according to ASTM 119 with two reference mixes (4286 and 11,429 psi) (30 and 80 MPa) and two mixes with silica fume (7143 and 14,286 psi) (50 and 100 MPa). The relative humidity in the concrete was monitored during the storage period of about six months, where a level of 71–74% RH was reached at the time of testing. No external spalling or internal damage to any of the specimens was observed.

Although the above test results appear to be contradictory, it seems that increasing moisture content leads to increased damage. Also, concrete with silica fume is more dense and dries out more slowly in such a way that a higher vapour pressure can build up internally and cause spalling.[58]

For offshore structures for oil and gas explorations where high-strength concrete may be exposed to the combination of a moist environment and fire, a passive fire-protection material may provide a safety precaution against concrete spalling.[56,59]

Intentionally induced relief for building up vapour pressure may be another approach for increasing the resistance to concrete spalling. Thus, some promising experiments have been carried out by incorporating polymer fibers[60,61] or polymer particles,[62] which melts already at a low temperature and thus provide relief channels for the vapour pressure.

5.7   Abrasion-erosion resistance

Although some building codes have recently increased their upper strength level for structural utilization of concrete to about 14,290 psi (100 MPa),[63] very few structures have so far been built with concrete strengths of more than 10,000 psi (70 MPa). At the same time, mineral aggregate-based concrete with very high strengths of up to approximately 33,000 psi (230 MPa) can now be produced.[1] There appears, therefore, to be a great potential for utilization of high-strength concrete to applications where high abrasion resistance is needed.

Mechanical abrasion is the dominant abrasion for pavements and bridge decks exposed to studded tires. This effect has been studied under controlled laboratory conditions; and accelerated load facility for full-scale testing of abrasion resistance of highway pavements exposed to heavy traffic by studded tires was built in Norway in 1985 (Fig. 5.12). Figure 5.13 presents some data from the test facility. By increasing the concrete strength from 7145 (50%) up to 14,290 psi (100 MPa) the abrasion of the concrete was reduced by roughly 50%. At 21,430 psi (150 MPa) the abrasion of the concrete was reduced to the same low level as that of high quality massive granite. Compared to an Ab 16t type of asphalt for a typical Norwegian highway traffic, this represents an increased service life of the highway pavement by a factor of approximately ten.

In the Scandinavian countries where studded tires are extensively used, even high quality asphalt pavements may have a service life of only two to

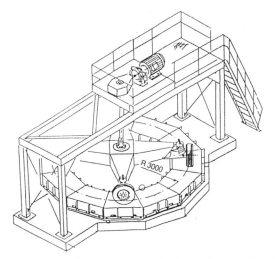

Fig. 5.12   Accelerated load facility for testing the abrasion resistance of highway pavements exposed to studded tires[64]

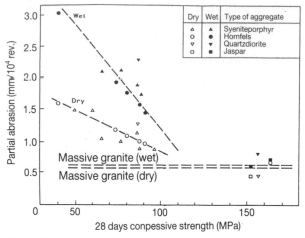

Fig. 5.13   Relationship between abrasion resistance and compressive strength[64]

three years or less. Encouraged by the above test results several high-strength concrete pavements have been completed over recent years.[65]

The wear of non-pavement surfaces such as concrete floors, sidewalks, stairs, etc. is caused primarily by foot traffic, light vehicular traffic, and the skidding, scraping, or sliding of objects on the surface. In some types of industrial operation, the use of steel wheeled vehicles, forks, buckets of lift trucks and loaders inflicts very severe damage to the concrete surfaces they operate on. The movement of abrasive granular material in and out of concrete storage facilities such as silos and bins also creates significant abrasion problems. In the Scandinavian countries the utilization of high-strength concrete for such problems has already started.

Hydraulic abrasion, or abrasion erosion is recognizable by the smooth, worn concrete surface in addition to cavitation erosion, where the surface is full of small holes and pits. Spillway aprons, stilling basins, sluiceways and tunnel linings are particularly susceptible to abrasion erosion.

Most concretes used in hydraulic structures in the past could not be classified as high-strength concrete. The concrete provided the mass, and strength was a secondary consideration only. Concerns about thermal cracking at early ages in the large concrete sections used in these structures have also led to low cement contents to minimize heat of hydration, with subsequent low concrete strengths. Where higher quality concrete is used, the resistance to high water velocities has been satisfactory for many years, but even these concretes do not fully resist the abrasive action, grinding or repeating impacts of the debris. Observations of abrasion-erosion of concrete surfaces in the stilling basins of several major USA dams have varied from 2 in. (50 mm) to 118 in. (3000 mm). At Dworshak Dam, 2000 yd^3 (1530 m^3) of concrete and bedrock were eroded from the stilling basin. In many of these instances, the abrasion-erosion is accelerated because of the impact forces of large rocks and boulders caught in turbulent flows. These forces weaken the concrete surface and make it more susceptible to removal.

In 1983 a major repair on the stilling basin of Kinzua Dam in Pennsylvania, USA, was carried out by using high-strength concrete.[66] The structure which has been in operation since 1967, had already had an extensive repair carried out in 1973–74. In 1983 approximately 1960 yd³ (1500 m³) of 10 in. (250 mm) slump concrete was placed, the 28 day compressive strength of which was 12,857 psi (90 MPa). Diver inspection of the concrete after one year of service including a period with very large volume of debris in the stilling basin, showed that the concrete was performing as intended.

For marine or offshore concrete structures exposed to 'ice abrasion', the actual mechanism that results in loss of surface is more complex than the simple act of ice rubbing on the concrete surface. Research work and field observations suggest that the concrete deterioration at or near the waterline is due to a combination of environmental causes plus the impact of loading of the concrete surface by repeated impacts from ice floes.

Pieces of ice, driven by wind and current, can possess significant kinetic energy, much of which is dissipated into the concrete when the ice collides with the concrete structure. Some of the energy is lost in the crushing of the ice. The frequency of the loading is dependent on the circumstances which occur at any given time at a particular structure and can vary from an occasional impact to repeated impacts every few seconds. A large ice floe in open waters will, upon initial contact with a structure, both load the structure and begin to crush itself at the point of contact with the structure. As the driving forces of the floe continue to move it forward against the structure, the resistance of the structure continues to increase to a point where the floe experiences a local failure in the ice, usually in flexure, some distance from the initial point of contact with the structure. This momentarily releases the load on the structure. The original ice, now damaged by crushing and cracking, is shunted away by the moving flow and new, undamaged ice in the floe then collides with the structure. The characteristics of the ice and the floe, combined with the dynamic response of the structure, will establish a 'ratcheting' effect on the concrete surface, repeatedly loading and unloading it. With time, this repetitive loading behavior can destroy the effectiveness of the aggregate bond near the surface of the concrete and both cause and propagate existing microcracks in the concrete.

Regardless of type of abrasion, both laboratory and field experience indicate that compressive strength is the single most important factor in determining the abrasion resistance of the concrete. Also, the abrasion resistance can be significantly improved by the use of hard and dense aggregate both in the upper and lower part of the grading curve. By replacing the 0.079 to 0.157 in. (2 to 4 mm) fraction of the natural sand with crushed high quality material, the compressive strength in the Norwegian highway pavement investigations[65] decreased from 23,471 to

21,900 psi (164.3 to 153.3 MPa), while the service life of the pavement increased by 50%. Generally, the abrasion of concrete is higher in wet than in dry condition, but the experience indicates that also this effect is reduced by increased strength level. From Fig. 5.13 it can be seen that at 7143 psi (50 MPa), the wet abrasion loss was approximately 100% higher than the dry abrasion loss, while at 14,286 psi (100 MPa) the wet abrasion loss was only 50% higher. At 214,429 psi (150 MPa), only a small difference between the wet and dry abrasion loss was observed. For such a dense concrete, it appears that the effect of moisture becomes more negligible.

5.8   Concluding remarks

In recent years more rapid developments in the field of concrete technology have taken place. Increasing construction challenges in combination with new innovations in materials and construction techniques have strengthened the stature of concrete as a major construction material.

There are several reasons for the above development. A rapid development in the general field of materials science has taken place. New cementitious materials and admixtures have been developed, and advancement in processing of aggregates has also taken place. Reinforced and prestressed concrete is being utilized in new areas, such as offshore structures for oil and gas explorations. From the first concrete structure produced for the North Sea operations in 1973 (the Ekofisk tank) to the most recent offshore structures currently under design, the strength of concrete has increased from 5715 to 11,430 psi (40 to 80 MPa). In North America high-strength concrete is being used in tall buildings, where concrete with strengths up to 18,570 psi (130 MPa) has been used in the columns.[67]

For highway pavements and bridge decks subjected to studded tires, experimental work has shown that utilization of high-strength concrete may increase the service life by a factor of up to ten, compared to that of a high-quality asphalt pavement. Industrial floors, hydraulic structures and structures exposed to ice abrasion also represent a great potential for utilization of high-strength concrete. In many areas, large quantities of resources are being spent on maintenance and rehabilitation of concrete structures due to lack of durability. There is a great challenge, therefore, for the engineering profession to utilize and further develop the technology of high-performance concrete for the benefit of society.

References

1   Division of Building Materials, The Norwegian Institute of Technology, NTH, Trondheim, Norway (unpublished data).
2   Elkem A/S Materials, Kristiansand, Norway (unpublished data).
3   Zhang, M.H. and Gjørv, O.E. (1990) Development of high-strength lightweight concrete. *High-Strength Concrete*, ACI SP-121, 667–81.

4   Sellevold, E.J., Bager, D.H., Klitgaard, J. and Knudsen, T. (1982) *Silica fume-cement pastes: hydration and pore structure*, Report BML 82.610. Division of Building Materials, the Norwegian Institute of Technology, NTH, 19–50.

5   Regourd, M., Mortureux, B. and Gautir, E. (1981) Hydraulic reactivity of various pozzolans. *Fifth International Symposium on Concrete Technology*, Monterey, Mexico.

6   Diamond, S. (1986) The microstructure of cement paste in concrete. *Proceedings, 8th Int. Congress on Chemistry of Cement, Vol. I*, Rio de Janeiro, 122–47.

7   Maso, J.C. (1980) The bond between aggregates and hydrated cement paste. *Proceedings, 7th Int. Congress on Chemistry of Cement, Vol. 1*, Paris, VII 1/3 to VII 1/15.

8   Monteiro, P.J.M. and Mehta, P.K. (1985) Ettringite formation on the aggregate-cement paste interface. *Cement and Concrete Research*, **15**, 2, 378–80.

9   Olliver, J.P. and Grandet, J. (1982) Sequence of formation of the aureole of transition. *Proceedings, RILEM Coloq. Liaisons Pates de Ciment Materiaux Associes*, Tolouse, A14 to A22.

10 Hoshino, M. (1988) Difference of the w/c ratio, porosity and microscopical aspect between the upper boundary paste and the lower boundary paste of the aggregate in concrete. *Materials and Structures*, RILEM, **21**, 175, 336–40.

11 Regourd, M. 'Microstructure of high strength cement paste systems,' Very High-Strength Cement-Based Materials. *Materials Research Society, Proceedings*, **42**, 3–17.

12 Regourd, M., Mortureux, B., Aitcin, P.C. and Pinsonneault, P. (1983) Microstructure of field concretes containing silica fume. *Proceedings, 4th Int. Symp. on Cement Microscopy*, Nevada, USA, 249–60.

13 Sarkar, S.L. and Aitcin, P.C. (1987) Comparative study of the microstructure of very high strength concretes. *Cement, Concrete and Aggregates*, **9**, 2, 57–64.

14 Aitcin, P.C. (1989) From gigapascals to nanometers. *Proceedings, Engr. Foundation Conf. on Advances in Cement Manufacture and Use*. The Engineering Foundation, 105–30.

15 Scrivener, K.L., Bentur, A. and Pratt, P.L. (1988) Quantitative characterization of the transition zone in high strength concretes. *Advances in Cement Research*, **1**, 2, 230–7.

16 Gjørv, O.E. (1983) Chemical processes related to concrete. *Proceedings, CEB-RILEM International Workshop on Durability of Concrete Structures*, Copenhagen, 341–4.

17 (1976) *Regulations for the structural design of fixed structures on the Norwegian continental shelf*. Norwegian Petroleum Directorate, Stavanger.

18 Zhang, M.H. and Gjørv, O.E. (1991) Effect of silica fume on pore structure and diffusivity of low porosity cement pastes. *Cement and Concrete Research*, Vol. 21, 800–8.

19 Diamond, S. (1983) Effects of microsilica (silica fume) on pore-solution chemistry of cement pastes. *Journal of the American Ceramic Society*, **66**, 5, C82–C84.

20 Page, C.L. and Vennesland, Ø. (1983) Pore solution composition and chloride binding capacity of silica-fume cement pastes. *Materials and Structures, RILEM*, **16**, 91, 19–25.

21 Nakamura, N., Sakai, M., Koibuchi, K. and Iijima, Y. (1986) Properties of high-strength concrete incorporating very finely ground granulated blast furnace slag. *ACI SP-91*, 1361–80.

22 Byfors, J., Hansson, C.M. and Tritthart, J. (1986) Pore solution expression as a method to determine influence of mineral additives on chloride binding. *Cement and Concrete Research*, **16**, 760–70.

23   Tuutti, K. (1982) *Corrosion of steel in concrete*, Report fo. 4-82. Swedish Cement and Concrete Research Institute, Stockholm.
24   (1990) *High-strength concrete, state-of-the-art-report*, FIP/CEB.
25   CEB Bulletin d'Information No. 182 (1989) *Durable concrete structures – CEB Design Guide*, Second edition.
26   Gjørv, O.E., Vennesland, Ø. and El-Busaidy, A.H.S. (1977) Electrical resistivity of concrete in the oceans. *Offshore Technology Conference, Proceedings*, OTC 2803, Houston, Texas, 581–8.
27   Vennesland, Ø. and Gjørv, O.E. (1983) Silica concrete – protection against corrosion of embedded steel. *ACI SP-79*, Vol. II, 719–29.
28   Danish Great Belt Link (unpublished data).
29   Gewertz, M.W., Tremper, B., Beaton, J.L. and Stratfull, R.F. (1958) Causes and repair of deterioration to a California bridge due to corrosion of reinforcing steel in a marine environment. *Highway Research Board, Bulletin 182*.
30   Browne, R.D. (1980) Mechanisms of corrosion of steel in concrete in relation to design, inspection, and repair of offshore and coastal structures. *ACI SP-65*, 169–204.
31   Houston, A. and Furguson, P.M. (1972) *Corrosion of reinforcing steel embedded in structural concrete*. Centre for Highway Research, The University of Texas at Austin, Research Report No. 112-1-F, 148 pp.
32   Schiessl, P. (1975) 'Admissible crack width in reinforced concrete structures', Contribution II 3-17, Inter-Association Colloquium on the Behaviour in Service of Concrete Structures. *Preliminary Reports, Vol. II*, Liege.
33   Vennesland, Ø. and Gjørv, O.E. (1981) Effect of cracks in submerged concrete sea structures on steel corrosion. *Materials Performance*, **20**, 49–51.
34   Gjørv, O.E. and Bathen, E. (1987) Quality control of the air-void system in hardened concrete. *Nordic Concrete Research*, 95–110.
35   Gjørv, O.E., Okkenhaug, K., Bathen, E. and Husevåg, R. (1988) Frost resistance and air-void characteristics in hardened concrete. *Nordic Concrete Research*, 89–104.
36   Siebel, E. (1989) Air-void characteristics and freezing and thawing resistance of superplasticized air-entrained concrete with high workability. *ACI SP 119*, 297–319.
37   Okada, E., Hisaka, M., Kazama, Y. and Hattori, K. (1981) Freeze-thaw resistance of superplasticized concretes. *Developments in the Use of Superplasticizers, ACI SP-68*, 269–82.
38   Foy, C., Pigeon, M. and Bauthia, N. (1988) Freeze-thaw durability and deicer salt scaling resistance of a 0.25 water-cement ratio concrete. *Cement and Concrete Research*, **18**, 604–14.
39   Gagne, R., Pigeon, M. and Aitcin, P.C. (1990) Durabilité au gel bétons de hautes performances mécaniques. *Materials and Structures, RILEM*, **23**, 103–9.
40   Malhotra, V.M., Painter, K. and Bilodeau, A. (1987) Mechanical properties and freezing and thawing resistance of high-strength concrete incorporating silica fume. *Proceedings, CANMET – ACI International Workshop on Condensed Silica Fume in Concrete*, Montréal, Canada, 25 p.
41   Hammer, T.A. and Sellevold, E.J. (1990) Frost resistance of high strength concrete. *ACI SP-121*, 457–87.
42   Petersson, P.-E. (1984) *Inverkan av salthaltiga miljøer på betongens frostbestandighet*, Technical Report SP-RAPP 1984: 34 ISSN 0280-2503. National Testing Institute, Borås, Sweden.
43   Biczók, I. (1972) *Concrete corrosion, concrete protection*. Akademiai Kiado, Budapest.

44 Mather, K. (1982) Current research in sulfate resistance at the Waterways Experiment Station. *George Verbeck Symposium on Sulfate Resistance of Concrete, ACI SP-77*, 63–74.

45 Cohen, M.D. and Bentur, A. (1988) Durability of portland cement – silica fume pastes in magnesium and sodium sulfate solutions. *ACI Materials Journal*, **85**, 148–57.

46 Mehta, P.K. (1985) Studies on chemical resistance of low water-cement ratio concretes. *Cement and Concrete Research*, **15**, 6, 969–78.

47 Gjørv, O.E. (1983) Durability of concrete containing condensed silica fume. *ACI SP-79, Vol. II*, 695–708.

48 Davis, G. and Oberholster, R.E. (1987) Use of the NBRI accelerated test to evaluate the effectiveness of mineral admixtures in preventing the alkali-silica reaction. *Cement and Concrete Research*, **16**, 2, 97–107.

49 Hooton, R. (1987) Some aspects of durability with condensed silica fume in concretes. *Proceedings, CANMET – ACI International Workshop on Silica Fume in Concrete*, Montréal, Canada.

50 Asgeirsson, H. and Gudmundsson, G. (1979) Pozzolanic activity of silica dust. *Cement and Concrete Research*, **9**, 249–52.

51 Sveinbjørnsson, S. (1987) *Alkali-aggregate demages in concrete with microsilica – field study*. Icelandic Building Research Institute, Reykjavik.

52 Diederichs, U., Jumppanen, U.M., Penttala, V. (1988) Material properties of high strength concrete at elevated temperatures. *IABSE 13th Congress*, Helsinki, June.

53 Pedersen, S. (1981) *Beregningsmetoder for varmepåvirkede betonkon-struktioner*. Institute of Building Design, Technical University of Denmark, Lyngby.

54 Shirley, S.I., Burg, R.G. and Fiorato, A.E. (1988) Fire endurance of high strength concrete slabs. *ACI Materials Journal*, **85**, 2, 102–8.

55 American Society for Testing and Materials, ASTM, Philadelphia, E119.

56 Jensen, J.J., Danielsen, U., Hansen, E.Aa. and Seglem, S. (1987) Offshore structures exposed to hydrocarbon fire. *Proceedings, First International Conference on Concrete for Hazard Protection*, Edinburgh, Sept, 113–25.

57 Williamson, R.B. and Rashed, A.I. (1987) *A comparison of ASTM E119 fire endurance exposure of two EMSAC concretes with similar conventional concretes*. Fire Research Laboratory, University of California, Berkeley, July.

58 Hertz, K. (1984) *Heat-induced explosion of dense concretes*. Institute of Building Design, Technical University of Denmark, Lyngby, Report no. 166.

59 Danielsen, U. (1989) Marine concrete structures exposed to hydrocarbon fires, Nordic Seminar on Fire Resistance of Concrete. *SINTEF Report STF65 A89036*, Trondheim, 56–76.

60 (1984) *Investigation of surface burning characteristics of fibermesh I fiber reinforcement in regular and lightweight concrete*. Southwest Research Laboratory, San Antonio, USA.

61 (1985) *Small scale fire tests of fiber-enhanced concrete*. Under-Writers Laboratories Inc., USA.

62 Chandra, S., Berntsen, L. and Anderberg, Y. (1980) Some effects of polymer addition on the fire resistance of concrete. *Cement and Concrete Research*, **10**, 367–75.

63 (1989) *Norwegian Standard NS 3473, Concrete structures, Design rules*. Oslo.

64 Gjørv, O.E., Bærland, T. and Rønning, H.R. (1990) Abrasion resistance of high-strength concrete pavements. *Concrete International*, **12**, 1, 45–8.

65 Helland, S. (1990) High-strength concrete used in highway pavements. *ACI SP-121*, 757–66.

66 Holland, T.C., Krysa, A., Luther, M.D. and Lin, T.C. (1986) Use of

silica-fume concrete to repair abrasion-erosion damage in the Kinzua dam stilling basin. *ACI SP-91*, 841–63.
67   Godfrey, K. (1987) Concrete strength record jumps 36%. *Civil Engineering*, 84–6 and 88.

6 Fracture mechanics

R Gettu and S P Shah

6.1 Introduction

Fracture mechanics is the study of crack propagation and the consequent structural response. Tremendous research interest in the 1980s led to fracture models that have been tailored to represent the quasi-brittle behavior of concrete. The validation of these approaches has opened two important avenues of application – materials engineering and structural analysis.

The mechanical behavior of concrete that is designed to have high strength is different in many aspects from that of normal concrete. These differences have yielded some characteristics that are not beneficial, such as brittleness. Fracture models can be used to understand the microstructural mechanics that control brittleness and crack resistance (toughness), and to provide reliable means of quantifying them. New high-performance concretes can then be engineered to possess higher toughness and lower brittleness. Increased resistance to cracking may also lead to better durability, long-term reliability and seismic resistance.

The application of fracture mechanics to structural analysis and design is motivated by the fact that the failure of concrete structures is primarily due to cracking, and several types of failures could occur catastrophically, especially in high-strength concrete structures. Certain aspects of such failures cannot be predicted satisfactorily by empirical relations obtained from tests, but can be explained rationally through fracture mechanics. In general, structural analysis based on fracture principles can lead to better estimates of crack widths and deformations under service loads, the safety factors under ultimate loads, and post-failure response during collapse.

The present chapter reviews the classical theory of fracture, its extensions for the modeling of concrete behavior, and the microstructural features that necessitate such models. Material characterization based on fracture mechanics and its implications are also discussed. Some examples of the applications are presented to demonstrate the scope of the fracture approach. More pertinent details can be found in recent publications such as the two RILEM reports,[1,2] the ACI report,[3] two detailed bibliographies[4,5] and several conference and workshop proceedings.[6–13]

6.2   Linear elastic fracture mechanics

Consider a panel (Fig. 6.1a) made of an ideal-elastic material. When it is uniformly stressed in tension, the stress-flow lines (imaginary lines showing the transfer of load from one loading-point to another are straight and parallel to the direction of loading. If the panel is cut as in Fig. 6.1b, the stress-flow lines are now forced to bend around the notch producing a stress-concentration (i.e., a large local stress) at the tip. The magnitude of the local stress depends mainly on the shape of the notch, and is larger when the notch is narrower. Also, since the stress-flow lines change direction a biaxial stress field is created at the notch-tip.

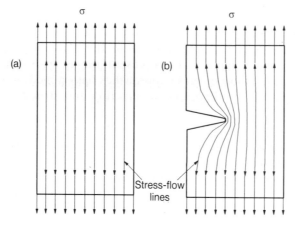

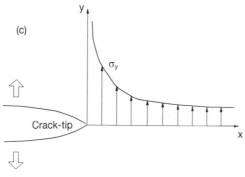

Fig. 6.1   Stress-flow lines for a panel in tension: (a) without crack; (b) with crack; and (c) stress concentration ahead of the crack

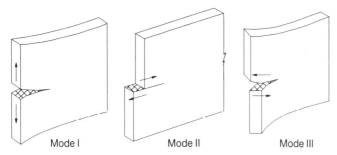

Fig. 6.2   Basic fracture modes

For a sharp notch or a crack, the stress (σ_y) in the loading direction and along the crack plane $(y = 0)$ is of the form:

$$\sigma_y = \frac{K_I}{\sqrt{2\pi x}} + C_1 + C_2 \sqrt{x} + C_3 x + \dots \tag{6.1}$$

where C_1, C_2, C_3, ... depend on geometry and loading. Note that the near-tip $(x \rightarrow 0)$ stresses depend only on the first term, and at the tip $(x = 0)$ this term and the theoretical stress become infinite (see Fig. 6.1c). The parameter K_I in Equation (6.1) characterizes the 'intensity' of the stress-field in the vicinity of the crack, and is called the stress intensity factor. It depends linearly on the applied stress (or load), and is a function of the geometry of the structure (or specimen) and the crack length:

$$K_I = \sigma F \sqrt{\pi a} \tag{6.2}$$

where σ = applied (nominal) stress as in Fig. 6.1b, a – crack length, and F = function of geometry and crack length. The subscript of K represents the mode of crack-tip deformation; the three basic modes (see Fig. 6.2) are Mode I – tensile or opening, Mode II – in-plane shear or sliding, and Mode III – anti-plane shear or tearing. When two or all three modes occur simultaneously, the cracking is known as mixed-mode fracture. Only tensile cracks are discussed in this chapter (unless mentioned otherwise) since crack propagation in concrete is dominated by Mode I fracture. Moreover, analytical treatments of the three modes are quite similar.

A more useful relation for K_I, in terms of the load and structural dimensions, can be obtained by writing Equation (6.2) as:

$$K_I = \frac{P}{b\sqrt{d}} f(\alpha) \tag{6.3}$$

where P = applied load, b = panel thickness, d = panel width (see Fig. 6.3a), $f(\alpha)$ = function of geometry, and $\alpha = a/d$ = relative crack length. When panel length $W = 2d$, the geometry function is (Chapter 3, Broek[14]):

$$f(\alpha) = \sqrt{\pi \alpha}(1.12 - 0.23\alpha + 10.56\alpha^2 - 21.74\alpha^3 + 30.42\alpha^4) \tag{6.4}$$

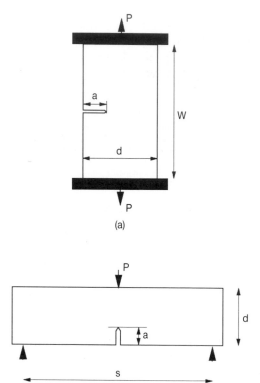

(a)

(b)

Fig. 6.3 Fracture specimens: (a) single-edge-notched panel in tension; and (b) single-edge-notched three points bend specimen

The advantage in Equation (6.3) is that it can also be used for several other geometries, such as the beam in Fig. 6.3b with span/depth $= s/d = 4$. Function f, however, is different for this beam:[15]

$$f(\alpha) = 6\sqrt{\alpha}\left\{\frac{1.99 - \alpha(1-\alpha)(2.15 - 3.93\alpha + 2.7\alpha^2)}{(1+2\alpha)(1-\alpha)^{3/2}}\right\} \qquad (6.5)$$

From Equation (6.1) it is obvious that a stress-based criterion for crack propagation is meaningless since the stress at the crack-tip is always infinite, irrespective of the applied load. Griffith,[16] in his landmark paper, suggested a criterion based on energy: Crack propagation occurs only if the potential energy of the structure is thereby minimized. This thermodynamic criterion forms the basis of fracture mechanics theory. Denoting the potential energy per unit thickness as U, the energy release rate is dU/da, which is usually designated as G (after Griffith). If energy (per unit length of crack-front) is consumed during fracture at the rate dW/da, denoted as R (for resistance of the material against fracture), the fracture criterion is:

$$\frac{dU}{da} = \frac{dW}{da} \quad \text{or} \quad G = R \qquad (6.6)$$

There is no fracture when $G<R$, and the fracture is unstable (i.e., sudden or catastrophic) when $G>R$. This criterion does not, however, imply that crack propagation is independent of the stress-field at the crack-tip. Irwin[17] showed that the energy release rate and the crack-tip fields are directly related by the equations suggested by Knott[18]:

$$G = K_I^2/E'   \qquad (6.7)$$

where $E' = E$ for plane stress, $E' = E/(1-v^2)$ for plane strain, E = modulus of elasticity, and v = Poisson's ratio. Using Equation (6.7), the fracture criterion of Equation (6.6) can be written as:

$$K_I = K_{IR}, \quad K_{IR} = \sqrt{E'R}   \qquad (6.8)$$

In the classical theory of linear elastic fracture mechanisms (LEFM), R is a constant (i.e., independent of the structural geometry and crack length), usually denoted as G (for critical strain energy release rate). Then, G and the associated critical stress intensity factor $K_{Ic}(=\sqrt{E'G_c}$; also known as fracture toughness) are material properties. When LEFM governs, the behavior is termed as Griffith or ideal-brittle fracture. More details of LEFM are given in textbooks such as Knott[18] and Broek.[14]

The LEFM parameter K_{Ic} (or G_c) can be determined from tests of specimens such as the single edge-notched tension specimen (Fig. 6.3a) and the three-point bend (3PB) specimen (Fig. 6.3b). Substitution of the experimentally determined maximum load (fracture load) for P in Equation (6.3) yields $K_I = K_{Ic}$. Geometry function $f(\alpha)$ can be determined using elastic analysis techniques including the finite element method.[13,14] For several common geometries, $f(\alpha)$ can be found in fracture handbooks such as Tada *et al.*[15] and Murakami.[19] Note that in the above-mentioned specimens, failure occurs immediately, i.e., the crack length is the notch length (a_0) and $\alpha = \alpha_0$ (= relative notch length = a/d), at failure.

The main features of LEFM can be summarized as:

1   The fracture criterion involves only one material property, which is related to the energy in the structure and the near-tip stress field.
2   The stresses near the crack-tip are proportional to the inverse of the square root of distance from the tip and become infinite at the tip.
3   During fracture, the entire body is elastic and energy is dissipated only at the crack-tip; i.e., fracture occurs at a point.

In reality, stresses do not become infinite, and some inelasticity always exists at the crack-tip. However, LEFM can still be applied as long as the inelastic region, called the fracture process zone, is of negligible size. Materials where such conditions exist include glass, layered silicate, diamond, and some high strength metals and ceramics (Chapter 4, Knott[18]). For other materials, the applicability of LEFM depends on the size of the cracked body relative to the process zone. In general, LEFM solutions can be used when the structural dimensions and the crack length are much larger than the process zone size. The process zone in concrete

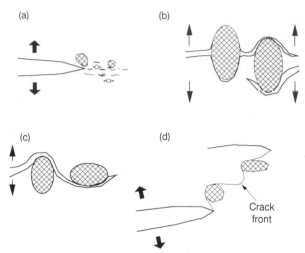

Fig. 6.4   Fracture process zone mechanisms: (a) microcracking; (b) crack bridging; (c) crack deflection; and (d) crack bowing

and the applicability of LEFM to concrete structures is discussed in the following sections.

6.3   The fracture process zone

Flaws, such as water-filled pores, air voids, lenses of bleed water under coarse aggregates and shrinkage cracks, exist in normal concrete even before loading. Due to the stress concentration ahead of a crack, microcracks could initiate at nearby flaws and propagate (Fig. 6.4a) forming a fracture process zone as suggested first by Glucklich.[20] Discrete microcracking at the tip of a propagating crack in hardened cement paste (hcp) has been observed by Struble *et al.*[21] with a scanning electron microscope. Also, acoustic waves generated during microcracking have been monitored using nondestructive techniques,[22,23,24] and used to determine the location, size and orientation of individual microcracks. Depending on the size and orientation of the microcracks, their interaction with the main crack could significantly decrease the K_I at its crack-tip.[25,26] This beneficial effect is known as crack-shielding or toughening. However, an amplification of K_I is also possible, especially if the main crack can propagate by coalescing with the microcracks. When the matrix is more compact and the flaws are less in number, as in high-strength concrete, the extent of microcracking is reduced.[27,28]

The most effective shielding mechanism in cement-based materials is crack-bridging (see Fig. 4b). A propagating crack is arrested when it encounters a relatively strong particle, such as an unhydrated cement grain, sand grain, gravel piece or steel fiber. Upon increase of load, the crack may be forced around and ahead of the particle which then bridges the crack-faces.[21,29] At this stage, crack-branching could also occur.[30,31]

Until the bridging particle is debonded (or broken), it ties the crack-faces together and thereby, decreases the K_I at the crack-tip.[32,33] In addition, energy is dissipated through friction during the detachment of the bridges and the separation of the crack-faces, which increases the fracture resistance R.

Microcracking, crack-bridging and frictional separation cause the process zone in concrete and other cementitious materials. Several investigations, using a wide range of techniques, have focused on whether and where these phenomena exist. However, results have been influenced significantly by methodology, and have been contradictory at times.[34] The various techniques used to detect and study the process zone have been reviewed by Mindess.[35] The most sensitive of the non-destructive methods are based on optical interferometry using laser light. Researchers using moiré interferometry[36,37] or holographic interferometry[38,39] observed well-defined crack-tips surrounded by extensive process zones with large strains. They concluded that strains behind the crack-tip (i.e., in the wake) decrease gradually to zero, and that energy dissipation near the crack-faces is most significant. Castro-Montero *et al.*[39] defined the process zone as the region where the experimentally determined strain-field deviates considerably from LEFM. Their results are shown in Fig. 6.5 for three stages in the fracture of a center-notched mortar plate. The zones labeled A have strains less than the LEFM solution, and appear to be of constant size. The zones B (in the wake) have strains greater than LEFM, and seem to increase with crack length and load. It was concluded, therefore, that most toughening occurs in the wake zone. Note that the dimensions of the crack are not constant through the thickness of the specimen,[40,41] and therefore the process zone has a complex three-dimensional character. Measurements of surface deformations should be interpreted with this feature in mind.

In some analytical fracture models, such as the micromechanics

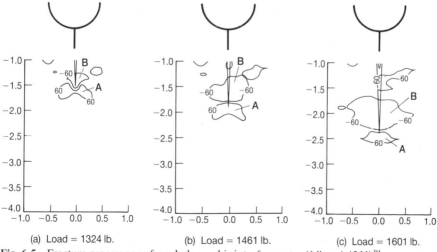

(a) Load = 1324 lb.   (b) Load = 1461 lb.   (c) Load = 1601 lb.

Fig. 6.5   Fracture process zone from holographic interferometry (1 lb = 4.45 N)[39]

approach of Horii,[42] each process zone mechanism is described individually. In general, however, the treatment of the process zone need not distinguish between the different mechanisms; i.e., the toughening effect of the bridging (wake) zone and the microcracked zone are equivalent.[43] Therefore, all the shielding mechanisms can be lumped into a conceptual fracture process zone that lies ahead of the traction-free zone of the crack or in the wake of the 'actual' crack-front. This zone can also include the toughening effects of other inelastic mechanisms such as crack-deflection (tortuosity of the crack path; Fig. 6.4c) and crack-bowing (unevenness of the crack front; Fig. 6.4d). Cracks in concrete follow paths of least resistance, and subsequently, are considerably tortuous with rough crack-faces. In the hardened cement paste, the crack usually passes around unhydrated cement grains and along calcium hydroxide cleavage planes.[21] The crack-face roughness appears to be less when the cement paste contains silica fume.[44] In mortar, the crack follows the interfaces between the sand grains and the hardened cement paste.[30] Since the weakest phase in normal concrete is the aggregate-mortar interface, cracks tend to avoid the aggregates and propagate through the interfaces.[45] In silica-fume concrete the interfaces and the mortar are much stronger, and therefore, cracks are less tortuous and sometimes pass through the gravel.[27,29,46] Also, cracks propagate through coarse aggregates that are weaker than the mortar, as in lightweight concrete.[46,47] The tortuosity of the crack gives rise to a higher R due to a larger surface area. Also, further shielding occurs since the non-planar crack experiences a lower K_I than the corresponding planar crack.[47] Similarly, shielding exists when the crack front is trapped by bridging particles and bows between the bridges until they break or slip. This crack-bowing effect (Fig. 6.4d) can be an important toughening mechanism for very-high strength concrete with strong interfaces where cracks propagate through the aggregates. The bowed crack-front has a lesser K_I than a straight front.[32,33]

In conclusion, a sizable fracture process zone occurs ahead of a propagating crack in cementitious materials. Energy is dissipated throughout this zone causing crack-shielding or toughening. Its size and effectiveness depend on the microstructure and inherent material heterogeneity; therefore, toughening in mortar is greater than in hardened cement paste and less than in concrete. The stress distribution (Fig. 6.6) in the process

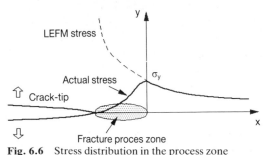

Fig. 6.6   Stress distribution in the process zone

zone differs considerably from the LEFM distribution (Fig. 6.1c). Within the process zone, the stress and strain undergo 'softening' with a gradual decrease to zero instead of an abrupt (brittle) drop. Models of fracture in cementitious materials should account for these factors in order to predict their behavior satisfactorily.

6.4   Notch sensitivity and size effects

In the 1960s and 1970s, several researchers examined the use of LEFM for concrete with tests of notched panels and beams.[4,48] They generally obtained K_{Ic} (and/or G_c) from LEFM relations by taking the initial notch length as the critical crack length at failure (i.e., at the observed maximum load). Early works[20,49,50] concluded that the critical parameters thus determined varied with notch length, and specimen size and shape. From these and later investigations[51–55] three significant trends can be identified: (1) the fracture behavior of materials with a finer microstructure is closer to LEFM (e.g., hcp vs. concrete); (2) for a given specimen shape and size, K_{Ic} depends on notch length; and (3) for a given specimen shape and relative notch length (α_0), K_{Ic} increases with specimen size. Also, the effects of specimen thickness and notch width on K_{Ic} are negligible within a certain range.[54,56–58]

One aspect of fracture is called notch-sensitivity, and is defined as the loss in net tensile strength due to the stress-concentration at a notch. Considering the 3PB specimen (Fig. 6.3b) with notch length a_0, the net tensile strength (σ_u) is calculated from bending theory using the failure load (P_u) and net depth ($d - a_0$), as $\sigma_u = 1.5 P_u s / b (d - a_0)^2$. The σ_u values corresponding to the 3PB ($s/d = 6.25$, $d = 64$ mm) tests of concrete by Nallathambi[59] are plotted in Fig 6.7a, along with the scatter. For determining the LEFM behavior, the above equation for σ_u is substituted in Equation (6.3) and K_I at $P = P_u$ is taken as K_{Ic}. The resulting curve is shown in Fig. 6.7a (for an assumed value of $K_{Ic} = 15.8$ MPa$\sqrt{\text{mm}}$). The observed response seems to be satisfactorily described by LEFM for $\alpha_0 = 0.1 \sim 0.5$. However, other tests do not yield such a correlation with LEFM. Data from tests of larger specimens ($s/d = 6$) by Nallathambi[59] are shown in Fig. 6.7b. It appears that for small notch lengths, σ_u is almost constant (notch insensitive) as in failure governed by limit stress criteria. Another important feature is the strong dependence of σ_u on specimen size. Notch sensitivity, therefore, is not a material property but depends on the structure and the flaw size. Also, decrease in σ_u with increase in crack length is seen only for short notches (Fig. 6.7a). Therefore, notch-sensitivity cannot always be used to judge the proximity of fracture behavior to LEFM.

The influence of notch (or flaw) length is illustrated better by studying the experimentally determined values of K_{Ic}. Data from four-point bend tests ($40 \times 40 \times 160$ mm beams) by Ohgishi et al.[57] are plotted in Fig. 6.8a.

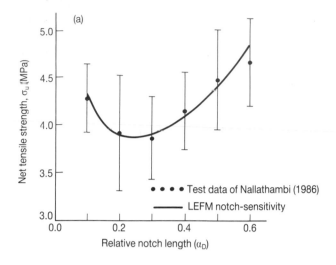

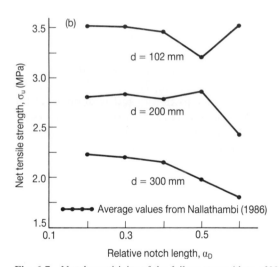

Fig. 6.7   Notch sensitivity of the failure stress (data of Nallathambi)

For hcp, practically constant values of K_{Ic} are obtained (as in other works such as Gjørv *et al.*[55]), but the values for mortar (with sand of maximum size 0.3 mm) increase with notch length. It appears that the behavior of hcp is much closer to LEFM than mortar. Unless very large specimens are tested, the observed values increase with notch length until a certain point and then decrease.[60] This trend can be seen in the tests of 3PB specimens $(50 \times 100 \times 40 \text{ mm})$ of high strength concrete (compressive strength $f_c = 124$ MPa) by Biolzi *et al.*[61] Their data, in Fig. 6.8b, shows the variation in LEFM-based K_{Ic} with increase in notch depth. It is obvious that concrete, even of high strength, does not behave according to LEFM in usual test specimens. It should be mentioned that the effect of notch length on K_{Ic} varies considerably with specimen geometry and size. From tests

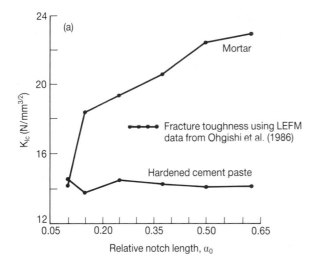

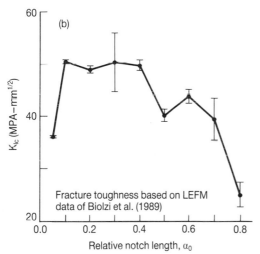

Fig. 6.8   Influence of notch length on K: (a) data of Ohgishi *et al.*[57]; and (b) data of Biolzi *et al.*[61]

conducted by Kesler *et al.*[62] on large wedge-loaded center-cracked plates of hcp, mortar and concrete, Saouma *et al.*[63] obtained K_{Ic}-values that were independent of crack length.

In LEFM, K_{Ic} does not depend on the size of the specimen or structure. However, the failure stress is size-dependent. This can easily be demonstrated by substituting $P = P_u$ and $K_I = K_{Ic}$ in Equation (6.3), which yields $K_{Ic} = P_u f(\alpha)/b\sqrt{d}$. Assuming that failure occurs with negligible crack extension ($\alpha = \alpha_0$), and defining the nominal failure stress as $\sigma_N = P_u/bd$,

$$\sigma_N = K_{Ic} f(\alpha_0) \frac{1}{\sqrt{d}} \tag{6.9}$$

Since K_{Ic} is constant in LEFM and $f(\alpha_0)$ is a shape-dependent constant, Equation (6.9) implies that $\sigma_N = (\text{constant}) x 1/\sqrt{d}$ in geometrically similar structures. This is the size effect according to LEFM where the failure stress decreases with increase in certain structural dimensions. Tests of concrete specimens exhibit a more complex behavior, yielding K_{Ic}-values that may be size-dependent. Therefore, the size effect on failure stress is not always that of LEFM. From tests of notched beams, Walsh[53] recognized a significant trend in this size effect. For small specimens, the failure stress is constant, as in failure criteria based on limit stress. For very large specimens, the size effect is the strongest; $\sigma_N \propto 1\sqrt{d}$ as in LEFM. Therefore, the structure should be greater than a certain size for LEFM to apply. Such a transition in failure mode has also been observed in hcp and mortar.[57] The data of Tian *et al.*[58] are shown in Figs. 6.9a and 6.9b for compact tension specimens of concrete (maximum aggregate size = 20 mm). It is clear that the failure stress and K_{Ic} are significantly size-dependent.

In summary, the fracture of cementitious materials that have a finer microstructure (e.g., hcp) is closer to LEFM than others (e.g., concrete, fiber-reinforced mortar). The mode of structural failure generally lies between two limits: the strength limit, and LEFM behavior. When the size of the initial crack or the size of the uncracked ligament is of the same order as the inherent material heterogeneity (i.e., much smaller than the structure), failure is governed by limit criteria. This failure mode also occurs when the characteristic structural dimensions are small (i.e., of the same order as the size of the heterogeneities). When the structure is very large, failure is governed by LEFM.

In order to model concrete failure, several methods have been developed for determining size-independent fracture properties using practical-size specimens. Also, it is essential that analysis techniques are able to simulate the observed geometry-effects on cracking and failure before they can be applied to predict actual structural behavior.

6.5 Fracture energy from work-of-fracture

The resistance of the material to fracture, R, was defined earlier (Equation (6.6)) as the energy needed to create a crack of unit area. When a notched specimen, such as the beam in Fig. 6.3b, is tested until failure, the total work done gives the total energy dissipated. Assuming that all the energy has been consumed in extending the crack, determination of the work done and the crack area would yield R. This value is usually called fracture energy, and is denoted here as G. Nakayama[64] tested beams with triangular notches and obtained the load versus load-point displacement curves. (The triangular or V-shaped notches prevented catastrophic failure.) The area under a curve provided the total energy. Taking the unnotched part of the cross-section (i.e., the ligament area) as the area of

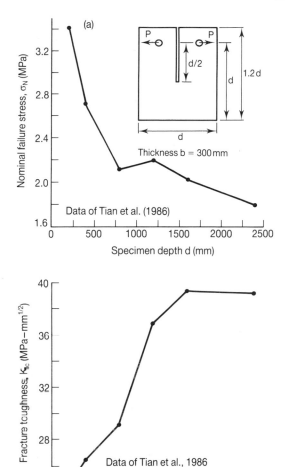

Fig. 6.9 Size effect on failure stress (data of Tian *et al.*[58])

the crack, the fracture energies for several brittle materials were determined.

For LEFM-type behavior, the fracture energy is a constant equal to the critical strain energy release rate, i.e. $G_f = G_c$. In practice however, G_f is not always constant. Tattersall and Tappin[65] found that the fracture energy increased with specimen size and decreased with a significant decrease in stiffness of the testing machine. In the first application of the work-of-fracture method to concrete, Moavenzadeh and Kuguel[66] showed that the determination of G_f was not straightforward. Since cracking in concrete is tortuous, they argued that the crack area should be determined exactly instead of simply using the ligament area. They obtained the crack area from microscopy techniques, and thereby computed fracture energies that were almost the same for hcp, mortar and concrete. Since the measurement of 'true' crack area is difficult, and to a certain extent subjective, later researchers have usually circumvented the problem by taking the ligament (or projected) area as the nominal area of the crack. Then, the fracture energy is also a nominal value. Moavenzadeh and Kuguel[66] used straight notches for their concrete beams, as have later researcher who have been able to obtain the entire load-displacement curve using stiff servo-hydraulic testing machines.

Petersson[67] proposed that the work-of-fracture from a notched beam test would provide a material constant for G_f if the following conditions are satisfied: (1) energy consumption outside the fracture process zone is negligible, (2) energy consumption is independent of specimen geometry, and (3) the fracture is always stable. Due to the effect of toughening or the increase in R with crack extension (see Section 6.6.2), condition (2) can never be satisfied exactly. However, as the sizes of ligament and specimen increase, G_f would tend asymptotically towards a constant value. Therefore, G_f approaches a material property when the fracture zone is negligible compared to the specimen size.[68] Accordingly, the work-of-fracture method was recommended by RILEM[69] with a lower limit on the specimen size. For concrete with a maximum aggregate size of 16–32 mm, the required beam depth is 200 mm and the length is 1.2 m. The notch length should be half the depth of the beam. If the total energy consumed (including the work done by the weight of the beam) is W_f, the fracture energy is:[69,70]

$$G_f = W_f / A_{\text{lig}} \tag{6.10}$$

where A_{lig} = area of the ligament. In the case of a beam with a straight notch, $A_{\text{lig}} = b(d - a_0)$. The corresponding value of fracture toughness can be calculated from Equation (6.7). Condition (3) is satisfied by using a testing machine with high stiffness and/or servo-control.

It has been shown that the specimen sizes recommended by RILEM do not always provide size-independent values of G_f. For concrete, Hillerborg (1983) concluded on the basis of the fictitious crack model (see Section 6.6.1) that the work-of-fracture method would provide constant values only at the LEFM limit when the beam was 2–6 m deep and has a notch longer than 200–400 mm. This would be the point at which condition (2) is practically satisfied. However, with increasing loads there is some crushing at the loading points and possible energy dissipation outside the process zone. Due to this, Petersson's condition (1) may not be satisfied in certain cases. Nallathabi *et al.*[71] conducted a detailed experimental study of concrete with a maximum aggregate size of 20 mm. The work-of-fracture of beams with depths ranging from 150 mm to 300 mm was determined. It was shown that G_f increased with an increase in beam depth, and decreased with an increase in notch depth. All the values obtained were much greater than G_c-values calculated from LEFM (Equation (6.3)). By comparing several studies, Hillerborg[72] concluded that in some cases the work-of-fracture method yielded size-independent values, but in others doubling the specimen size resulted in a 20% increase in G_f. A 20% increase in G was also observed for high strength concrete beams by Gettu *et al.*[73] by increasing the beam depth from 200 mm to 400 mm. Such a size effect on G_f has been observed in compact tension specimens by Wittmann *et al.*[74] and was attributed to the increase in the width of the fracture zone with increase in the ligament length. Other possible sources of size effects and errors in the work-of-fracture method have been described in detail by Planas and Elices,[75] and Brameshuber and Hilsdorf.[76]

Though the method yields size-dependent values, it has widely been used due to its simplicity. Availability of extensive data has led to empirical 'code-type' relations linking fracture energy to conventional design properties. One such equation is provided by the CEB-FIP Model Code:[77,78]

$$G_f = x_F f_{cm}^{0.7} \qquad (6.11)$$

where x_F is a tabulated coefficient that depends on the aggregate size (e.g., for a maximum aggregate size $= 16$ mm, $x_F = 6$), $f_{cm} =$ mean compressive strength of the concrete in MPa, and G_f is obtained in N/m. The commentary to the code cautions that Equation (6.11) gives values that may be size dependent with deviations up to $\pm 20\%$. Nevertheless, the relation is useful when experimental data is lacking.

It should be emphasized that G_f, by itself, is not a reliable measure of toughness or ductility, and that using G_f as the sole fracture parameter could lead to erroneous conclusions. If one were to conclude from the observed increase in G_f with compressive strength (as suggested by Equation (6.11)) that ductility increases with the strength, this would be wrong. With the use of additional parameters (see Section 6.7), the higher brittleness in high-strength concrete can be adequately characterized.

6.6 Nonlinear fracture mechanics of concrete

Since LEFM (or any theory with only one fracture parameter) cannot adequately characterize the cracking and failure of concrete, several nonlinear techniques have been proposed. These approaches are, in general, modifications of LEFM, with fracture criteria based on two or more material parameters that account for the process zone. Based on whether the fracture zone is modeled explicitly or implicitly, the approaches may be categorized as cohesive crack models and effective crack models, respectively. All the models represent the actual three-dimensional fracture zone through an 'equivalent' line crack which simulates only a certain part of the actual process. Since the relation between the model and reality is different in each approach, the various models are not equivalent in all aspects.

Cohesive crack models

In the cohesive crack models, the fracture zone is simulated by a set of line forces imposed on the crack faces that tend to close the crack. The magnitude and distribution of these forces, or cohesive stresses, are related to material parameters in a prescribed manner. The cohesive stresses cause toughening by decreasing the G at the crack tip. The stresses have also been taken to represent the bridging action of aggregates and fibers. The fictitious crack model, the crack band model and several other closing pressure models that are discussed here can be classified as cohesive crack models. All cohesive crack models have to be used along with an appropriate constitutive model for the uncracked concrete. Normally, it is reasonable to assume that the uncracked concrete is elastic.

Fictitious crack model (FCM)

Hillerborg *et al.*[79] proposed a discrete-crack approach for the finite element analysis of fracture in concrete, which became known as the fictitious crack model. It is similar in principle to the Barenblatt model where there is a zone at the crack-tip with varying stresses. According to the FCM, a crack is initiated when the tensile stress reaches the material strength f_t, and it propagates in a direction normal to the stress. As the crack-opening (w) increases, the stresses (f) in the fracture zone behind the crack-tip gradually decrease from f_t to zero. The stress-opening relation is assumed to be a property of the material. The fracture criteria of the FCM are similar to the final stretch model of Wnuk[80] for elastic-plastic fracture.

The fracture criteria of the FCM (see Fig. 6.10) can be summarized as:

$$w = 0, \quad f = f_t \tag{6.12a}$$

$$w = w_c, \quad f = 0 \tag{6.12b}$$

$$\int_0^{w_c} f\,dw = G_f \tag{6.12c}$$

where w_c is the critical crack opening. The conditions in Equations (6.12a) and (6.12b) occur at the tips of the fictitious crack (a_f) and the traction-free crack (a_0), respectively. Note that Equation (6.12c) gives the energy absorbed for producing unit area of a traction-free crack. If the shape of $f(w)$ is known, there are only two independent material parameters which usually are f_t and G_f. Several different shapes have been assumed for the $f(w)$ relation including linear, bilinear and smoothly varying functions. The experimental determination and the significance of the $f(w)$ curve are discussed in on Constitutive Relations below.

In a parametric study of the FCM, Carpinteri *et al.*[81] used a linear $f(w)$ function in the finite element analysis of a notched beam. They showed that with an increase in G_f, the peak load and the deformation at the peak increase, and the steepness of the post-peak load-deformation curve

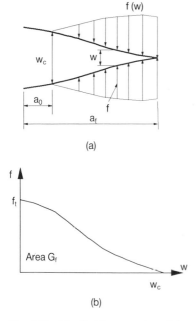

(a)

(b)

Fig. 6.10   The fictitious crack model

decreases. Increasing f_t caused an increase in the peak load but no change in the final part of the load-deformation curve. Increasing the specimen size produced an effect identical to that obtained by decreasing G_f.

The FCM has been extended to include non-planar fracture[82-84] where the direction of crack extension is defined at each step by remeshing. Other discrete crack models with mixed-mode considerations have also been proposed.[85-87] The effects of suatained and cyclic loading have been predicted with a discrete crack model by Akutagawa *et al.*[88]

Crack band model (CBM)

A convenient approach for the finite element analysis of fracture in concrete structures is to assume that cracking is distributed throughout the element. Crack propagation is then represented by a stress-strain relation that exhibits softening. The crack band model of Bazant differs from earlier smeared crack models due to the use of fracture mechanics criteria.[89,90] This eliminates spurious mesh dependence that is obtained when the criterion for crack propagation is a stress or strain limit.

In the analysis with the CBM, finite elements of width equal to that of the fracture zone have to be used in the region where cracking is expected. Excluding this aspect, analyses with the CBM and the FCM are practically equivalent. The strain due to cracking, ε_c, in the constitutive relation of the CBM can be related to the crack opening of the FCM (see Equation (6.12)) as:

$$\varepsilon_c = w/w_h \qquad (6.13)$$

where w_h is the element width. Note that w_h is not the actual process zone width but an effective width of the softening region needed for the numerical analysis. To allow the use of finite elements that are wider than w_h, which is necessary in the analysis of large structures, certain modifications have to be made as described by Bazant.[91]

When cracking is non-planar (i.e., not parallel to the usually square mesh lines), the crack pattern can be roughly predicted with the CBM.[91] For obtaining accurate results step-wise remeshing is needed as in the FCM. Recently, smeared-cracking models that are more general have been proposed[92,93] to analyze cases where mixed-mode cracking occurs with the transfer of tensile and shear stresses across the process zone. However, in a comparative study of smeared and discrete crack models, Rots and Blaauwendraad[85] suggested that, in non-planar crack problems, the smeared approach should be used as a qualitative predictor, followed by a corrector analysis with the discrete approach.

Singular closing pressure models

The process zone of concrete is treated in several approaches like the plastic zone in metals. Accordingly, the stresses at the crack-tip are taken to be finite (i.e., non-singular), and therefore, $K_I = 0$. Another class of cohesive crack models include a singular crack-tip (i.e., $K_I \neq 0$) as in LEFM, with the criterion for crack extension as $K_I = K_{Ic}$. Here K_{Ic} represents the fracture toughness of the matrix phase (hcp or mortar).

The use of a singular closing pressure models is justified by the fact that, in concrete, the crack profile at the tip is parabolic[39,94] as in the case of LEFM. On the other hand, a non-singular model with $K_I = 0$ produces a crack profile having zero slope at the tip which does not match the experimentally observed profile (see Fig. 6.11a). Singular models have been proposed by Jenq and Shah,[95] Foote et al.,[96] and Yon et al.[97] A closing pressure model has also been proposed for mixed-mode fracture by Tasdemir et al.[98] In these models, toughening is simulated by closing stresses or tractions imposed behind the singular crack-tip (see Fig. 6.11b). Consequently, the strain energy release rate has the following form until the critical state:

$$G = \frac{K_i^2}{E'} + \int_0^{\text{CTOD}} f(w)\,dw \qquad (6.14)$$

where CTOD = opening at the intial crack-tip. The first term in Equation (6.14) is due to the singularity and the second term gives the energy dissipated by the cohesive stresses. The use of a singular model will lead to realistic predictions of crack profiles and near-tip deformations. However, in the analysis of global structural behavior, singular and non-singular models give practically the same results.

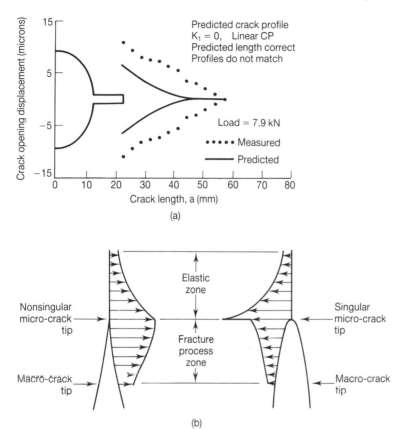

Fig. 6.11   Singular cohesive crack models

Constitutive relations for the cohesive crack

For the experimental determination of the tensile stress-strain relation or stress-opening relation for a crack, a uniaxial tension test seems to be the ideal method. However, in a concrete specimen under direct tension, the stress field and the cracking, especially after the peak, are not always uniform. Also, the transverse compression applied for gripping the specimen, and the dimensions of the specimen can influence the results considerably. In spite of these problems, reliable $f(w)$ relations have been obtained from tension tests by some investigators.[99] Under specific test conditions, Gopalaratnam and Shah,[100] Cornelissen *et al.*,[101] Guo and Zhang,[102] and Hordijk[99] could determine the $f(w)$ relation for monotonic and cyclic loading. Their results show that the envelope of the cyclic behavior is not affected by the cycling of the load, and is the same as that obtained for monotonic loading.

A generalizing stress-operating relation, that is applicable to several cement-based materials including normal and high-strength concrete, has been proposed by Wecharatana.[103] From a detailed study of the tensile softening response of notched dog-bone and end-tapered specimens, he obtained an equation of the form:

$$\left(\frac{f}{f_t}\right)^m + \left(\frac{w}{w_c}\right)^{2m} = 1 \qquad (6.15)$$

where m is a material parameter; for 24 MPa concrete, $m = 0.27$ and for 83 MPa concrete, $m = 0.2$.

Several other experimental techniques have been proposed for determining the stress-opening relation. The method of Li,[104] based on the J-integral (Chapter 4 of Broek[14]), requires the load, displacement and crack-tip opening of two fracture specimens to be monitored until failure. Using this data, the $f(w)$ curve including its parameters, f_t, w_c and G_f, can be obtained.[104]

The ability to measure deformations continuously over a wide field with sufficient accuracy by means of optical interferometry has led to experimental-numerical approaches for determining the constitutive relations of cracking. In these investigations, deformations near the propagating crack in a test specimen were obtained using moiré interferometry[37,105] or sandwich-hologram interferometry.[38,39] From these data, the parameters of the $f(w)$ curve having a prescribed shape were obtained. In the works of Castro-Montero et al.[39] and Miller et al.,[106] holograms of an area near the notches of center-cracked plates were made after each loading step, with a laser light source. The interference fringes observed in a sandwich of two consecutive holograms, with the same illumination direction, represent one component of the relative displacement undergone during the load step. To obtain the correct displacement vectors three holograms have to be made at each step with different illumination directions. An image analysis system was used to analyze the fringes objectively. The crack-opening and the strain field were thereby computed at each load step. Using finite element analysis with the FCM, the parameters of the $f(w)$ relation were determined such that the calculated crack-openings matched those computed from the holograms.

Other experimental-numerical approaches (e.g. Roelfstra and Wittmann[107]) use the global load-deformation response of fracture specimens to obtain the parameters of the $f(w)$ relation. Wittmann et al.[108] assumed a bilinear form for $f(w)$ characterized by four constants f_t, f_c, f_1 and w_1, with a change of slope at (f_1, w_1). By fitting the load-displacement curve from tests of compact tension specimens, they found that several $f(w)$ relations could produce the same global behavior. However, the computed fracture energy G_f (as in Equation (6.12c)) was almost the same. They proposed that, in order to get unique results, the tensile strength f_t should be determined independently and the ratio f_t/f_1 should be set at a certain value (in the range of 3–5). The resulting stress-opening relations seem to be independent of specimen size and loading rate.

The global structural behavior that is predicted by fracture analysis can be strongly influenced by the shape of the $f(w)$ relation.[107,109,110] From available data, several code-type formulations have been proposed for the stress-opening curve. The CEB-FIP Model Code[77,78] recommends a bilinear relation whose parameters are f_t, G_f, w_c and x_f, where w_c and x_F (see Equation (6.11)) have tabulated values that depend on the aggregate size. For a maximum aggregate size of 16 mm, $w_c = 0.15$ mm. Values for G_f can be obtained from Equation (6.11). The change in slope is at $f = 0.15f_t$, and the mean tensile strength can be estimated from:

$$f_t = 0.30f_c^{2/3} \tag{6.16}$$

where f_c is the characteristic compressive strength in MPa. A similar relation has also been formulated by Liaw *et al.*[111]

Some recent works have focused on establishing rational relations between the microstructure of the concrete and the stress-opening behavior of the crack. The relation of Li and Huang[112] for $f(w)$ is formulated in terms of the aggregate content, maximum aggregate size, fracture toughness of the hcp and the characteristics of the aggregate-hcp interface. Their model assumes that microcracking, frictional debonding of the aggregates and crack-deflection are the main toughening mechanisms. Another model proposed by Duda[113] incorporates also the aggregate size distribution through a probabilistic formulation. Note that these models are only valid for concrete with weak interfaces, and with aggregates that are stronger than the hcp.

Effective crack models

Several models have been proposed where the fracture analysis is performed on an elastic structure that is geometrically identical to the actual structure. The length of the effective (LEFM) crack is initially the same as the crack or flaw in the actual structure. Further equivalence between the actual and effective cracks is prescribed by the model. The nonlinearity of the fracture process is usually represented by fracture criteria that are extensions of LEFM.

R-curve models

Consider a concrete specimen with an initial crack or flaw. As the load increases, a process zone at the tip will form and increase in size until the traction-free crack propagates. An effective crack can be defined such that its compliance is equal to that of the actual crack including its fracture zone. Accordingly, this effective crack will extend as the process zone grows. Since toughening increases with process zone sizes, its effect can now be modeled very simply by assuming that LEFM relations can be applied to the effective crack, and that the crack resistance varies with the effective crack extension. Then, the fracture criterion is $G = R$ as in Equation (6.6) (or $K_I = K_{IR}$ as in Equation (6.8), where R is not constant

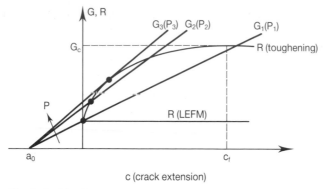

Fig. 6.12   The R-curve model

(as in LEFM) but is an increasing function of the effective crack growth c, which is denoted as $R(c)$ or the R-curve.

R-curve models can also be motivated by other concepts. Clarke and Faber[114] suggested that increase in crack resistance is due to statistical variability of the material microstructure. They argued that as the crack lengthens, the probability of encountering features of largest resistance increases. Therefore, the overall fracture resistance increases until the crack attains a length beyond which the probability is unity. Subsequently, the R-curve rises gradually only when there is a wide distribution of microstructural resistance. When the mortar is as strong as the aggregates, as in several high strength concretes, the R-curve may be expected to rise very sharply. This also occurs when the aggregates are of constant size and equi-spaced, or when the material is practically homogeneous.

The implications of the R-curve can be seen from Fig. 6.12, where the variation of G is shown for a specimen such as a beam (Fig. 6.3b) with a notch of length a_0. (The discussion would be identical if R and G were replaced by K_{IR} and K_I, respectively.) For a certain load P, G can be obtained as a function of crack length (with $a = a_0 + c$) from LEFM. Two R-curves are donsidered in Fig. 6.12: the 'LEFM' R-curve for an ideal-brittle material where R is practically constant; and the 'Toughening' R-curve which rises monotonically to a horizontal asymptote, for a material such as concrete. For constant R, fracture occurs when $G = R$ at $c = 0$, and subsequent crack propagation (i.e., $c>0$) is unstable or catastrophic since $G>R$. For the rising $R(c)$, fracture is again initiated when $G = R$ at $c =$ but subsqently fracture cannot occur at the same load since $G<R$. With increasing P, stable fracture occurs until the slope of $R(c)$ is less than that of $G(a)$. Therefore, with a rising $R(c)$ catastrophic failure occurs only after some stable fracture, and the load capacity increases even after fracture is initiated, i.e., the structure is flaw tolerant. Note that increasing $G(a)$ functions (as in Fig. 6.12) are exhibited by several geometries, called positive geometries. For negative geometries, the slopes maybe negative over a certain range of a. Discussion in this work

is limited to positive geometries, and the reader is referred to Jenq and Shah,[115] Planas and Elices,[116] Ouyang et al.[117] and Bazant et al.[118] for details on negative geometries.

Usually R is taken to be a constant after a certain amount of crack extension. In Fig. 6.12, R asymptotically reaches a constant value G_c at $c = c_f$, where G_c is the critical strain energy release rate in LEFM. This reiterates that when R is constant, the R-curve model is identical to LEFM. It is, however, possible that after a horizontal plateau, R may decrease, especially due to the interaction of the crack with the specimen boundary.[119] Also, the plateau value may depend on structural dimensions.

The R-curve behavior of concrete has been studied by several investigators. From fracture tests, Brown[51] found that for mortar the crack resistance increased with crack length, while R was constant for hcp. Wecharatana and Shah[120] obtained R-curves from load-displacement curves of test specimens and the corresponding crack lengths measured optically. They concluded that mortar and concrete exhibited rising R-curves that were geometry-dependent. Generally, it is not possible to accurately measure the crack extension in a concrete specimen. Therefore, indirect procedures such as the compliance methods have been used. In the multi-cutting method of Hu and Wittmann,[121] the notch of a fractured specimen is extended several times and the compliance is determined at each step. The total crack length is that beyond which the compliance is the same as that of a virgin specimen. The traction-free crack length is that where notch extension does not change the compliance. Based on these crack lengths, Hu[122] obtained R-curves that decrease after reaching a maximum.

R-curves can also be derived from stress-opening relations of the cohesive crack models. For a given $f(w)$ relation, Foote et al.[96] obtained geometry-dependent K_{IR}-curves using the Green's function approach. Their curves were independent of the initial crack length for a given specimen but the values of c_f (see Fig. 6.12) varied with specimen geometry and size. Also, in their approach, c_f is approximately equal to the length of the fully developed cohesive zone. It was shown later that these R-curves reach a plateau only for large specimens, and that for smaller specimens the slope may become constant and even increase.

Planas et al.[124] proposed an $R(\text{CTOD})$ curve that is directly related to $f(w)$. CTOD (for crack tip opening displacement) is the opening of the effective crack at the location of its initial tip. In their model,

$$R(\text{CTOD}) = \sum_0^{\text{CTOD}} f(w)\, dw \qquad (6.17)$$

The $R(\text{CTOD})$ curve is geometry-independent unlike the $R(c)$ curve. The CTOD can be related to c through LEFM for obtaining a conventional $R(c)$ relation.

Bazant and Kazemi[125] considered c_f and G_c (see Fig. 6.12) to be material properties, and derived an *R*-curve given in parametric form as:

$$\text{for } c < c_f, \quad R\left(c = G_c \frac{g'(\gamma)}{g'(\alpha_0)} \frac{c}{c_f}, \quad \frac{c}{c_f} = \frac{g'(\alpha_0)}{g(\alpha_0)} \left(\frac{g(\gamma)}{g'(\gamma)} - \gamma + \alpha_0 \right) \right) \tag{6.18a}$$

$$\text{for } c \geq c_f, \quad R(\iota) - G_c \tag{6.18b}$$

where $g(\alpha) = \{f(\alpha)\}^2$ (see Equation (6.3)), $g'(a) =$ derivative of $g(\alpha)$ with respect to α, G_c and c_f are the fracture energy and process zone size defined by the size effect model (see below), and γ is a dummy parameter. Equation (6.18) was derived for an infinitely large specimen where the process zone develops without restrictions. For finite structures, the maximum load occurs at $c < c_f$ and $R < G_c$. Bazant *et al.*[118] suggested later that for finite size, Equation (6.18) is valid until the maximum load and subsequently the *R*-curve is constant. With this modification, Gettu *et al.*[73] satisfactorily predicted load-deflection curves of high strength concrete, using Equation (6.18).

Ouyang *et al.*[117] proposed an *R*-curve that is the envelope of several specimens that are geometrically similar and different in size, but with the same initial notch length. For infinite size, their *R*-curve is given as:

$$R = \xi \left\{ 1 - \left(\frac{d_2 \kappa - \kappa + 1}{d_1 \kappa - \kappa + 1} \right) \left(\frac{\kappa a_0 - a_0}{a - a_0} \right)^{d_2 - d_1} \right\} (a - a_0)^{d_2} \tag{6.19a}$$

$$d_{1,2} = \frac{1}{2} + \frac{\kappa - 1}{\kappa} \pm \left\{ \frac{1}{4} + \frac{\kappa - 1}{\kappa} - \left(\frac{\kappa - 1}{\kappa} \right)^2 \right\}^{1/2} \tag{6.19b}$$

where κ and ξ are functions of a_0, E, K_{Ic}, $COTD_c$ and specimen geometry.[126] K_{Ic} and $COTD_c$ are the critical stress intensity factor and critical CTOD, respectively, defined according to the two parameter fracture model (see the next Section). For finite specimens, the plateau of the *R*-curve begins at the maximum load. Using this model, the load-deformation response of several fracture specimens have been predicted.[117]

The *R*-curve can be easily used in nonlinear fracture analysis of structures when the $R(c)$ and $G(a)$ functions are available. However, the *R*-curve approach has some limitations: (1) $R(c)$ is not a true material property but also depends on structual geometry, and (2) $G(a)$ functions cannot be determined without knowing the crack pattern.

Two parameter fracture model (TPFM)

The TPFM of Jenq and Shah[115] proposes an effective crack that is equivalent in compliance to the elastic component of the actual crack. As shown in Fig. 6.13a, two independent material parameters K_{Ic}^s and $CTOD_c$ are defined in terms of the critical state of the effective crack. The critical state is usually at the maximum load and corresponds to the

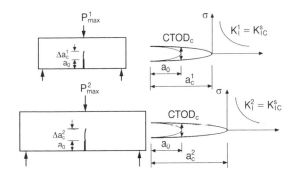

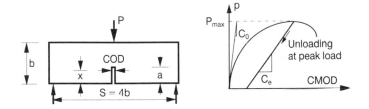

(a) Fracture criteria: $K_I = K_{IC}^s$ and $CTOD = CTOD_c$
(Superscripts 1 and 2 correspond to small and large
specimens, respectively)

(b) Determination of K_{IC} and $CTOD_c$ from C_0 and C_e obtained in
a notched-beam test (C_0 and C_e are the initial compliance and
the unloading compliance at the peak load, respectively)

Fig. 6.13 The two parameter fracture model

effective crack length a_c. The associated critical stress intensity factor is $K_{Ic}{}^s$, and the critical CTOD is $CTOD_c$. The fracture criteria of the model are:

$$K_I = K_{Ic} \text{ and } CTOD = CTOD_c \tag{6.20}$$

These criteria have to be satisfied simultaneously for fracture to occur (see Fig. 6.13a). The TPFM has also been extended to mixed mode fracture.[127]

A RILEM[128] recommendation describes the procedure for obtaining the parameters of the TPFM from the load-CMOD response of a 3PB specimen, where CMOD (for crack-mouth opening displacement) is the opening of the notch mouth. As shown in Fig. 6.13b, the unloading compliance C_e is determined just after the peak load P_{max} (within 95% of P_{max}), and used along with the initial compliance C_i in LEFM relations to get a_c. The parameters $K_{Ic}{}^s$ and $CTOD_c$ are then computed for the effective crack at load P_{max}. It has been demonstrated that this method yields parameters that are practically size-independent.[110] Though several specimen geometries can be used, the RILEM recommendation proposes the 3PB specimen since the method has been extensively verified with this geometry. Typical values of the TPFM parameters are given in Section 6.7.

An effective crack model has been proposed by Karihaloo and Nallathambi[129,130] that is similar in principle to the TPFM. In their model, the critical length of the effective crack is determined such that the deflection of the effective specimen is the same as that of the actual specimen, under peak load P_{max}. The K_{Ic}-values determined from this method are comparable to those of the TPFM; typical values are about 31 MPa$\sqrt{mm}$ for normal concrete with $f_c = 27$ MPa, and about 58 MPa$\sqrt{mm}$ for 78 MPa high strength concrete.[131] See Section 6.7 for more details.

Size effect method (SEM)

In Section 6.4, the size effect observed in fracture tests of concrete was discussed. As a result of this phenomenon, tests on geometrically similar specimens (with same shape and relative notch length) of different sizes yield failure stresses that vary with specimen size. This trend has been modeled by Bazant[132] with the relation:

$$\sigma_N = \frac{B}{\sqrt{1+\beta}}, \quad \beta = \frac{d}{d_0} \tag{6.21}$$

where σ_N = nominal failure stress (see Equation (6.9)), d = characteristic structural dimension (for the 3PB specimen, d = depth), and B and d_0 are empirical parameters. Equation (6.21) has the form shown graphically in Fig. 6.14, which implies that for small sizes, the failure is governed by limit stress criteria (no size effect), and that for large sizes, the failure is governed by LEFM ($\sigma_N \propto 1/\sqrt{d}$).

Since Equation (6.21) relates the failure stress of small specimens to LEFM behavior, Bazant proposed that the LEFM asymptote be used to define fracture parameters unambiguously for an infinitely large size. Test data could then be extrapolated to the limit of $d \to \infty$, where effects of the specimen geometry are theoretically absent. This approach was used to determine fracture parameters for concrete and mortar by Bazant and

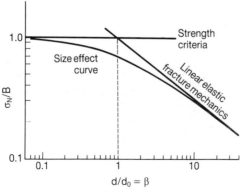

Fig. 6.14   The size effect model

Pfeiffer.[133] This also lead to a reformulation of Equation (6.21) in terms of two fracture parameters:[125]

$$\sigma_N = \frac{K_{Ic}}{\sqrt{g'(\alpha_0)c_f + g(\alpha_0)d}} \tag{6.22}$$

where $K_{Ic}(=\sqrt{E'G_c})$ and c_f are the fracture toughness and the maximum process zone length, for the LEFM limit of finite size. (See Fig. 6.12 and Equation (6.18).) The functions g and g' are the same as in Equation (6.18). In this model, the effective crack is similar to that of the TPFM. The SEM has been used to model the fracture of several types of specimens and structures.[73,125]

The procedure for obtaining the fracture parameters of the SEM has been proposed as a RILEM[134] recommendation. The peak loads of at least three sizes of 3PB specimens are needed to calibrate Equations (6.21) and (6.22), and consequently determine K_{Ic}, G_c and c_f. Typical values for these parameters are given in Section 6.7.

Comments on the effective crack models

Planas and Elices[135,136] conducted comparative studies of several effective crack models, and have concluded that their fracture criteria and their performance within the practical size range are indistinguishable. The models, however, may exhibit differences in their asymptotic behavior when they are used in the analysis of structures far larger than those used for determining their parameters (see also Karihaloo and Nallathambi[130]).

It should be emphasized that consistency in the choice of a model is of utmost importance; i.e., the same model should be used for both the material characterization and the analysis. Also, results from the same model should be used for comparing the characteristics of different materials or structures.

It should also be noted that G_c used in the effective crack models is not equal to G_f obtained from the work-of-fracture method (see Section 6.5) unless the fracture is governed by LEFM.

6.7   Material characterization

For the characterization of concrete, fracture mechanics provides parameters that are complementary to traditional measures such as strength and limit strain. In some applications where cracking governs the response, fracture parameters may become the only data that are needed for analysis. In general, two aspects of fracture behavior can be identified: the resistance against cracking, and the brittleness of crack propagation. This section deals with the quantification of both these aspects and the review of certain factors that influence them.

Crack resistance

In most of the nonlinear fracture models, one of the parameters quantifies resistance of the material against cracking, and the other denotes the brittleness or ductility of the material (see below). In cohesive crack models such as the FCM and CBM, the crack resistance is measured by the tensile strength f_t. It should be noted that when f_t is determined from a tensile test it could vary with the size and shape of the specimen. In general, f_t increases with an increase in f (see Equation (6.16)). The other parameter of the FCM, the fracture energy G_f determined from the work-of-fracture method, depends on both the crack resistance and the process zone deformation (see Equation (6.11)).

The parameters of the effective crack models that quantify crack resistance are K_{Ic} and G_c. The K_{Ic}^s of the TPFM varies between 30–50 MPa$\sqrt{mm}$ for normal concrete and between 25–30 MPa$\sqrt{mm}$ for mortar.[130] The variation of K_{Ic}^s with strength, as determined by Nallathambi and Karihaloo,[131] is shown in Table 6.1. The values of K_{Ic}

Table 6.1   Variation of K_{Ic}^s with compressive strength

f_c (MPa)	27	39	49	68	68
K_{Ic}^s (MPa$\sqrt{mm}$)	31	40	44	48	58

obtained from their effective crack model are almost the same. The increase in K_{Ic}^s with strength has also been demonstrated by John and Shah.[137] Lange[138] determined K_{Ic}^s for several compositions of hcp and mortar, with and without silica fume and the results are given in Table 6.2.

K_{Ic} from the SEM usually varies between 25–60 MPa$\sqrt{mm}$ for normal concrete, and its value compares well with K_{Ic}^s from the TPFM.[130] From an analysis of the tests of Bazant and Pfeiffer,[133] values of 36 MPa$\sqrt{mm}$ for concrete $f_c = 34$ MPa and 27 MPa$\sqrt{mm}$ for mortar ($f_c = 48$ MPa) were obtained by Bazant and Kazemi[125] using Equation (6.22). For 86 MPa high-strength concrete, Gettu et al.[73] obtained K_{Ic} of 30 MPa$\sqrt{mm}$. Chern and Tarng[139] conducted size effect tests on fiber-reinforced concrete to determine G_c. They used 20 mm long steel fibers and crushed limestone aggregates. Their results are summarized in Table 6.3.

Table 6.2 K_{Ic}^s for hcp and mortar, with and without silica fume

Material	f_c (MPa)	K_{Ic}^s (MPa$\sqrt{mm}$)
hcp	62	14
hcp + 5% sf*	65	14
hcp + 10% sf*	65	15
coarse mortar 1:1	59	20
coarse mortar 1:2	47	22
fine mortar 1:1	63	23

*silica fume

Table 6.3 G_c of fiber-reinforced concrete

Max. aggregate size (mm)	Fiber content (%)	f_c (MPa)	G_c (N/m)
25	0	33	39
	1	37	59
	2	40	112
13	0	31	29
	1	40	67
	2	44	118
5	0	40	20
	1	45	75
	2	48	124

K_{Ic} of high-strength concretes has also been determined by other means. Using a compliance method, de Larrard *et al.*[140] found that by increasing the concrete strength from 54 to 105 MPa, the K_{Ic} increased by about 30%. A general conclusion is that the crack resistance increases with increase in the conventional compressive strength, but at a lesser rate. However, when fibers are added to concrete, the crack resistance increases much more than the strength.

Brittleness

The failure of plain concrete is generally brittle, but not as brittle as that of glass. This ductility, or rather the 'pseudo-ductility', that concrete possesses can be quantified through fracture mechanics. In this work (as in others; cf. Hucka and Das[141]), ductility is taken to be the inverse of brittleness.

In almost all of the nonlinear fracture models, the brittleness of the material can be related to parameters that depend on the dimensions or the deformations of the (effective) fracture process zone. Some brittleness quantifiers have been defined by combining G_f with other properties. Hillerborg[68] defined a characteristic length l_{ch} that is proportional to the process zone length:

$$l_{ch} = \frac{EG_f}{f_t^2} \qquad (6.23)$$

The relation is similar to that used by Irwin for the size of the plastic zone, which is of significant importance with respect to the ductile-brittle transition in the fracture of metals. In the context of concrete, a smaller l_{ch} implies that the material is more brittle. Typical values for l_{ch} are given in Table 6.4.

A code-type relation has also been proposed for l_{ch} by Hilsdorf and Brameshuber[78] that is valid for concrete with f_c in the range of 10–100 MPa:

$$l_{ch} = 600 x_F f_{cm}^{-0.3} \qquad (6.24)$$

Table 6.4 Typical values for l_{ch}

Material	l_{ch}	Reference
glass	1 micron	Bache[142]
dense silica cement paste	1 mm	Bache[142]
hcp	5–15 mm	Hillerborg[68]
mortar	100–200 mm	Hillerborg[68]
high-strength concrete (50–100 MPa)	150–300 mm	Hillsdorf and Brameshuber[78]
normal concrete	200–55 mm	Hillerborg[68]
dam concrete max. aggregate size = 38 mm	0.7 m	Brühwiler *et al.*[143]
glass fiber reinforced mortar	0.5–3 m	Hillerborg[68]
steel fiber reinforced concrete	2–20 m	Hillerborg[68]

where x_F and f_{cm} are defined below Equation (6.11).

The brittleness of the material can also be judged from the shape of the $f(w)$ curve (Fig. 6.10). A material with a steep incline in f is more brittle than one that exhibits a more gradual decline. This aspect is quantified by the model of Wecharatana[103] in Equation (6.15), where parameter m is a brittleness index. The critical separation w_c of the FCM also reflects the ductility of a material, but its value is affected significantly by the shape chosen for the $f(w)$ curve (see Section 6.6).

The R-curve has traditionally represented the brittleness of a toughening material. A gradually rising R-curve (see Section 6.6) implied that the material was less brittle than one where the R-curve rises rapidly. Accordingly, when two materials have almost the same G_c, that with a smaller c_f would be more brittle (see Fig. 6.12).

Three quantities derived from the TPFM have been used as measures of ductility. The higher the pre-peak crack extension, the lower is the brittleness of a quasi-brittle material. Therefore, the critical effective crack length a_c is an indicator of the material ductility. It was used by John and Shah[137] to compare the brittleness of high-strength mortar to that of normal concrete. It can be seen in Fig. 6.15 that a_c decreases with an

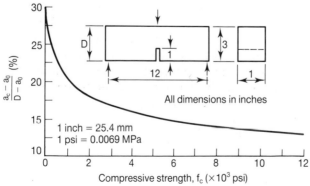

Fig. 6.15 Decrease in ductility with the increase of strength (results of John and Shah[137])

Table 6.5 Typical values for CTOD$_c$ and Q

Material	Q (mm)	CTOD$_c$ (mm)	Reference
hcp	25	0.0035	Lange[138]
hcp + 5% sf *	10	0.0025	Lange[138]
hcp + 10% sf *	5	0.0017	Lange[138]
coarse mortar 1:2	57	0.0046	Lange[138]
concrete	300	0.02	Jenq and Shah[115]
high strength mortar (110 MPa)	90	0.011	John and Shah[137]

* silica fume

increase in compressive strength demonstrating that brittleness increases with strength. Since a_c depends on the size and shape of the specimen tested, two other size-independent quantifiers of ductility have also been defined (Jenq and Shah[115]) – one is CTOD$_c$ and the other is a length parameter Q:

$$Q = \left(\frac{E\,\text{CTOD}_c}{K_{Ic}^s}\right)^2 \qquad (6.25)$$

Both CTOD$_c$ and Q are smaller for a more brittle material. Typical values of these parameters are given in Table 6.5 (see Table 6.2 also).

In the SEM of Bazant, the process zone size, c_f, defined in Equation (6.22) is a measure of the material ductility. Bazant and Kazemi[125] obtained c_f-values of 10–25 mm for concrete ($f_c = 34$ MPa) and 6–15 mm for mortar ($f_c = 48$ MPa). For a high strength concrete ($f_c = 86$ MPa), Gettu *et al.*[73] found that c_f was about 3–6 mm at 14 days.

From a review of the test data, it may be concluded that brittleness increases with increase in the strength of concrete. For normal concrete, brittleness decreases with increase in aggregate size and with the addition of fibers. The presence of silica fume seems to increase the brittleness significantly.

Previous discussion in this section has dealt with the brittleness of the material. This can be described simply as the lack of flaw-tolerance or the inability of the material to prevent unstable and catastrophic crack propagation. The brittleness of structural failure, however depends on both the material and the structural geometry. Hillerborg[68] and Bache[142] have taken the quantity d/l_{ch} to indicate structural brittleness. Carpinteri[48] considered the square-root of this quantity as a brittleness number.

Carpinteri *et al.*[81] proposed a dimensionless brittleness number that increases with decreasing structural brittleness. This is based on the parameters of the FCM (see Section 6.6), as given as:

$$s_E = \frac{G_f}{f_t d} \qquad (6.26)$$

For a linear $f(w)$relation, $s_E = w_c/2d$.

In the brittleness numbers s_E and d/l_{ch}, the structural brittleness increases with increase in the material brittleness characterized by w_c and l_{ch}, respectively. They also imply that the structural brittleness increases with the dimensions of the structure. The second effect causes the characteristic increase in steepness of the post-peak part of load-deflection curve with increase in size (see Bosco et al.[144]). Bazant has argued that the effects of specimen geometry and relative length of the initial flaw should also be included in the structural brittleness number. Accordingly, the brittleness number β of the SEM (see Equations 6.21 and 6.22) is defined as:[125, 133]

$$\beta = \frac{g(\alpha_0)}{g'(\alpha_0)} \frac{d}{c_f} \tag{6.27}$$

where the first term accounts for the shape of the structure, and the second accounts for the structural dimensions and the material brittleness. β can be directly determined from size effect tests by fitting Equation (6.21). The size effect curve in Fig. 6.14 shows the implications of β. When the size or material brittleness is large (i.e., β is large), the behavior is close to LEFM, and when β is small the behavior is governed by limit criteria. The position of the data on the size effect curve also gives an idea of the brittleness. Chern and Tarng[139] found that with the addition of fibers, the behavior shifted away from LEFM implying a decrease in structural brittleness.

Factors influencing the fracture parameters

Discussion in previous sections assumed that the concrete was subjected to normal conditions of temperature, loading rate, humidity, etc., and that its composition did not vary significantly. In this section, the effects of these factors on the fracture parameters will be briefly reviewed.

Concrete composition

The presence of aggregates influences the fracture properties of a cement-based material considerably (as already mentioned above). This was first studied by Moavenzadeh and Kuguel,[66] and Naus and Lott[145] using LEFM. It was demonstrated that K_{Ic} for concrete was larger than that of mortar, and K_{Ic} for mortar was larger than that of hcp. In normal concrete, where cracks are arrested and deflected by the aggregates, the crack resistance generally increases with increase in the size of the aggregates.[71,146] This trend is, at times, opposite to that observed for the compressive strength. From studies based on the FCM, Roelfstra and Wittmann,[107] and Wittmann et al.[108] found that increase in the aggregate size caused the increase of G_f and w_c. These trends are reflected in the relations proposed by Hilsdorf and Brameshuber[78] for G_f and l_{ch} (see Equations (6.11) and (6.24)). The increase in crack resistance and the decrease in brittleness, with an increase in aggregate size, is due to the larger fracture process which arises when there is a wider range of heterogeneities, as explained in Section 6.3. Mihashi *et al.*[23] observed this effect in a study of the acoustic emissions recorded during fracture. It should also be mentioned that in a study of dam concrete, Saouma *et al.*[147] concluded that the crack resistance of concrete was higher for angular aggregates than for round aggregates, but was practically independent of the aggregate size.

In some high-strength concretes where the aggregates rupture during fracture, aggregate size is not expected to influence the fracture parameters. However, the presence of fibers will increase both the resistance and the ductility.

It has also been found that K_{Ic} decreases with an increase in water-cement ratio[71,145,148] and air content.[71,149] This implies that a denser microstructure would exhibit higher crack resistance. This also explains the increase in K_{Ic} with longer curing periods,[66,71] and with aging in the first few weeks.[54,148–150] Consequently, high strength concrete has a higher crack resistance than normal concrete, but also a higher brittleness due to fewer inhomogeneities.

Loading rate

Under dynamic and impact loading, the failure stress of concrete increases considerably with an increase in loading rate.[151] Several investigators have studied rate effects on fracture parameters of concrete in order to explain this characteristic increase in strength. By applying the TPFM, John and Shah[152,153] showed that there is a small increase in K_{Ic}^s with increase in loading rate, but a_c and $CTOD_c$ decrease considerably. Consequently, it was concluded that the brittleness of concrete increases with loading rate in the dynamic regime. Using work-of-fracture methods, increase in G_f with rate has been observed by Wittmann *et al.*[108] and Oh.[154]

Rate effects in the static range involve a strong interaction of creep and fracture.[155,156] At slow rates, Wittmann *et al.*[148] found that while f_t increases with loading rate, G_f and w_c seem to decrease. From size effect tests, Bazant and Gettu[157,158] obtained a shift in failure behavior towards LEFM, implying an increase in brittleness, when the deformation rate was decreased. Their results show a decrease in K_{Ic} and c_f with a decrease in rate. They also showed from relaxation tests that the effect of creep is much stronger in the presence of cracks.

The effect of loading rate on the fracture of high-strength concrete has not yet been investigated thoroughly. However, it appears that time-dependent effects are lesser in high strength concrete than in normal concrete (e.g., Banthia *et al.*[159]).

Temperature and humidity

The crack resistance of concrete generally increases with the decrease of temperature.[151] Using the SEM, Bazant and Prat[160] showed that at high temperatures, G_c decreases more when the concrete is wet than when it is dry. This effect was observed on G_f, even at sub-zero temperatures by Planas *et al.*[161] and Ohlsson *et al.*[162] At very low temperatures, Maturana *et al.*[163] also found that the freezing of the water in the concrete gave rise to a considerable decrease in the brittleness in terms of l_{ch}.

The detrimental effect of free water on the crack resistance of concrete was attributed to stress-corrosion by Shah and Chandra.[164] Michalske and Bunker[165] suggested that this phenomenon arises due to the weakening of strained silicate bonds at the crack-tip by water. Rossi[166] has proposed that free water is also the cause of rate effects in concrete, and therefore, there were lesser effects of loading rate in dry concrete and high-strength concrete due to the absence of moisture.

The effect of water in the curing stage is, however, different. As in the case of compressive strength, proper curing with water results in K_{Ic} that is significantly higher than for concrete cured in air.[149]

Loading history

The influence of fatigue loading on fracture parameters is of importance since cracks can propagate under repeated loads to cause failure.[167] Pons *et al.*[168] found that under low-frequency cyclic loading, G_f and LEFM-based K_{Ic} decreased considerably with increase in the number of cycles and the amplitude of the loading. They concluded that characterization of concrete through its fracture parameters is not independent of the loading history. Using the SEM, Schell[169] found that fatigue fracture of high-strength concrete was similar to that of normal concrete, and that it could be modeled adequately through nonlinear fracture mechanics coupled with the Paris' law.

Another important aspect of loading history is the effect that prior compressive loads may have on subsequent fracture. Tinic and Brühwiler[170] showed that when concrete is subjected to repeated compressive loading, the tensile strength could decrease by about 50%. Such losses of crack resistance seem to be lesser in high-strength concrete, but still significant.[171]

6.8   Other aspects of fracture in concrete

Other aspects of fracture characterization in concrete include the mixed mode fracture, interfacial fracture, field and core-based fracture tests, and stochastic models for fracture. These are summarized in this section.

Mixed-mode fracture

Since the crack resistance of concrete in Mode I is much lower than in other modes, several investigators have doubted whether mixed- mode fracture can exist in concrete.[172] In this work, a distinction is made between non-planar Mode I cracking, and mixed-mode cracking where the crack undergoes both opening and sliding at its tip. It appears that mixed-mode fracture occurs only when normal and shear loads are simultaneously applied on the crack.[173,174] Otherwise, the crack chooses a non-planar path where K_I dominates. Its behavior is, however, influenced by crack-face friction.[127,175,176] In either case, the fracture is less brittle than planar Mode I, especially for concrete where the crack path is tortuous.

Interfacial fracture

The bond between aggregates and hcp dictates several aspects of the behavior of concrete (*cf.* Mindess and Shah[177]). In normal concrete, the crack resistance of the interface is lower than that of the hcp and the aggregates,[178] and therefore, fracture occurs through these interfaces. This also causes crack tortuosity and a wider fracture process zone. In some high-strength concretes, the interfaces are stronger, and consequently, the brittleness is higher. Due to its importance in materials engineering and micromechanical modeling, researchers[179,180] are focusing more attention on the fracture mechanics and characterization of interfaces.

Field and core-based fracture tests

Fracture tests may sometimes have to be performed on existing structures, such as dams, for characterizing the actual material. For this purpose, in situ methods are being validated (e.g., by Saouma *et al.*[181,182]). Core-based fracture tests for concrete are also being explored[183,184] since concrete is usually extracted from structures in the form of cores. It should be noted that core-based tests have been used extensively for the characterization of rocks.[185]

Stochastic models for fracture

As concrete is a hectrogeneous material, the randomness in its microstructure plays an important role in its behavior. However, most of the prevalent fracture models for concrete are deterministic. In future these models may be extended to also include stochastic effects. Nevertheless, there have been several other studies that account for the randomness in concrete; for example, the random crack model of Zaitsev and Wittmann[186] which is applicable for normal and high-strength concrete, and the stochastic theory of Mihashi.[187]

6.9   Applications

Fracture mechanics can be used as a tool in the materials engineering of concrete and in structural analysis. The motivations for such applications and some illustrative examples are presented in this section.

Materials engineering

The engineering of cement-based materials based on fracture mechanics has been emphasized by several researchers. The introduction of weaker inclusions, which function as flaws causing irregular crack surfaces and high dissipation of energy, has been suggested as a means of improving crack resistance and brittleness. For example, Mai *et al.*[188] showed that the resistance of cement mortar could be increased more than 10 times by embedding pieces of paper. The resulting loss in bending strength was less than 50%. Beaudoin[189] found that when a small volume fraction of mica flakes were added to high alumina cement paste, the fracture toughness and the bending strength increased considerably. The compressive strength, however, decreased with increase in mica content. Not much work has been done on the engineering of structural concrete using fracture principles, but the ideas discussed in the section establish the basis for such approaches.

Some of the flaws that are usually present in concrete are pores. They influence fracture behavior significantly, especially in the hcp. Alford *et al.*[190] conducted fracture tests on macro-defect free (MDF) hcp plates that were prepared by pressing a mixture of cement paste and a water soluble polymer. Their results show that while the compressive strength and crack resistance increase appreciably, the brittleness also increases due to the reduced porosity.[191] Several other methods for modifying the microstructure of high-strength hcp to enhance the fracture performance have been suggested by Beaudoin and Feldman.[192]

Usually, high-strength concretes are materials with higher strength and good workability. Unfortunately, they are also highly brittle. This increase in brittleness with strength has been discussed in Section 6.7 and is illustrated in Fig. 6.15.[193] The results from a study[73] where concretes with f_c of 86 MPa and 33 MPa were tested, are shown in Fig. 6.16. It can be seen for a 160% increase in f_c, K_{Ic} increases only 25% and the ductility decreases by about 60% (see Section 6.6). Generally, the problem of brittleness is handled by confining the high-strength concrete with steel; for example, to avoid catastrophic failures of columns under seismic loading.

The best available method for increasing the ductility of concrete is through the use of fibers. Naaman[194] showed that an efficient high-performance material with superior strength and ductility is obtained when a high-strength matrix is reinforced with fibers. It appears that when a large volume (8–12%) of fibers is used the tensile strength of the matrix also

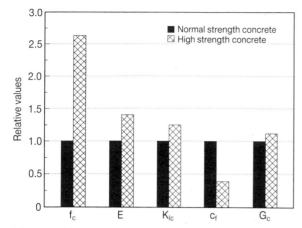

Fig. 6.16   Fracture parameters for normal and high-strength concrete (data of Gettu *et al.*[73])

increases significantly.[195] The fracture behavior of fiber-reinforced concrete involves several concepts discussed in this work, and has been treated more thoroughly elsewhere.[196–198]

The use of fracture mechanics principles in the design of materials with higher toughness and ductility has produced encouraging results in the ceramics industry,[199] and a similar approach is needed for high-performance concretes. Other aspects of concrete behavior will also benefit from an increase in crack resistance and a decrease in brittleness. These include the bond between steel and concrete, which is more brittle in silica fume concrete than normal concrete.[200] The durability, long-term reliability and thermal performance, which are greatly affected by cracking, would also be enhanced.

Structural analysis and design

The application of fracture models is most straightforward to concrete structures with negligible reinforcement, such as dams,[201–203] plinths of bridge piers,[204] and thin-walled pipes and shells.[205] Fracture analysis has been used successfully to study the collapse mechanisms of such structures and to make their design safer. In usual structures, fracture analysis is more difficult due to the influence of the steel reinforcement and distributed cracking. Nevertheless, several aspects of structural behavior have been identified where fracture mechanics would lead to rational and conservative design.[206,207] Some of these are discussed in this section.

In the calculation of beam and slab deflections, the modulus of rupture is usually taken to be a material property. However, it has been shown that the flexural strength decreases considerably with an increase in the depth, more so for high-strength concrete.[208] A similar trend has also been observed for the shear strength (see below). In the presence of shrinkage stresses this effect is even greater, and could result in a large underestimation of the deflections. Fracture analysis accounts for the dependence of failure stress on brittleness, and can be used to determine the value of the strength that should be used in the design. Hillerborg[209,210] has also argued that the compressive stress-strain diagram used in flexural design should depend on the brittleness of the concrete and on the beam depth. This could lead to better estimates of the rotational capacity, the ultimate moment of over-reinforced beams, and the balanced reinforcement ratio.

Fracture principles have also been used to study failure mechanisms of reinforced concrete. For example, the bond-slip behavior of reinforcing steel was simulated with a nonlinear mixed-mode model by Ingraffea *et al.*[211] They concluded that secondary radial cracks that emanate from the ribs of the bar allow stable bond-slip before the occurrence of primary debonding cracks. Also, failure mechanisms in the pullout of anchor bolts have been modeled using fracture mechanics.[212,213]

Bosco *et al.*[144,214] have proposed that the minimum flexure reinforcement should be determined based on fracture mechanics principles. Their approach is based on a structural brittleness number N_p derived by Carpinteri:[215]

$$N_p = \frac{f_y \sqrt{d}}{K_{Ic}} \frac{A_s}{A}$$ (6.28)

where A_s/A = steel reinforcement percentage, and f_y = yield strength of the steel. The moment-rotation response is similar for beams with the same N_p. The condition for design is that $N_p = N_{p_c}$, where N_{p_c} corresponds to the critical case when the steel yielding moment is equal to the first cracking moment. From tests on various sizes of reinforced concrete beams with different steel ratios, they obtained $N_{p_c} = 0.14$ for concrete with $f_c = 30$ MPa, and $N_{p_c} = 0.26$ for high-strength concrete with $f_c = 76$ MPa. The minimum steel percentage increases with an increase in concrete strength and a decrease in beam depth. Since design codes specify constant values, they may be unconservative for smaller beams, especially at higher concrete strengths.

Shear failure

One type of failure that deserves more discussion is shear failure. Diagonal tension and torsional failure of beams, and punching shear failure of slabs occur in a brittle manner. To avoid such catastrophic collapses, reinforcement is provided across potential crack locations. In the design against such failures, the resistance is taken to be the sum of the contributions of steel and concrete. However, since failure is due to fracture, there is a size effect on the shear strength of concrete.[205,206,216] The effect of beam depth on shear failure has been studied extensively by Bazant using the SEM.[217,218] He has proposed that nonlinear fracture mechanics be used to model the shear failure, and that the design shear strength of concrete should decrease with increase in structural brittleness. Similar conclusions have been made by Gustafsson and Hillerborg,[219] Shah,[193] and Thorenfeldt and Drangsholt.[220] Bazant and Kazemi[218] also found that the brittleness decreases when there is bond-slip. Walraven[221] showed from tests on normal and lightweight concrete that the size effect is important in both slender and short beams.

Shear failure has been analyzed numerically using several fracture models. Saouma and Ingraffea[222] employed a discrete crack approach based on LEFM, with aggregate interlock and a nonlinear constitutive model for the uncracked concrete. A crack band model was used by de Borst and Nauta[223] who obtained load-displacement behavior and crack patterns that matched the experimentally obtained results. The same model was also used to analyze the punching shear failure of reinforced concrete slabs. The approach of Wang and Blaauwendraad[224] involves two stages. First, a predictor analysis is performed using a crack band model with shear transfer across the crack. The resulting crack pattern is utilized in the choice of possible crack paths for discrete analyses from which the load-displacement response is determined. The path that gives the lowest load corresponds to the actual failure pattern. They concluded that shear failure can be modeled with just Mode I fracture, and that the size effect was caused only by the fracture of concrete.

Jenq and Shah[225] proposed an approximate method, based on mixed-mode fracture criteria and the TPFM, to calculate the shear failure loads of beams. They assume that Mode I cracks initiate due to flexure at the tensile face and propagate up to the longitudinal rebars. Beyond the rebars, one of the cracks propagates towards the compression face in mixed-mode. The fracture criteria are the same as in Equation (6.20), except that K_I and $CTOD_c$ are replaced by vector sums of the stress intensity factors and crack-tip displacements for the decoupled Modes I and II, respectively. An experimentally calibrated relation is used for modeling the bond-slip behavior of the rebars. The analyses is carried out for different crack paths, and that which gives the lowest failure load is taken to be the critical case. The predicted dependence of the failure load on beam depth and steel ratio (ρ) is given in Fig. 6.17 for normal and high-strength concretes.

The decrease of shear strength of concrete with increase in brittleness is of significant importance to high-strength concrete. Several works have shown that empirical relations based on tests of normal concrete can be unconservative for concretes of higher strengths.[226-228] The increased brittleness could also pose problems in other types of shear failure, such as the punching of slabs[229] and the failure of moment-resistant joints.[230] Such failures are potential applications where fracture mechanics can lead to rational design that accounts for the brittleness of the structure.

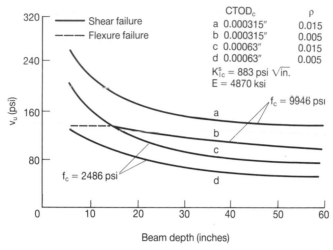

Fig. 6.17   Influence of beam size and compressive strength on shear failure loads. v = shear strength; r = reinforcement ration (1 in. = 25.4 mm), 1 psi = 0.0069 MPa, 1 ksi = 6.9 MPa) (results of Jenq and Shah[225])

References

1   Elfgren, L. (ed.) (1989) *Fracture mechanics of concrete structures: from theory to application*, Report of RILEM Technical Committee 90-FMA. Chapman and Hall, London.

2   Shah, S.P. and Carpinteri, A. (eds) (1991) *Fracture mechanics test methods for concrete*, Report of RILEM Technical Committee 89-FMT Fracture mechanics of concrete: test methods. Chapman and Hall, London.

3   ACI Committee 446 (1990) *Fracture mechanics of concrete: concepts, models and determination of material properties*, Report 446.1R, ACI, Detroit; also in *Fracture mechanics of concrete structures*, Bazant, Z.P. (ed.); Abstract in *Concrete International* ACI, **12**, 67–70.

4   Mindess, S. (1983) The cracking and fracture of concrete: an annotated bibliography, 1928–1981, in Wittmann, F.H. (ed.) *Fracture mechanics of concrete*, Elsevier Science, Amsterdam, 539–661.

5   Mindess, S. (1986) The cracking and fracture of concrete: an annotated bibliogrphy, 1982–1985, in Wittmann, F.H. (ed.) *Fracture toughness and fracture energy of concrete* (International Conference, Lausanne). Elsevier Science, Amsterdam, 539–661.

6   Shah, S.P. and Swartz, S.E. (eds) (1989) *Fracture of concrete and rock*, SEM-RILEM International Conference, Houston, 1987. Springer Verlag, New York.

7   Mihashi, H., Takahashi, H. and Wittmann, F.H. (eds) (1989) *Fracture toughness and fracture energy: test methods for concrete and work*, International Workshop, Sendai, Japan, 1988. A.A. Balkema, Rotterdam.

8   Mazars, J. and Bazant, Z.P. (eds) (1989) *Cracking and damage: strain localization and size effect*, France-USA Workshop, Cachan, France, 1988. Elsevier Applied Science, London.

9   Rossmanith, H.P. (ed.) (1990) *Fracture and damage of concrete and rock*, International Conference, Vienna, 1988. Pergamon Press, Oxford.

10 Elfgren, L. and Shah, S.P. (eds) (1991) *Analysis of concrete structures by fracture mechanics*, International RILEM Workshop, Abisko, Swden, 1989. Chapman and Hall, London.

11 Shah, P. (ed.) (1991) *Toughening mechanisms in quasi-brittle materials*, NATO Workshop, Evanston, USA, 1990. Kluwer Academic, Dordrecht, Netherlands.

12 van Mier, J.G.M., Rots, J.G. and Bakker, A. (eds) (1991) *Fracture processes in concrete, rock and ceramics*, International RILEM/ESIS Conference, Noordwijk, Netherlands. Spon, London.

13 Bazant, Z.P. (ed.) (1992) *Fracture mechanics of concrete structures*, International Conference, Breckenridge, USA. Elsevier Applied Science, London.

14 Broek, D. (1989) *The practical use of fracture mechanics*. Kluwer Academic, Dordrecht.

15 Tada, H., Paris, P.C. and Irwin, G.R. (1985) *The stress analysis of cracks handbook*. Paris Productions, St Louis, Missouri, USA.

16 Griffith, A.A. (1920) The phenomena of rupture and flow in solids. *Phil. Trans. Roy. Soc.*, **221A**, 163.

17 Irwin, G.R. (1957) Analysis of stresses and strains near the end of a crack traversing a plate. *J. Applied Mechanics*, **24**, 361–4.

18 Knott, J.F. (1979) *Fundamentals of fracture mechanics*. Butterworth, London.

19 Murakami, Y. (editor-in-chief) (1990) *Stress intensity factors handbook*. Pergamon Press, Oxford.

20 Glucklich, J. (1963) Fracture of plain concrete. *J. Engineering Mech. Div. (ASCE)*, **89**, EM6, 127–38.

21 Struble, L.J., Stutzman, P.E. and Fuller, E.J., Jr. (1989) Microstructural aspects of the fracture of hardened cement paste. *J. American Ceramic Society*, **72**, 12, 2295–9.

22 Maji, A.K., Ouyang, C. and Shah, S.P. (1990) Fracture mechanisms of brittle materials based on acoustic emission. *J. Materials Research*, **5**, 207–17.

23 Mihashi, H., Nomura, N. and Niiseki, S. (1991) Influence of aggregate size on fracture process zone of concrete detected with three dimensional acoustic emission technique. *Cement and Concrete Research*, **21**, 737–44.

24 Ouyang, C., Landis, E. and Shah, S.P. (1991) Damage assessment in concrete using quantitative acoustic emission. *J. Engineering Mechanics*, **117**, 11, 2681–98.

25 Kachanov, M. (1986) Interaction of a crack with some microcrack systems, in Wittmann, F.H. (ed.) *Fracture toughness and fracture energy of concrete*, International Conference, Lausanne, 1985. Elsevier Science, Amsterdam.

26 Hutchinson, J.W. (1987) Crack tip shielding by micro-cracking in brittle solids. *Acta. Metall.*, **35**, 1605–19.

27 Carrasquillo, R.L., Slate, F.O. and Nilson, A.H. (1981) Microcracking and behavior of high strength concrete subject to short-term loading. *ACI Journal*, **78**, 3, 179–86.

28 Mihashi, H., Nomura, N., Izumi, M. and Wittmann, F.H. (1991) Size dependence of fracture energy of concrete, in van Mier, J.G.M., Rots, J.G. and Bakker, A. (eds) *Fracture processes in concrete, rock and ceramics*. Spon, London.

29 van Mier, J.G.M. (1991) Crack face birdging in normal, high strength and Lytag concrete, in van Mier, J.G.M., Rots, J.G. and Bakker, A. (eds) *Fracture processes in concrete, rock and ceramics*. Spon, London.

30 Mindess, S. and Diamond, S. (1982) The cracking and fracture of mortar. *Materials Structure*, **15**, 86, 107–13.

31 Diamond, S. and Bentur, A. (1985) On the cracking in concrete and fiber reinforced cements, in Shah, S.P. (ed.) *Application of fracture mechanics to*

cementitious composites, NATO Workshop, Evanston, USA. 1984. Martinus Nijhoff, Dordrecht, Netherlands, 87–140.

32   Li, V.C. and Huang, J. (1990) Crack trapping and bridging as toughening mechanisms in high strength concrete, in Shah, S.P., Swartz, S.E. and Wang, M.L. (eds) *Micromechanisms of failure of quasi-brittle material*, International Conference, Albuquerque, USA. Elsevier Applied Science, London, 579–88.

33   Bower, A.F. and Ortiz, M. (1991) Three-dimensional analysis of crack trapping and bridging, in van Mier, J.G.M., Rots, J.G. and Bakker, A. (eds) *Fracture processes in concrete, rock and ceramics*. Spon, London, 110–28.

34   Mindess, S. (1991) The fracture process zone in concrete, Shah, S.P. (ed.) *Toughening mechanisms in quasi-brittle materials*. Kluwer Academic, Dordrecht, Netherlands, 271–86.

35   Mindess, S. (1991) Fracture process zone detection, in Shah S.P. and Carpinteri, A. (eds) *Fracture mechanics test methods for concrete*. Chapman and Hall, London, 231–61.

36   Cedolin, L., Dei Poli, S. and Iori, I. (1983) Experimental determination of the fracture process zone in concrete. *Cement and Concrete Research*, **13**, 557–67.

37   Cedolin, L., Dei Poli, S. and Iori, I. (1987) Tensile behavior of concrete. *J. Engineering Mechanics*, **113**, 3, 431–49.

38   Miller, R.A., Shah, S.P. and Bjelkhagenm, H.I. (1988) Crack profiles in mortar measured by holographic interferometry. *Experimental Mechanics*, **28**, 4, 388–94.

39   Castro-Montero, A., Shah, S.P. and Miller, R.A. (1990) Strain field measurement in fracture process zone. *J. Engineering Mechanics*, **116**, 11, 2463–84.

40   Swartz, S.E. and Go, C.G. (1984) Validity of compliance calibration to cracked concrete beams in bending. *Experimental Mechanics*, **24**, 2, 129–34.

41   Bascoul, A., Kharchi, F. and Maso, J.C. (1989) Concerning the measurement of the fracture energy of microconcrete according to the crack growth in a three-points bending test on notched beans, in Shah, S.P. and Swartz, S.E. (eds) *Fracture of concrete and rock*. Springer-Verlag, New York, 396–408.

42   Horii, H. (1991) Mechanisms of fracture in brittle disordered materials, in van Mier, J.G.M., Rots, J.G. and Bakker, A. (eds) *Fracture processes in concrete, rock and ceramics*. Spon, London, 95–110.

43   Thouless, M.D. (1988) Bridging and damage zones in crack growth. *J. American Ceramic Society*, **71**, 6, 408–13.

44   Diamond, S. and Mindess, D. (1992) SEM investigations of fracture surfaces using stereo pairs: I Fracture surfaces of rocks and of cement paste. *Cement and Concrete Research*, **22**, 67–78.

45   Hsu, T.T.C., Slate, F.O., Sturman, G.M. and Winter, G. (1963) Microcracking of plain concrete and the shape of the stress-strain curve. *ACI Journal*, **60**, 209–24.

46   Bentur, A. and Mindess, S. (1986) The effect of concrete strength on crack patterns. *Cement and Concrete Research*, **16**, 59–66.

47   Faber, K.T. and Evans, A.G. (1983) Crack deflection processes – I: Theory. *Acta Metall.*, **31**, 4, 565–76.

48   Carpinteri, A. (1982) Application of fracture mechanics to concrete structures. *J. Struct. Div. (ASCE)*, **108**, No. ST4, 833–48.

49   Kaplan, M.F. (1961) Crack propagation and the fracture of concrete. *J. ACI*, **58**, 5, 591–610.

50   Romualdi, J.P. and Batson, G.B. (1963) Mechanics of crack arrest in concrete. *J. Engineering Mech. Div. (ASCE)*, **89**, EM3, 147–68.

51   Brown, J.H. (1972) Measuring the fracture toughness of cement paste and mortar. *Magazine of Concrete Research*, **24**, 81, 185–96.
52   Shah, S.P. and McGarry, F.J. (1971) Griffith fracture criterion and concrete. *J. Engineering Mech. Div. (ASCE)*, **97**, EM6, 1663–76.
53   Walsh, P.F. (1972) Fracture of plain concrete. *Indian Concrete Journal* (Bombay), **46**, 11, 469–70; 476.
54   Higgins, D.D. and Bailey, J.E. (1976) Fracture measurements on cement paste. *J. Materials Science*, **11**, 11, 1995–2003.
55   Gjørv, O.E., Sorensen, S.I. and Arnesen, A. (1977) Notch sensitivity and fracture toughness of concrete. *Cement and Concrete Research*, **7**, 3, 333–44.
56   Mindess, S. and Nadeau, J.S. (1976) Effect of notch width on *K* for mortar and concrete. *Cement and Concrete Research*, **6**, 4, 529–34.
57   Ohgishi, S., Ono, H., Takatsu, M. and Tanahashi, I. (1986) Influence of test conditions on fracture toughness of cement paste and mortar, in Wittmann, F.H. (ed.) *Fracture toughness and fracture energy of concrete*, International Conference, Lausanne, 1985. Elsevier Science, Amsterdam, 281–90.
58   Tian, M., Huang, S., Liu, E., Wu, L., Long, K. and Yang, Z. (1986) Fracture toughness of concrete, in Wittmann, F.H. (ed.) *Fracture toughness and fracture energy of concrete*, International Conference, Lausanne, 1985. Elsevier Science, Amsterdam, 281–90.
59   Nallathambi, P. (1986) 'A Study of Fracture of Plain Concrete', Doctoral thesis, University of Newcastle, NSW, Australia.
60   Carpinteri, A. (1982) Sensitivity and stability of progressive cracking in plain and reinforced cement composites. *International Journal of Cement Composites and Lightweight Concrete*, **4**, 1, 47–56.
61   Biolzi, L., Cangiano, S., Tognon, G. and Carpinteri, A. (1989) Snap-back softening instability in high strength concrete beams. *Materials Structure*, **22**, 429–36.
62   Kesler, C., Naus, D. and Lott, J. (1972) Fracture mechanics – its applicability to concrete. *Proceedings of the 1971 International Conference on Mechanical Behavior of Materials* (Japan), IV, 113–24.
63   Saouma, V.E., Ingraffea, A.R. and Catalona, D.M. (1982) Fracture toughness of concrete: K revisited. *J. Engineering Mech. Div. (ASCE)*, **108**, EM6, 1152–66.
64   Nakayama, J. (1965) Direct measurement of fracture energies of brittle heterogeneous materials. *J. American Ceramic Society*, **48**, 11, 583–7.
65   Tattersall, H.G. and Tappin, G. (1966) The work of fracture and its measurement in metals, ceramics and other materials. *J. Materials Science*, **1**, 299–301.
66   Moavenzadeh, F. and Kuguel, R. (1969) Fracture of concrete. *J. Materials (ASTM)*, **4**, 3, 497–519.
67   Petersson, P.E. (1980) Fracture energy of concrete: method of determination. *Cement and Concrete Research*, **10**, 78–89.
68   Hillerborg, A. (1983) Analysis of one single crack, in Wittmann, F.H. (ed.) *Fracture mechanics of concrete*. Elsevier Science, Amsterdam, 223–49.
69   RILEM Committee of Fracture Mechanics of Concrete (1985) Determination of the fracture energy of mortar and concrete by means of three-point bend tests on notched beams, RILEM Draft Recommendation. *Materials Structures*, **18**, 106, 285–90.
70   Hillerborg, A. (1985) The theoretical basis of a method to determine the fracture energy *G* of concrete. *Materials Structures*, **118**, 106, 291–6.
71   Nallathambi, P., Karihaloo, B.L. and Heaton, B.S. (1984) Effect of specimen and crack sizes, water/cement ratio and coarse aggregate texture upon fracture toughness of concrete. *Magazine of Concrete Research*, **36**, 129, 227–36.

72   Hillerborg, A. (1989) Existing methods to determine and evaluate fracture toughness of aggregative material – RILEM recommendation on concrete, in Mihashi, H., Takahashi, H. and Wittmann, F.H. (eds) *Fracture toughness and fracture energy, test methods for concrete and rock*. A.A. Balkema, Rotterdam, 145–52.

73   Gettu, R., Bazant, Z.P. and Karr, M.E. (1990) Fracture properties and brittleness of high-strength concrete. *ACI Materials Journal*, **87**, 6, 608–18.

74   Wittmann, F.H., Mihashi, H. and Nomura, N. (1990) Size effect on fracture energy of concrete. *Engineering Fracture Mechanics*, **35**, 1, 2, 3, 107–15.

75   Planas, J. and Eliccs, M. (1989) Conceptual and experimental problems in the determination of the fracture energy of concrete, in Mihashi, H., Takahashi, H. and Wittmann, F.H. (eds) *Fracture toughness and fracture energy, test methods for concrete and rock*. A.A. Balkema, Rotterdam, 165–82.

76   Brameshuber, W. and Hilsdorf, H.K. (1990) Influence of ligament length and stress state on fracture energy of concrete. *Engineering Fracture Mechanics*, **35**, 1, 2, 3, 95–106.

77   CEB (1990) *CEB-FIP model code*, Chapter 2 – Material properties, Bulletin No. 195. Ciomite Euro-International du Béton, Lausanne.

78   Hilsdorf, H.K. and Bramshuber, W. (1991) Code-type formulation of fracture mechanics concepts for concrete. *International Journal of Fraction*, **51**, 61–72.

79   Hillerborg, A., Modwer, M. and Petersson, P.-E. (1976) Analysis of crack formation and crack growth in concrete by means of fracture mechanics and finite elements. *Cement and Concrete Research*, **6**, 773–82.

80   Wnuk, M.P. (1974) Quasi-static extension of a tensile crack contained in a viscoelastic-plastic solid. *Journal of Applied Mechanics*, **41**, 1, 234–42.

81   Carpinteri, A., Di Tommaso, A. and Fanelli, M. (1986) Influence of material parameters and geometry on cohesive crack propagation, in Wittmann, F.H. (ed.) *Fracture toughness and fracture energy of concrete*, International Conference, Lausanne, 1985. Elsevier Science, Amsterdam, 117–35.

82   Bocca, P., Carpinteri, A. and Valente, S. (1991) Mixed mode fracture of concrete. *International Journal of Solids Structure*, **27**, 9, 1139–53.

83   Gerstle, W.H. and Xie, M. (1992) FEM modeling of fictitious crack propagation in concrete. *J. Engineering Mechanics*, **118**, 2, 416–34.

84   Bittencourt, T.N., Ingraffea, A.R. and Llorca, J. (1992) Simulation of arbitrary, cohesive crack propagation, in Bazant, Z.P. (ed.) *Fracture mechanics of concrete structures*. Elsevier Applied Science, London, 339–50.

85   Rots, J.G. and Blaauwendraad, J. (1989) Crack models for concrete: discrete or smeared? Fixed, multi-directional or rotating? *HERON*, **34**, 1.

86   Schellekens, J.C.J. (1990) *Interface elements in finite element analysis*, TU-Delft report 25-2-90-5-17. Stevin Laboratory, Delft University of Technology, Delft, Netherlands.

87   Feenstra, P.H., de Borst, R. and Rots, J.G. (1990) Stability analysis and numerical evaluation of crack-dilatancy models, in Bicanic, N. and Mang, H. (eds) *Computer aided analysis and design of concrete structures*, Second International Conference, Zell am See, Austria. Pineridge Press, Swansea, UK, 987–99.

88   Akutagawa, S., Jeang, F.L., Hawkins, N.M., Liaw, B.M., Du, J. and Kobayashi, A.S. (eds) (1991) Effects of loading history on fracture properties of concrete. *ACI Materials Journal*, **88**, 2, 170–80.

89   Bazant, Z.P. and Cedolin, L. (1979) Blunt crack band propagation in finite element analysis. *J. Engineering Mech. Div. (ASCE)*, **105**, EM2, 297–315.

90   Bazant, Z.P. and Oh, B.H. (1983) Crack band theory for fracture of concrete. *Materials Structure*, **16**, 93, 155–77.

91 Bazant, Z.P. (1985) Mechanics of fracture and progressive cracking in concrete structures, in Sih, G.C. and Di Tommaso, A. (ed.) *Fracture mechanics of concrete: structural application and numerical calculation.* Martinus Nijhoff, Dordrecht, Netherlands, 1–93.

92 Rots, J.G. and de Borst, R. (1987) Analysis of mixed-mode fracture in concrete. *J. Engineering Mechanics*, **113**, 11, 1739–58.

93 Carol, I. and Prat, P.C. (1990) A statically constrained microplane model for the smeared analysis of concrete cracking, in Bicanic, N. and Mang, H. (eds) *Computer aided analysis and design of concrete structures*, Second International Conference, Zell am See, Austria. Pineridge Press, Swansea, UK, Vol. 2, 919–30.

94 Du, J., Yon, J.H., Hawkins, N.M. and Kobayashi, A.S. (1990) Analysis of the fracture process zone of a propagating crack using moire interferometry, in Shah, S.P., Swartz, S.E. and Wang, M.L. (eds) *Micromechanics of a failure of quasi-brittle materials*, International Conference, Albuquerque, USA. Elsevier Science, London, 146–55.

95 Jenq, Y.S. and Shah, S.P. (1985) A fracture toughness criterion for concrete. *Engineering Fracture Mechanics*, **21**, 5, 1055–69.

96 Foote, R.M.L., Mai, Y.-W. and Cotterell, B. (1986) Crack growth resistance curves in strain-softening materials. *J. Mech. Phys. Solids*, **34**, 6, 593–607.

97 Yon, J.-H., Hawkins, N.M. and Kobayashi, A.S. (1991) Numerical simulation of Mode I dynamic fracture of concrete. *J. Engineering Mechanics*, **117**, 7, 1595–610.

98 Tademir, M.A., Maji, A.K. and Shah, S.P. (1990) Crack propagation in concrete under compression. *J. Engineering Mechanics*, **116**, 5, 1058–76.

99 Hordijk, D.A. (1991) 'Local Approach to Fatigue of Concrete', Doctoral Thesis, Delft University of Technology, Delft, Netherlands.

100 Gopalaratnam, V.S. and Shah, S.P. (1985) Softening response of plain concrete in direct tension. *ACI Journal*, **82**, 3, 310–23.

101 Cornelissen, H.A.W., Hordijk, D.A. and Reinhardt, H.W. (1986) Experimental determination of crack softening characteristics of normalweight and lightweight concrete. *HERON*, **31**, 2, 45–56.

102 Guo, Z. and Zhang, X. (1987) Investigation of complete, stress-deformation curves for concrete in tension. *ACI Materials Journal*, 4, **84**, 278–85.

103 Wecharatana, M. (1990) Britteleness index of cementitious composites, in Suprenant, B.A. (ed.) *Serviceability and durability of construction material*, First Materials Engineering Congress, Denver, USA. ASCE, New York, 966–75.

104 Li, V.C., Chan, C.-M. and Leung, C.K.Y. (1987) Experimental determination of the tension-softening relations for cementitious composites. *Cement and Concrete Research*, **17**, 441–52.

105 Du, J., Yon, J.H., Hawkins, N.M. and Kobayashi, A.S. (1990) Analysis of the fracture process zone of a propagating crack using moire interferometry, in Shah, S.P., Swartz, S.E. and Wang, M.L. (eds) *Micromechanics of a failure of quasi-brittle materials*, International Conference, Albuquerque, USA. Elsevier Science, London, 146–55.

106 Miller, R.A., Castro-Montero, A. and Shah, S.P. (1991) Cohesive crack models for cement mortar examined using finite element analysis and laser holographic measurement. *J. American Ceramics Society*, **74**, 130–8.

107 Roelfstra, P.E. and Wittmann, F.H. (1986) Numerical method to link strain softening with failure of concrete, in Wittmann, F.H. (ed.) *Fracture toughness and fracture energy of concrete*, International Conference, Lausanne, 1985. Elsevier Science, Amsterdan, 163–75.

108 Wittmann, F.H., Rokugo, K., Brühwiler, E., Mihashi, H. and Simonin, P. (1988) Fracture energy and strain softening of concrete as determined by

means of compact tension specimens. *Materials Structure*, **21**, 21–32.

109   Alvaredo, A.M. and Torrent, R.J. (1987) The effects of the strain-softening diagram on the bearing capacity of concrete beams. *Materials Structure*, **20**, 448–54.

110   Ratanalert, S. and Wecharatana, M. (1989) Evaluation of the fictitious crack and two-parameter fracture models, in Mihashi, H., Takahashi, H. and Wittmann, F.H. (eds) *Fracture toughness and fracture energy, test methods for concrete and rock*. A.A. Balkema, Rotterdam, 345–66.

111   Liaw, B.M., Jeang, F.L., Du, J.J., Hawkins, N.M. and Kobayashi, A.S. (1990) Improved nonlinear model for concrete fracture. *J. Engineering Mechanics*, **116**, 2, 429–45.

112   Li, V.C. and Huang, J. (1990) Relation of concrete fracture toughness to its internal structure. *Engineering Fracture Mechanics*, **35**, 1, 2, 3, 39–46.

113   Duda, H. (1991) Grain-model for the determination of the stress-crack-width relation, in Elfgren, L. and Shah, S.P. (eds) *Analysis of concrete structures by fracture mechanics*. Chapman and Hall, London, 88–96.

114   Clarke, D.R. and Faber, K.T. (1987) Fracture of ceramics and glasses. *J. Physical Chemistry of Solids*, **48**, 11, 1115–57.

115   Jenq, Y.S. and Shah, S.P. (1985) A two parameter fracture model for concrete. *J. Engineering Mechanics*, **111**, 4, 1227–41.

116   Planas, J. and Elices, M. (1990) Anomalous size effect in cohesive materials like concrete, in Suprenant, B.A. (ed.) *Serviceability and durability of construction materials*, First Materials Engineering Congress, Denver, USA. ASCE, New York, Vol. 2, 1345–56.

117   Ouyang, C. and Shah, S.P. (1991) Geometry-dependent R-curve for quasi-brittle materials. *J. American Ceramics Society*, **74**, 11, 2831–36.

118   Bazant, Z.P., Gettu, R. and Kazemi, M.T. (1991) Identification of nonlinear fracture properties from size effect tests and structural analysis based on geometry-dependent R-curves. *International Journal of Rock Mechanics and Mineral Science*, **28**, 1, 43–51; corrigenda: **28**, 2, 3, 233.

119   Sakai, M. and Bradt, R.C. (1986) Graphical methods for determining the nonlinear fracture parameters of silica and graphite refractory composites, in Bradt, R.C., Evans, A.G., Hasselman, D.P.H. and Lange, F.F. (eds) *Fracture mechanics of ceramics*. Plenum, New York, Vol. 7, 127–42.

120   Wecharatana, M. and Shah, S.P. (1983) Predictions of nonlinear fracture process zone in concrete. *J. Engineering Mechanics*, **109**, 5, 1231–46.

121   Hu, X. and Wittmann, F.H. (1990) Experimental method to determine extension of fracture-process zone. *J. Materials in Civil Engineering*, **2**, 1, 15–23.

122   Hu, X. (1990) *Fracture process zone and strain-softening in cementitious materials*, Research report 1, Institute of Building Materials. Swiss Federal Institute of Technology, Zurich, 1990.

123   Cottrell, B. and Mai, Y.-W. (1987) Crack growth resistance curve and size effect in the fracture of cement paste. *J. Materials Science*, **22**, 2734–8.

124   Planas, J., Elices, M. and Toribio, J. (1989) Approximation of cohesive crack models by R-CTOD curves, in Shah, S.P., Swartz, S.E. and Barr, B. (eds) *Fracture of concrete and rock: recent development*, International Conference, Cardiff. Elsevier Applied Science, London, 203–12.

125   Bazant, Z.P. and Kazemi, M.T. (1990) Determination of fracture energy, process zone length and brittleness number from size effect, with application to rock and concrete. *International Journal of Fracture*, **44**, 111–31.

126   Ouyang, C and Shah, S.P. (1991) Geometry-dependent R-curve for quasi-brittle materials. *J. American Ceramics Society*, **74**, 11, 2931–6.

127   Jenq, Y.S. and Shah, S.P. (1988) Mixed-mode fracture of concrete. *International Journal of Fracture*, **38**, 123–42.

128 RILEM Committee on Fracture Mechanics of Concrete – Test methods (1990) Determination of fracture parameters (K and CTOD) of plain concrete using three-point bend tests, RILEM Draft Recommendation. *Mater. Struct.*, **23**, 457–60.
129 Karihaloo, B.L. and Nallathambi, P. (1989) An improved effective crack model for the determination of fracture toughness of concrete. *Cement and Concrete Research*, **19**, 603–10.
130 Karihaloo, B.L. and Nallathambi, P. (1991) Notched beam test: Mode I fracture toughness, in Shah, S.P. and Carpinteri, A. (eds) *Fracture mechanics test methods for concrete*. Chapman and Hall, London, 1–86.
131 Nallathambi, P. and Karihaloo, B.L. (1990) Fracture of concrete: application of effective crack model, *Proceedings of Ninth International Conference on Experimental Mechanics*, Lyngby, Denmark, Vol. 4, 1413–22.
132 Bazant, Z.P. (1984) Size effect in blunt fracture: concrete, rock, metal. *J. Engineering Mechanics*, **110**, 518–35.
133 Bazant, Z.P. and Pfeiffer, P.A. (1987) Determination of fracture energy from size effect and brittleness number. *ACI Materials Journal*, **84**, 463–79.
134 RILEM Committee on Fracture Mechanics of Concrete – Test methods (1990) Size-effect method for determining fracture energy and process zone size of concrete, RILEM Draft Recommendation. *Materials Structure*, **23**, 461–65.
135 Planas, J. and Elices, M. (1990) Fracture criteria for concrete: mathematical approximations and experimental validation. *Engineering Fracture Mechanics*, **35**, 1, 2, 3, 87–94.
136 Planas, J. and Elices, M. (1990) The approximation of a cohesive crack by effective elastic cracks, in Firrao, D. (ed.) *Fracture behaviour and design of materials and structures*, Eighth European Conference on Fracture, Turin, Italy. Engineering Materials Advisory Services, Warley, UK, Vol. 2, 605–11.
137 John, R. and Shah, S.P. (1989) Fracture mechanics analysis of high-strength concrete. *J. Materials in Civil Engineering*, **1**, 4, 185–98.
138 Lange D.A. (1991) 'Relationship Between Microstructure, Fracture Surfaces and Material Properties of Portland Cement', Ph.D. Thesis, Northwestern University, Evanston, USA.
139 Chern, J.-C. and Tarng, K.-M. (1990) Size effect in fracture of steel fiber reinforced concrete, in Shah, S.P., Swartz, S.E. and Wang, M.L. (eds) *Micromechanics of failure of quasi-brittle materials*, International Conference, Albuquerque, USA. Elsevier Applied Science, London, 244–53.
140 de Larrard, F., Boulay, C. and Rossi, P. (1987) Fracture toughness of high-strength concretes, in Holand, I., Helland, S., Jakobsen, B. and Lenschow, R. (eds) *Utilization of high strength concrete*, Symposium, Stavanger, Norway. Tapir, Trondheim, Norway, 215–23.
141 Hucka, V. and Das, B. (1974) Brittleness determination of rocks by different methods. *Int. J. Rock Mech. Min. Sci.*, **11**, 3289–392.
142 Bache, H.H. (1986) Fracture mechanics in design of concrete and concrete structures, in Wittman, F.H. (ed.) *Fracture toughness and fracture energy of concrete*, International Conference, Lausanne, 1985. Elsevier Science, Amsterdam, 577–86.
143 Brühwiler, E., Broz, J.J. and Saouma, V.E. (1991) Fracture model evaluation of dam concrete. *J. Materials in Civil Engineering*, **3**, 4 235–51.
144 Bosco, C., Carpinteri, A. and Debernardi, P.G. (1990) Minimum reinforcement in high-strength concrete. *J. Structural Engineering*, **116**, 2, 427–37.
145 Naus, D.J. and Lott, J.L. (1969) Fracture toughness of Portland cement concretes. *J. ACI*, **66**, 6, 481–9.
146 Strang, P.C. and Bryant, A.H. (1979) The role of aggregate in the fracture of concrete. *J. Materials Science*, **14**, 1863–8.

147 Saouma, V.E., Broz, J.J., Brühwiler, E. and Boggs, H.L. (1991) Effect of aggregate and specimen size on fracture properties of dam concrete. *J. Materials in Civil Engineering*, **3**, 3, 204–18.

148 Wittmann, F.H., Roelfstra, P.E., Mihashi, H., Huang, Y.-Y., Zhang, X.-H. and Nomura, N. (1987) Influence of age of loading, water-cement ratio and rate of loading on fracture energy of concrete. *Materials Structure*, **20**, 103–10.

149 Bascoul, A., Detriche, C.H. and Ramoda, S. (1991) Influence of the characteristics of the binding phase and the curing conditions on the resistance to crack propagation of mortar [in French]. *Materials Structure*, **24**, 129–36.

150 Ojdrovic, R.P., Stojimirovic, A.L. and Petroski, H.J. Effect of age on splitting tensile strength and fracture resistance of concrete. *Cement and Concrete Research*, **17**, 70–76, 150.

151 Reinhardt, H.W. (1991) Loading rate, temperature and humidity effects, in Shah, S.P. and Carpinteri, A. (eds) *Fracture mechanics test methods for concrete*. Chapman and Hall, London, 199–230.

152 John, R. and Shah, S.P. (1986) Fracture of concrete subjected to impact loading. *Cement and Concrete Aggregates* (ASTM), **8**, 1, 24–32.

153 John, R., Shah, S.P. and Jenq, Y.-S. (1987) A fracture mechanics model to predict the rate sensitivity of Mode I fracture of concrete. *Cement and Concrete Research*, **17**, 249–62.

154 Oh, B.H. (1990) Fracture behavior of concrete under high rates of loading. *Engineering Fracture Mechanics*, **35**, 1, 2, 3, 327–32.

155 Liu, Z.-G., Swartz, S.E., Hu, K.K. and Kan, Y.-C. (1989) Time-dependent response and fracture of plain concrete beams, in Shah, S.P., Swartz, S.E. and Barr, B. (eds) *Fracture of concrete and rock: recent developments*, International Conference, Cardiff. Elsevier Applied Science, London, 577–86.

156 Zhou, F. and Hillerborg, A. (1992) Time-dependent fracture of concrete: testing and modelling, in Bazant, Z.P. (ed.) *Fracture mechanics of concrete structures*. Elsevier Applied Science, London, 906–11.

157 Bazant, Z.P. and Gettu, R. (1990) Size effect in concrete structures and influence of loading rate, in Suprenant, B.A. (ed.) *Serviceability and durability of construction materials*, First Materials Engineering Congress, Denver, USA. ASCE, New York, Vol. 2, 1113–23.

158 Bazant, Z.P. and Gettu, R. (1992) Rate effects and load relaxation in static fracture of concrete. *ACI Materials Journal* (in press).

159 Banthia, N.P., Mindess, S. and Bentur, A. (1987) Impact behaviour of concrete beams. *Materials Structure*, **20**, 293–302.

160 Bazant, Z.P. and Prat, P.C. (1988) Effect of temperature and humidity on fracture energy of concrete. *ACI Materials Journal*, **85**, 262–71.

161 Planas, J., Maturana, P., Guinea, G. and Elices, M. (1989) Fracture energy of water saturated and partially dry concrete at room and at cryogenic temperatures, in Salama, K., Ravi-Chandar, K., Taplin, D.M.R. and Rama Rao, P. *Advances in fracture research*, Seventh International Conference on Fracture, Houston, USA. Pergamon Press, Oxford, Vol. 2, 1809–17.

162 Ohlsson, U., Daerga, P.A. and Elfgren, L. (1990) Fracture energy and fatigue strength of unreinforced concrete beams at normal and low temperatures. *Engineering Fracture Mechanics*, **35**, 1, 2, 3, 195–203.

163 Maturana, P., Planas, J. and Elices, M. (1990) Evolution of fracture behaviour of saturated concrete in the low temperature range. *Engineering Fracture Mechanics*, **35**, 4, 5, 827–34.

164 Shah, S.P. and Chandra, S. (1970) Fracture of concrete subjected to cyclic and sustained loading. *ACI Journal*, **67**, October, 816–25.

165 Michalske, T.A. and Bunker, B.C. (1984) Slow fracture model based on strained silicate structures. *Journal of Applied Physics*, **56**, 10, 2686–93.

166 Rossi, P. (1991) Influence of cracking in the presence of free water on the mechanical behaviour of concrete. *Magazine of Concrete Research*, **43**, 154, 53–7.

167 Swartz, S.E., Hu, K.-K. and Jones, G.L. (1978) Compliance monitoring of crack growth in concrete. *J. Engineering Mech. Div. (ASCE)*, **104**, EM4, 789–800.

168 Pons, G., Ramoda, S.A. and Maso, J.C. (1988) Influence of the loading history on fracture mechanics parameters of microconcrete: effects of low-frequency cyclic loading. *ACI Materials Journal*, **85**, 341–6.

169 Schell, W.F. (1992) 'Fatigue Fracture of High Strength Concrete under High Frequency Loading', M.S. thesis, Northwestern University, Evanston, IL, USA; also Bazant, Z.P. and Schell, W.F. Fatigue fracture of high strength concrete and size effect. *ACI Materials Journal* (submitted).

170 Tinic, C. and Brühwiler, E. (1985) Effect of compressive loads on the tensile strength of concrete at high strain rates. *Int. J. Cem. Comp. Lightweight Conc.*, **7**, 103–8.

171 Gettu, R., Oliveira, M.O.F., Carol, I. and Aguado, A. (1992) Influence of transverse compression on Mode I fracture of concrete, in Bazant, Z.P. (ed.) *Fracture mechanics of concrete structures*. Elsevier Science, London, 193–7.

172 Carpinteri, A. and Swartz, S. (1991) Mixed-mode crack propagation in concrete, in Shah, S.P. and Carpinteri, A. (eds) *Fracture mechanics test methods for concrete*. Chapman and Hall, London, 129–90.

173 Reinhardt, H.W., Cornelissen, H.A.W. and Horijk, D.A. (1989) Mixed mode fracture tests on concrete, in Shah, S.P. and Swartz, S.E. (eds) *Fracture of concrete and rock*. Springer-Verlag, New York, 117–30.

174 van Mier, J.G.M., Nooru-Mohamed, M.B. and Timmers, G. An experimental study of shear fracture and aggregate interlock in cement-based composites. *HERON*, **36**, 4.

175 Swartz, S.E. and Taha, N.M. (1990) Mixed mode crack propagation and fracture in concrete. *Engineering Fracture Mechanics*, **35**, 1, 2, 3, 137–44.

176 Ballatore, E., Carpinteri, A., Ferrara, G. and Melchiorri, G. (1990) Mixed mode fracture energy of concrete. *Engineering Fracture Mechanics*, **35**, 1, 2, 3, 145–57.

177 Mindess, S. and Shah, S.P. (eds) (1988) *Bonding in cementitious composites*, Symposium, Boston, 1987. Materials Research Society, Pittsburgh, USA.

178 Hollemeier, B. and Hilsdorf, H.K. (1977) Fracture mechanics studies on concrete compounds. *Cement and Concrete Research*, **7**, 5, 523–36.

179 Buyukozturk, O. and Lee, K.M. (1992) Interface fracture mechanics of concrete composites, in Bazant, Z.P. (ed.) *Fracture mechanics of concrete structures*. Elsevier Applied Science, London, 163–8.

180 Maji, A.K., Wang, J. and Cardiel, C.V. (1992) Fracture mechanics of concrete, rock and interface, in Bazant, Z.P. (ed.) *Fracture mechanics of concrete structures*. Elsevier Applied Science, London, 413–18.

181 Saouma, V.E., Broz, J.J. and Boggs, H.L. (1991) In situ testing for fracture properties of dam concrete. *J. Materials in Civil Engineering*, **3**, 3, 219–34.

182 Saouma, V.E., Brühwiler, E., Keating, S., Ryan, J. and Shulz, J. (1991) Innovative fracture testing techniques for dam engineering, in Saouma, V.E., Dungar, R. and Morris, D. (eds) *Dam fracture*, Proceedings, International Conference, Boulder, USA. Electric Power Research Institute, Palo Alto, USA, 459–75.

183 Brühwiler, E. (1991) Determination of fracture properties of dam concrete from core samples, in Saouma, V.E., Dungar, R. and Morris, D. (eds) *Dam*

fracture, Proceedings, International Conference, Boulder, USA. Electric Power Research Institute, Palo Alto, USA, 427–43.

184 Linsbauer, H.N. (1991) Fracture mechanics material parameters of mass concrete based on drilling core tests – review and discussion, in van Mier, J.G.M., Rots, J.G. and Bakker, A. (eds) *Fracture processes in concrete, rock and ceramics*. Spon, London, Vol. 2, 661–3.

185 Ouchterlony, F. (1990) Fracture toughness testing of rock with core based specimens. *Engineering Fracture Mechanics*, **35**, 1, 2, 3, 351–66.

186 Zaitsev, Y.B. and Wittmann, F.H. (1981) Simulation of crack propagation and failure of concrete. *Materials Structure*, **14**, 83, 357–65.

187 Mihashi, H. (1985) Stochastic approach to study the fracture and fatigue of concrete, in Eggwertz, S. and Lind, N.C. (eds) *Probabilistic methods in the mechanics of solids and structures*, IUTAM Symposium, Stockholm, 1984. Springer-Verlag, Berlin, 307–17.

188 Mai, Y.-W., Hakeem, M. and Cotterell, B. (1982) Imparting fracture resistance to cement mortar by intermittent interlaminar bonding. *Cement and Concrete Research*, **12**, 661–3.

189 Beaudoin, J.J. (1982) Properties of high alumina cement paste reinforced with mica flakes. *Cement and Concrete Research*, **12**, 157–66.

190 Alford, N.M.N., Groves, G.W. and Double, D.D. (1982) Physical properties of high strength cement pastes. *Cement and Concrete Research*, **12**, 349–58.

191 Mai, Y.-W. and Cotterell, B. (1985) Porosity and mechanical properties of cement mortar. *Cement and Concrete Research*, **15**, 995–1002.

192 Beaudoin, J.J. and Feldman, R.F. (1985) High strength cement pastes – a critical appraisal. *Cement and Concrete Research*, **15**, 105–16.

193 Shah, S.P. (1990) Fracture toughness for high-strength concrete. *ACI Materials J.*, **87**, 3, 260–65.

194 Naaman, A.E. (1985) High strength fiber reinforced cement composites, in Young, J.F. (ed.) *Very high strength cement-based materials*, Materials Research Society Symposium Proceedings, Vol. 42. Materials Research Society, Pittsburgh, USA, 219–29.

195 Shah, S.P. (1991) Do fibers increase the tensile strength of cement-based matrices? *ACI Materials Journal*, **88**, 6, 595–602.

196 Gopalaratnam, V. and Shah, S.P. (1987) Failure mechanisms and fracture of fiber reinforced concrete, in Shah, S.P. and Batson, G.B. (eds) *Fiber reinforced concrete properties and applications*, SP-105. ACI, Detroit, USA, 1–25.

197 Shah, S.P. and Ouyang, C. (1991) Mechanical behavior of fiber-reinforced cement-based composites. *J. American Ceramics Society*, **74**, 11, 2727–38 and 2947–53.

198 Balaguru, P.N. and Shah, S.P. (1992) *Fiber-reinforced cement composites*. McGraw Hill, New York.

199 Becher, P.F. (1991) Microstructural design of toughened ceramics. *J. American Ceramics Society*, **74**, 2, 255–69.

200 Ezeldin, A.S. and Balaguru, P.N. (1989) Bond behavior of normal and high-strength fiber reinforced concrete. *ACI Materials Journal*, **86**, 6, 515–24.

201 Linsbauer, H.N. (1991) Design and construction of concrete dams under consideration of fracture mechanics aspects, in Elfgren, L. and Shah, S.P. (eds) *Analysis of concrete structures by fracture mechanics*. Chapman and Hall, London, 160–8.

202 Saouma, V.E., Brühwiler, E. and Boggs, H.L. (1982) A review of fracture mechanics applied to concrete dams. *Dam Engineering*, **1**, 1, 41–57.

203 Martha, L.F., Llorca, J., Ingraffea, A.R. and Elices, M. (1991) Numerical simulation of crack initiation and propagation in an arch dam. *Dam Engineering*, **2**, 3, 193–213.

204 Swenson, D.V. and Ingraffea, A.R. (1990) The collapse of the Schoharie Creek Bridge: a case study in the fracture mechanics of concrete, in Bicanic, N. and Mang, H. (eds) *Computer aided analysis and design of concrete structures*, Second International Conference, Zell am See, Austria. Pineridge Press, Swansea, Vol. 1, 403–24.

205 Gustafsson, P.J. and Hillerborg, A. (1985) Improvements in concrete design achieved through the application of fracture mechanics, in Shah, S.P. (ed.) *Application of fracture mechanics to cementitious composites*, NATO Workshop, Evanston, USA, 1984. Nijhoff, Dordrecht, Netherlands, 639–80.

206 Hawkins, N.M. (1985) The role for fracture mechanics in conventional reinforced concrete design, in Shah, S.P. (ed.) *Application of fracture mechanics to cementitious composites*, NATO Workshop, Evanston, USA, 1984. Nijhoff, Dordrecht, Netherlands, 639–80.

207 Bazant, Z.P., Ozbolt, J. and Eligehausen, R. (1992) Fracture size effect: I. Review of evidence for concrete structures. *J. Structural Engineering* (submitted).

208 Hillerborg, A. (1989) Fracture mechanics and the concrete codes, in Li, V.C. and Bazant, Z.P. (eds) *Fracture mechanics: application to concrete*, SP-118. ACI, Detroit, 157–69.

209 Hillerborg, A. (1990) Fracture mechanics concepts applied to moment capacity and rotational capacity of reinforced concrete beams. *Engineering Fracture Mechanics*, **35**, 1, 2, 3, 233–40.

210 Hillerborg, A. (1991) Size dependency of the stress-strain curve in compression, in Elfgren, L. and Shah, S.P. (eds) *Analysis of concrete structures by fracture mechanics*. Chapman and Hall, London, 171–8.

211 Engaffea, A.R. Gerstle, W., Gergely, P. and Saouma, V. (1984) Fracture mechanics of bond in reinforced concrete. *J. Structural Engineering*, **110**, 4, 871–90.

212 Ballarini, R. and Shah, S.P. (1991) Fracture mechanics based analyses of pull-out tests and anchor bolts, in Elfgren, L. and Shah, S.P. (eds) *Analysis of concrete structures by fracture mechanics*. Chapman and Hall, London, 245–80.

213 Ohlsson, U. and Elfgren, L. (1991) Anchor bolts analyzed with fracture mechanics, in van Mier, J.G.M., Rots, J.G. and Bakker, A. (eds) *Fracture processes in concrete, rock and ceramics*. Spon, London, 887–97.

214 Bosco, C., Carpinteri, A. and Debernardi, P.G. (1991) Use of the brittleness number as a rational approach to minimum reinforcement design, in Elfgren, L. and Shah, S.P. (eds) *Analysis of concrete structures by fracture mechanics*. Chapman and Hall, London, 133–51.

215 Carpinteri, A, A fracture mechanics model for reinforced concrete collapse, in *Advanced mechanics of reinforced concrete* (IABSE Colloquium, Delft, Netherlands), Reports of the Working Commissions, Vol. 34, IABSE, Zurich, 17–30, 1981

216 Shioya, T., Iguro, M., Nojiri, Y., Akiyama, H. and Okada, T. (1989) Shear strength of large reinforced concrete beams, in Li, V.C. and Bazant, Z.P. (eds) *Fracture mechanics: applications to concrete*, SP-118. ACI, Detroit, USA, 259–79.

217 Bazant, Z.P. and Kim, J.-K. (1984) Size effect in shear failure of longitudinally reinforced beams. *ACI Journal*, **81**, 456–68.

218 Bazant, Z.P. and Kazemi, M.T. (1991) Size effect on diagonal shear failure of beams without stirrups. *ACI Structural Journal*, **88**, 3, 268–76.

219 Gustafsson, P.J. and Hillerborg, A. (1988) Sensitivity in shear strength of longitudinally reinforced concrete beams to fracture energy of concrete. *ACI Structural Journal*, **85**, 3, 286–94.

220 Thorenfeldt, E. and Drangsholt, G. (1990) Shear capacity of reinforced high

strength concrete beams, in Hester, W.T. (ed.) *High-strength concrete*, Second International Symposium, SP-121. ACI, Detroit, USA, 129–54.

221   Walraven, J. (1990) Scale effects in beams with unreinforced webs, loaded in shear, in *Progress in concrete research*, Annual Report. Delft University of Technology, Netherlands, Vol. 1, 103–12.

222   Saouma, V.E. and Ingraffea, A.R. Fracture mechanics analysis of discrete cracking, in *Advanced mechanics of reinforced concrete* (IABSE Colloquium, Delft, Netherlands), Reports of the Working Commissions, Vol. 34, IABSE, Zurich, 413–36, 1981

223   de Borst, R. and Nauta, P. (1984) Smeared crack analysis of reinforced concrete beams and slabs failing in shear, in Damjanic, F., Hinton, E., Owen, D.R.J., Bicanic, N. and Simonvic, V. (eds) *Computer-aided analysis and design of concrete structures*, International Conference, Split, Yugoslavia. Pineridge Press, Swansea, Vol. 1, 261–73.

224   Wang, Q.B. and Baauwendraad, J. (1991) Consequence of concrete fracture: brittle failure and size effect of RC beams under shear, in van Mier, J.G.M., Rots, J.G. and Bakker, A. (eds) *Fracture processes in concrete, rock and ceramics*. Spon, London, 909–18.

225   Jenq, Y.-S. and Shah, S.P. (1989) Shear resistance of reinforced concrete beams – a fracture mechanics approach, in Li, V.C. and Bazant, Z.P. (eds) *Fracture mechanics: application to concrete*, SP-118. ACI, Detroit, USA, 237–58.

226   Mphonde, A.G. and Frantz, G.C. (1984) Shear tests of high- and low-strength concrete beams. *ACI Journal*, **81**, 350–7.

227   Elzanaty, A.H., Nilson, A.H. and Slate, F.O. (1986) Shear capacity of reinforced concrete beams using high-strength concrete. *ACI Journal*, **83**, 2, 290–6.

228   Johnson, M.K. and Ramirez, J.A. (1989) Minimum shear reinforcement in beams with higher strength concrete. *ACI Structural Journal*, **86**, 4, 376–82.

229   Marzouk, H. and Hussein, A. (1991) Experimental investigation on the behavior of high-strength concrete slabs. *ACI Structural Journal*, **88**, 6, 701–13.

230   Ehsani, M.R. and Alameddine, F. (1991) Design recommendations for Type 2 high-strength reinforced concrete connections. *ACI Structural Journal*, **8**, 3, 277–91.

7 Structural members

Arthur H Nilson

7.1 Introduction

High performance concretes with higher strengths, having mix proportions quite different from ordinary concretes and modified by several additives, have mechanical properties differing in important ways from ordinary concretes.

It has been well established, for example, that for high-strength concrete the internal microcracking that occurs in concrete as load is applied is delayed until a higher fraction of ultimate load is reached. One result is that the range of linear elastic response to compression is extended. On the other hand, the extensive ductility exhibited by lower strength concretes after maximum stress, as microcracks spread to form an interconnected network, is not characteristic of higher strength concretes, and material ductility is less.[1-3] The response to sustained loading is also different. The much lower creep coefficient of high-strength concretes has also been correlated with differences in internal microcracking.[4-9]

Given the empirical nature of many of the equations in current use for the design of reinforced concrete members and structures, it is clear that a re-examination of those equations is imperative, in order to insure the safe and economical use of high-strength concrete. It is remarkable that the use of the material has preceded a full knowledge of its properties and behavior.

The emphasis in this chapter will be placed on the practical design of members for short-term and sustained load. While much work remains to be done, great progress has been made in the past decade in developing the needed information on material and member performance.[10-17] Design procedures have been checked and, in the USA, provisions of the American Concrete Institute (ACI) Code reviewed. Recommendations for change or for additional research have been presented where appropriate.

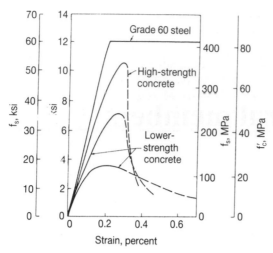

Fig. 7.1   Typical concrete and steel stress-strain curves

7.2   Axially loaded columns

Columns subject to purely axial loads are rare in practice. Generally moments exist concurrently, due to rigid frame action or load eccentricity. Most design codes recognize this, requiring that column moments be accounted for in design. However, in examining the differences in performance of high-strength concrete members, it is helpful to treat axially loaded columns first.

Contribution of steel and concrete

In most present design practice, the nominal strength of a column is calculated using the direct addition law, i.e., summing the individual strength contributions of the concrete and steel. The justification for this is evident in Fig. 7.1, which superimposes typical stress-strain curves in compression for three concretes with that for reinforcing steel having 60,000 psi (414 MPa) yield strength. The curve for reinforcing steel is drawn to a different vertical scale for convenience.

For low-strength concrete, when the concrete reaches the range of significant nonlinearity (about 0.001 strain), the steel is still in the elastic range and consequently starts to pick up a larger share of the load. When the strain is close to 0.002, the slope of the concrete curve is nearly zero, and there is little change in stress. The steel reaches its yield point at about this strain. Both materials deform plastically, and the strength of the column is accurately predicted by

$$P_n = 0.85f_c'A_c + f_yA_s \tag{7.1}$$

where f_c' = cylinder compressive strength of concrete
f_c' = yield strength of steel
A_c = area of concrete section
A_s = total area of steel

The factor 0.85 is used to account for the observed difference in the strength of concrete in a column compared with concrete of the same mix in a standard test cylinder.

A similar analysis holds for high-strength concrete, except that the steel will yield somewhat before the concrete reaches its peak stress. However, the steel will yield at constant stress until the concrete reaches its maximum stress. Prediction of strength may therefore still be based on Equation (7.1). Test evidence confirms that use of the factor 0.85 is still appropriate.

Lateral confinement

Lateral steel, preferably in the form of continuous spirals, has two beneficial effects on column behaviour: (a) it greatly increases the strength of the core concrete through confinement against lateral expansion, and (b) it increases the axial strain capacity of the concrete, permitting a more gradual and ductile failure. Equating the increase in strength due to the spiral confinement and the loss of capacity in spalling and incorporating a suffery factor of 1.2, the following minimum spiral reinforcement ratio ρ is specified in the ACI 318–89 Code:

$$\rho = 0.45[A_g/A_c - 1]f_c'/f_y \tag{7.2}$$

where ρ = ratio of volume of spiral reinforcement to volume of con-
   $A_g$crete core
   A_c = gross area of concrete cross section
   f_c' = area of concrete core (measured to outside of spiral)
   f_y = cylinder compressive strength of concrete
   = yield strength of spiral steel

The increase in compressive strength of columns provided with spiral steel is based on an experimentally derived relationship for strength gain:

$$f_c - f_c'' = 4.0f_2' \tag{7.3a}$$

where f_c = compressive strength of spirally reinforced concrete column
   f_c'' = compressive strength of unconfined concrete column
   f_2' = concrete confinement stress produced by spiral

Equation (7.3a) can be shown to lead directly to Equation (7.2). The concrete confinement stress produced by the spiral is calculated on the basis that the spiral steel has yielded, using the familiar hoop tension equation, from which

$$f_2' = [2A_{sp}f_y]/d_c s \tag{7.4}$$

where A_{sp} = area of spiral steel
   d_c = diameter of concrete core
   s = pitch of spiral

and other terms are as already defined.

Research by Ahmad and Shah[18] has shown that spiral reinforcement is less effective for high-strength concrete columns, and that the stress in the spiral steel at maximum column load is often significantly less than yield stress, as assumed above. These results are consistent with research by Martinez.[19,20] Martinez *et al.*, used an 'effective' confinement stress $f_2(1 - s/d_c)$ in evaluating their experimental results, where f_2 is the confinement stress in the concrete, calculated using the *actual* stress (not necessarily the yield stress) in the spiral steel. The term $(1 - s/d_c)$ reflects the reduction in effectiveness of spirals associated with increased spacing of the wires. Thus an improved version of Equation (7.3a) is:

$$f_c - f_c'' = 4.0f_2(1 - s/d_c) \qquad (7.3b)$$

Figure 2, from Martinez *et al.*,[20] shows results ot tests of columns having different concrete strengths. Clearly the strength gain predicted by Equation (7.3b) is valid for normal-weight concrete of all strengths for confinement stress up to about 300 psi (21 MPa). A similar plot based on Equation (7.3a) would show a somewhat unconservative prediction for higher confinement stresses, but it can be shown that confinement stresses for practical column spirals are seldom more than about 1000 psi (7 MPa). For this range, the ACI Code basis derived using Equation (7.3a) gives satisfactory results. It follows that Equation (7.2) can be used without change for high-strength normal-weight concrete columns as well as for columns using lower strength concrete.

It is important to note, however, that a spiral has much less confining effect in lightweight concrete columns, probably due to the crushing of the lightweight concrete under the spirals at heavy loads. This results in a greatly reduced strength gain in the core, compared with that for normal density concrete (see Fig. 7.2). Lightweight columns would require about

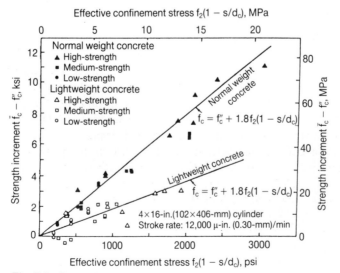

Fig. 7.2   Strength increment provided by spiral reinforcement action

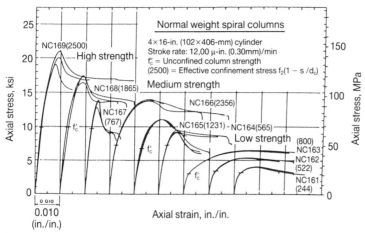

Fig. 7.3   Experimental stress-strain curves for spirally reinforced columns

2.5 times more spiral steel than corresponding normal-weight concrete columns. Whether or not such heavy spirals are practical is questionable.

There is not yet general agreement on the effectiveness of spirals for improving the *ductility* of high-strength concrete columns, that is, for increasing the strain limit and flattening the negative slope of the stress-strain curve past peak stress. Ahmad and Shah[18] conclude that confining spirals are about as effective in flattening the negative slope for high-strength as for low-strength concrete columns. This conclusion was reached on the basis of the results from their analytical model. The analytical model employs the basic constitutive relationship of concrete and the confining spiral steel and assume perfect compatibility in the lateral direction. However, the experimental results of Martinez *et al.*[20] showed significant differences, summarized in Fig. 7.3. Three groups of curves are identical by the three concrete strengths. Each of these groups consists of three sets of curves corresponding to three different amounts of lateral reinforcement. (Indicated in each set of curves with a short horizontal line is the average unconfined column strength corresponding to that particular set of confined columns.) Different behavior for different confinement stress is evident. Not only is the strain at peak stress much less for higher strength concrete, but the stress falls off sharply after peak value. This is true even for high-strength columns with very large confinement stress.

It is concluded that normal weight concrete columns with spiral steel show strength gain resulting from spirals that is well-predicted by present equations, but past peak stress they are likely to show much less ductility.[18–24] Furthermore, spirals in lightweight concrete columns are much less effective than predicted by present equations, and reserve strength past spalling of the concrete shell may be lacking.[20,25]

A separate issue is the relative merit of column ties versus spirals.[26,27] Ties are relatively ineffective in developing confinement pressure on the core concrete in columns of all strengths, and tied columns for all concrete strengths typically show less ductility than spirally-reinforced columns. In applications where severe overloads are likely, and where column toughness is critical, as in seismic zones, spirally-reinforced columns are much preferred.

Long-term loads

In concrete structures, particularly tall buildings, the larger part of the column load may be dead load, sustained continuously over the life of the structure. Significant creep occurs, which not only results in column shortening but may also modify all moments, thrusts and shears in the building if differential shortening occurs. Creep is conveniently described in terms of the creep coefficient:

$$C_c = \text{creep strain/initial elastic strain} \qquad (7.5)$$

There is general agreement that the coefficient of creep for high-strength concrete is significantly less than that of low-strength concrete.[4–9] The exact values of the creep coefficients for concretes of varying strengths depend on many factors other than compressive strength, including mix components and proportions, water/cementitious ratio, additives, temperature and humidity, age at first loading, and stress level with respect to strength. For common mixes, for sustained stress levels not more than about one-half the ultimate strength, and for normal ambient conditions, the values of Table 7.1 are representative.

It would seem, therefore, that the use of high-strength concrete in tall building columns, for example, would provide a means for reducing long-term column shortening. On the other hand, the sustained load stress for the high strength concrete columns would normally be higher, hence the initial elastic strain higher, tending toward creep strains that may be as high as before, even though the creep coefficient is less. Russell and Corley[28] present results of a study of time-dependent column behavior for a tall cast-in-place concrete structure.

Cyclic loading and fatigue

High-strength concrete may be relatively free of internal microcracking

Table 7.1 Variation of creep coefficient with compressive strength

Compressive strength psi (MPa)	Creep coefficient
3000 (21)	3.1
4000 (28)	2.9
6000 (41)	2.4
8000 (55)	2.0
10,000 (69)	1.6
12,000 (83)	1.4

even up to about 75% of ultimate load when loaded monotonically. On the other hand, it is known to be more brittle than low-strength concrete, lacking much of the ductility that accompanies progressive crack growth. Some experimental research indicates that fatigue strength is essentially independent of compressive strength.[29] Other research indicates that failure of concrete subject to repeated loading can be predicted approximately by the concept of the envelope curve, directly related to the short-term monotonic stress-strain curve.[24,30,31] For high-strength concrete, each load application causes relatively less incremental damage. However, the number of cycles to failure may not necessarily be larger because of the greater slope of the post-peak envelope curve.

7.3 Flexure in beams

The special mechanical properties of high-strength concrete affect the behavior of beams. In most ways, high-strength concrete beams behave according to the same rules that provide the basis for design of beams using lower strength concrete, but some differences have been found, mostly relating to deflection calculations, to be discussed in the following section. Topics pertaining to flexural calculations will be treated here.

Compressive stress distribution

The shape of the flexural compressive stress distribution in beams is directly related to (although not necessarily identical to) the shape of the compressive stress-strain curve in uniaxial compression. Considering the differences in the uniaxial test curves shown in Fig. 7.1, it is reasonable to expect corresponding differences in the flexure stress block shape.

Figure 7.4a shows the generally parabolic stress block typical of lower strength concrete. Three stress block parameters, k_1, k_2, and k_3 are sometimes used to define the characteristics of the block, where k_1 = ratio of average to maximum stress, k_2 = ratio of depths of compressive resultant to neutral axis, and k_3 = ratio of maximum stress in beam to maximum stress in axial compression test. In ordinary design in the USA and many countries, it is recognized that the exact shape of the stress block is unimportant provided that one knows (a) the magnitude of the compression result, and (b) the level at which it acts. These may be defined using an 'equivalent rectangular distribution' shown in Fig. 7.4b, with only two parameters: the first, β_1, defining the ratio of the rectangular block depth to the actual neutral axis depth, and the second (shown as a constant 0.85 in Fig. 7.4b) defining an equivalent constant stress that produces the proper value of the compressive resultant. The two parameters of Fig. 7.4b are easily related to the three parameters of Fig. 7.4a.[32]

For high strength concrete, which responds to compression in a more linear way, and to a higher fraction of ultimate stress, the actual compress-

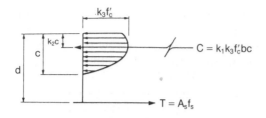

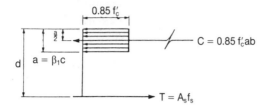

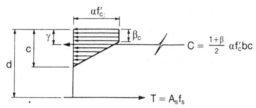

Fig. 7.4 Concrete flexural compressive stress distributions: (a) parabolic; (b) equivalent rectangular; (c) trapezoidal

ive stress distribution in the beam might be more closely represented by the trapezoidal distribution of Fig. 7.4c. However, for practical beams, which are invariably design as underreinforced with failure triggered by yielding of the tensile steel, comparative studies[33–42] indicate that the usual rectangular stress block is perfectly satisfactory. This holds even for compressive strength of 12,000 psi (83 MPa) and above, with differences of only a few percentage points between strengths predicted by mathematically continuous representation of stress, the trapezoidal block and the rectangle. It follows that the nominal flexural strength of an underreinforced high-strength concrete beam can be calculated by the usual equations:

$$M_n = A_s f_y (d - a/2) \tag{7.6}$$

$$a = A_s f_y / 0.85 f_c'' \tag{7.7}$$

where A_s = total tension steel area, f_y = steel yield stress, f_c'' = concrete compressive strength, d = effective depth to tensile steel, b = width of compression zone.

Limiting compressive strain and balanced steel ratio

While high-strength concrete reaches its maximum stress at a strain that is somewhat higher than that for lower-strength concrete, its ultimate strain is lower. However, for concretes in the present-day strength range of interest, say from 4000 psi to 12,000 psi (28 MPa to 83 MPa) the differences are not great, and tests indicate that the assumption of limiting strain of 0.003, as found in the 1989 ACI Code, is satisfactory. This strain limit, coupled with the yield strain of the tensile steel, provides the basis for calculating a balanced steel ratio, for which steel yielding and concrete crushing would theoretically occur simultaneously. In USA design practice, the maximum steel ratio is set at 0.75 times the balanced value. Tests have confirmed that this provides adequate ductility for high-strength concrete beams as well as beams of normal concrete. Thus the maximum steel ratio is:

$$\rho_{max} = 0.75 \times 0.85\beta_1 \left(f_c'/f_y\right) \times \left(\varepsilon_{cu}/\varepsilon_{cu} + \varepsilon_y\right) \tag{7.8}$$

in which $\beta_1 = a/c$, $c = $ depth to the actual neutral axis, $\varepsilon_{cu} = $ ultimate concrete strain $= 0.003$, and $\varepsilon_y = $ steel yield strain.

Minimum tensile steel ratio

In lightly reinforced beams, if the flexural strength of the newly cracked section is less than the moment that produced cracking, the beam may fail immediately and without warning of distress upon formation of the first flexural crack. To avoid this, a lower limit steel ratio is established. This is done by equating the cracking moment, computed from the concrete modulus of rupture, to the flexural strength of the cracked section. The minimum steel ratio clearly depends upon the modulus of rupture, which in USA practice is related to the square root of the compressive strength. It can be shown that, for rectangular cross section beams, the minimum steel ratio should be:

$$\begin{aligned} \rho_{min} &= 1.8\sqrt{f_c'}/f_y && \text{in psi units} \\ \rho_{min} &= 0.149\sqrt{f_c'}/f_y && \text{in MPa units} \end{aligned} \tag{7.9a}$$

and for T beams of normal proportions with flanges in compression

$$\begin{aligned} \rho_{min} &= 2.7\sqrt{f_c'}/f_y && \text{in psi units} \\ \rho_{min} &= 0.224\sqrt{f_c'}/f_y && \text{in MPa units} \end{aligned} \tag{7.9b}$$

In the 1989 ACI code, concrete strength is not included as a variable, and the minimum steel ratio continues to be set at

$$\begin{aligned} \rho_{min} &= 200/f_y && \text{in psi units} \\ \rho_{min} &= 1.4/f_y && \text{in MPa units} \end{aligned} \tag{7.9c}$$

It is easily confirmed that the ACI minimum steel ratio is conservative for rectangular beams for all but very high concrete strengths, and conservative for T beams for strengths to about 5000 psi (35 MPa). For other cases, use of Equation (7.9a) or (7.9b) is preferable.

7.4   Beam deflections

Immediate deflections

The main uncertainties in predicting elastic deflections of reinforced concrete beams are (a) elastic modulus E_c, (b) modulus of rupture f_r, and (c) effective moment of inertia I_e, which depends on the extent of cracking of the beam.

For the elastic modulus, research studies indicate the following expression may be used for concrete strengths to 12,000 psi (83 MPa):

$$E_c = 40,000\sqrt{f_c'} + 1,000,000 \text{ psi} \qquad (7.10)$$
$$E_c = 3320\sqrt{f_c'} + 6900 \text{ MPa}$$

Equation (7.10) should be multiplied by $(\omega_c/145)^{1.5}$ in psi units and by $(\omega_c/2300)^{1.5}$ in MPa units for concrete densities other than 145 lb/ft^3 or 2320 kg/m^3. The ACI Code value for elastic modulus, $E_c = 57,000\sqrt{f_c'}$ psi or $E_c = 4700\sqrt{f_c'}$ MPa, gives values that may be as much as 15% too high for high-strength concrete.[43]

For deflection prediction, a value for modulus of rupture of $7.5\sqrt{f_c'}$ psi or $0.7\sqrt{f_c'}$ MPa may be used. Test evidence confirms that the ACI Code equation for effective moment of inertia gives good results for high-strength beams as well as normal-strength beams:

$$I_e = [M_{cr}/M_a]^3 I_g + [1 - (M_{cr}/M_a)^3] I_{cr} \qquad (7.11)$$

in which M_{cr} = cracking moment, M_a = maximum moment in the span, I_g = gross moment of inertia of section, and I_{cr} = moment of inertia of cracked transformed section.

Deflection under sustained loads

It is the general practice to calculate time-dependent deflections of beams due to creep and shrinkage by applying multipliers to computed elastic deflections. This procedure is valid for high-strength concrete beams, but because high-strength concrete has a creep coefficient that is significantly below that of ordinary concrete, it follows that creep deflections of high-strength concrete beams should be less, relative to initial elastic deflections, than for otherwise identical low-strength beams. It can further be expected that compression steel, which reduces time-dependent beam deflections by limiting creep displacement of the concrete on the compression side of a beam, would be less important in high-strength than in

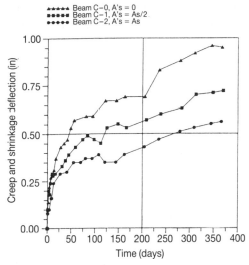

Fig. 7.5 Creep and shrinkage deflection of beams using 5400 psi (37 MPa) concrete

low-strength concrete beams. Experimental research confirms both observations.[44–46]

One-year data for tensile-reinforced beams of concrete compressive strength 5400 psi and 13,000 psi, are summarized in Figs 7.5 and 7.6, respectively.[45,46] First comparing beams with no compressive steel, Beam C-0 versus Beam A-0, it is seen that use of high strength concrete reduced creep and shrinkage deflection by about 32%. Comparing beams with compressive steel area half the tensile area, Beam C-1 versus Beam A-1, using high-strength concrete produced a deflection reduction of about 17%. For Beam C-2 versus Beam A-2, both with compressive steel area

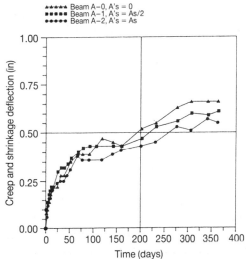

Fig. 7.6 Creep and shrinkage deflection of beams using 13,100 psi (90 MPa) concrete

equal to the tensile steel area, use of high-strength concrete produced no reduction in creep and shrinkage deflection.

From these data it is concluded that use of high-strength concrete is an effective means to reduce time-dependent deflections of beams with no compression steel. On the other hand, if compressive steel is present, there is little additional advantage gained, with regard to deflection reduction, through the use of high-strength concrete.

According to the 1989 ACI Code, additional long-term deflection resulting from creep and shrinkage is found by multiplying the immediate deflection caused by the sustained load by the factor:

$$\lambda = \zeta/(1+50\rho') \tag{7.12}$$

where ρ' is the compression steel ratio A_s'/bd at midspan for simple and continuous spans and ζ is a time-dependent factor to be taken as follows: time 5 years or more $= 2.0$; 12 months $\zeta = 1.4$; 6 months $\zeta = 1.2$; and 3 months $\zeta = 1.0$. These factors were based on tests where concrete strength was mostly in the range from 3000 psi to 4000 psi (21 to 28 MPa) and consequently do not recognize concrete compressive strength as a variable. With high-strength concrete beams now having strength of 12,000 psi (83 MPa) or more, a revised equation is needed.

Review of available test data for low-, medium- and high-strength concrete beams suggests the following equation,[46] recommended as a replacement for Equation (7.12):

$$\lambda = \mu\zeta/(1+50\mu\rho') \tag{7.13}$$

in which:

$$\mu = 1.4 - f_c'/10,000 \tag{7.14}$$

with the limits that: $0.4 < \mu < 1.0$. It is evident that including the additional factor μ in the numerator of Equation (7.13) accounts for the lower creep coefficient of high-strength concrete, and including it in the second term of the denominator accounts for the reduced influence on deflections of compression steel in high-strength beams. Analysis of test data indicates that while two separate factors might be used, the differences are small, and this resulted in recommendation of the single factor of Equation (7.14). Note that for concrete strength of 4000 psi (28 MPa) or lower, Equation (7.13) gives a long-term deflection multiplier identical with that resulting from use of Equation (7.12) from the 1989 ACI Code.

7.5   Shear in beams

In current practice in the USA, total shear resistance is made up of two parts: V_s provided by the stirrups and V_c, nominally the 'concrete contribution'. The concrete contribution includes, in an undefined way, the contributions of the still uncracked concrete at the head of a hypothetical diagonal crack, the resistance provided by aggregate interlock along the diagonal crack face, and the dowel resistance provided by the main reinforcement. The 1989 (and earlier) ACI Code includes two equations for, V_c, either one of which may be used at the designer's option. The simple Code Eq. (11-3):

$$V_c = 2\sqrt{f_c'}\,b_w d \qquad\qquad (7.15)$$

is widely used in practice. For greater refinement, Code Eq. (11-6) is available:

$$V_c = [1.9\sqrt{f_c'} + 2500\,\rho_w V_u/M_u]\,b_w d \qquad\qquad (7.16)$$

In Equations (7.15) or (7.16), V_u and M_u are, respectively, the factored shear and moment at the section, b_w is the web width, d the effective depth to the flexural steel, and ρ the longitudinal flexural steel ratio $A_s/b_w d$.

High-strength concrete loaded in uniaxial compression fractures suddenly and, in doing so, may form a failure surface that is smooth and nearly a plane.[1-3] This is in contrast to the rugged failure surface characteristic of low-strength concrete. In beams controlled by shear strength, the state of stress is biaxial, combining diagonal compression in the direction from the load point to the support with diagonal tension in the perpendicular direction. Diagonal tension cracks formed in high-strength concrete beam tests have been found to have relatively smooth surfaces.[47-52] Tests confirm that aggregate interlock decreases as concrete strength increases. Thus a shear strength deficiency may exist. Data from tests at the University of Connecticut[47,48] indicated that the value of V_c calculated using the 1983 ACI Code equation should be reduced for high-strength concrete. Data from Cornell University tests[49,50] confirmed that those equations do not properly account for the influence of the several parameters. Further research at North Carolina State University[51,52] confirmed that 1983 ACI Code provisions for shear overestimated the benefits of increasing concrete strength, and may be unconservative, particularly for beams in the normal range of a/d ratios, and normal (not heavy) amounts of flexural tension steel.

A comparison of some test results with predicted concrete shear contribution from the ACI Code Eqs. (11-3) and (11-6), for beams without web steel, is shown in Figs. 7.7, 7.8 and 7.9.[50] Figure 7.7 clearly shows that for beams with a moderate amount of flexural tensile steel, Code Eq. (11-6) is unconservative above about 6000 psi (41 MPa). Increasing the flexural steel ratio improves the value of V_c, as seen in Fig. 7.8, but the predictions remain unconservative for normally-proportioned beams with shear span ratio a/d of 4 to 6, except for very heavy flexural steel ratios. The influence of increasing shear span is further seen in Fig. 7.9.

On the other hand, for beams with web reinforcement, ACI Code Eq. (11-6) gives conservative results, at least for strengths tested, up to 9500 psi (66 MPa), as shown in Fig. 7.10. From this it is concluded that the provision of extra vertical stirrups may make up, in some way, for

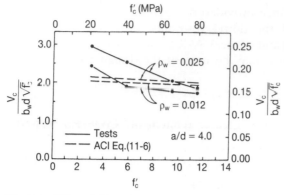

Fig. 7.7   Effect of f_c' on shear strength of beams without stirrups

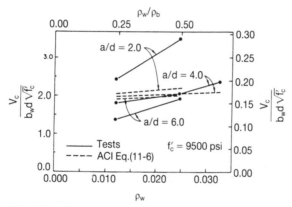

Fig. 7.8   Effect of ρ_ω on shear strength of beams with f_c' of 9500 psi (66 MPa)

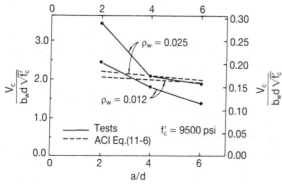

Fig. 7.9   Effect of shear span ratio a/d on shear strength of beams with $\sqrt{f_c'}$ of 9500 psi (66 MPa)

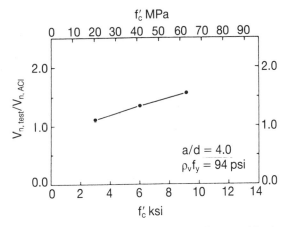

Fig. 7.10   Effect of f_c' on shear strength of beams with stirrups.

deficiencies in the contribution that is called V_c in beams using high-strength concrete.

While important progress has been made in recent years toward developing a rational method for design of web reinforcement that would take account of the actual behavior of reinforced concrete beams,[53] changes were made in the 1989 ACI Code to account for some of the deficiencies mentioned above on an entirely ad hoc basis.[54] The earlier shear design methodology was modified as follows:

1   The values of $\sqrt{f_c'}$ used in the shear design equations shall not exceed 100 psi (0.69 MPa), except that,
2   Values of $\sqrt{f_c'}$ greater than 100 psi (0.69 MPa) are permitted in calculating V_c for reinforced concrete beams in which the minimum web reinforcement equals $f_c'/5000$ times ($f_c'/35$ times), but not more than three times, the amount normally required.

Clearly these provisions affect only members with concrete strengths greater than 10,000 psi (69 MPa). While comparison with the test data of Roller and Russell[54] indicates that these changes are appropriate, it should be noted that many of the beams tested in that investigation contained extraordinarily large amounts of flexural tension steel (a feature that was established in other investigations to greatly enhance shear strength), and could not be considered practical beams.

7.6   Bond and anchorage

Present methods of design for development length and anchorage of tensile steel are based on tests, generally using concrete with compressive strength not greater than about 4000 psi (28 MPa). Although some information has recently become available for high strength concrete,[55] not enough data have been obtained to permit a definitive statement. Pending further research, a new limitation was introduced with the 1989 ACI Code to the effect that values of $\sqrt{f_c'}$ used with the normal bond and anchorage equations must not be taken greater than 100 psi (0.69 MPa). Clearly this limit affects only members with concrete strength greater than 10,000 psi (69 MPa). It is to be hoped that current research will provide the information needed, on the one hand, to verify the safety of the present ad hoc approach, and on the other hand, to permit optimum use of the undoubtedly greater bond strength of the new high-performance concretes.

7.7   Flexural and shear cracking

The modulus of rupture, which is an appropriate measure of concrete tensile strength for use in predicting flexural cracking load, has been reported to be in the range from 8 to $12\sqrt{f_c'}$ psi (0.66 to $0.99\sqrt{f_c'}$ MPa) for normal density concrete, and from 6 to $8\sqrt{f_c'}$ psi (0.50 to $0.66\sqrt{f_c'}$ MPa) for lightweight concrete.[25] Thus it appears that the values stated or implied in the 1989 ACI Code, $7.5\sqrt{f_c'}$ psi ($0.62\sqrt{f_c'}$ MPa) for normal density and $5.6\sqrt{f_c'}$ psi ($0.46\sqrt{f_c'}$ MPa) for lightweight concretes are quite conservative for higher concrete compressive strengths. However, curing conditions in the field are seldom as ideal as in the laboratory, and the ACI values may be closer to actual tensile strength of the concrete in beams on the job site.

It may be observed that the assumption of modulus of rupture lower than the actual value is neither conservative nor unconservative, but simply will result in an inaccurate prediction of cracking load. This results, in turn, in an inaccurate prediction of both initial and time-dependent deflections.

The direct tensile strength is seldom measured, but is of interest in studying web-shear cracking, mainly in prestressed concrete beams. This form of shear cracking initiates in the web, spreading both upward and downward to form the complete diagonal tension crack, and contrasts with flexure-shear cracking, initiating with a flexure crack and diagonally upward through the web. The second type of crack is typical in reinforced concrete beams, and often occurs in prestressed beams as well.

In USA practice, web-shear strength is predicted by empirical equations involving a number of terms, only one of which relates to the direct tensile strength. As noted above, the prediction of shear shrength of high-strength concrete reinforced beams is not firmly based on theory; shear strength for prestressed beams is even less so. While significant experimental research has been done,[47–54] the development and adoption of rational code provisions, perhaps based on the concepts of Schlaich *et al.*,[53] is still ahead.

7.8   Prestressed concrete beams

Characteristics of high-strength concrete, discussed previously in this chapter in the context of axially loaded members and reinforced concrete beams, affect the behavior of prestressed concrete beams in corresponding ways. There appears to be no reason why the normal procedures for flexural design cannot be applied to prestressed concrete beams using high-strength concrete, as they were to reinforced concrete beams. In fact, it is interesting to note that use of relatively high-strength concrete, with compressive strength in the range from 8000 psi to 10,000 psi (55 MPa to 69 MPa) was almost routine in the precast prestressed concrete industry long before such concretes were proposed for cast-in-place construction. No special difficulties have been encountered, although with strengths trending rapidly upward, questions do arise, mainly in connection with shear strength prediction.

The tests of high-strength concrete beams upon which shear provisions of the 1989 ACI Code are based were of reinforced, not prestressed, members. However, the code provision permitting values of $\sqrt{f_c'}$ greater than 100 psi (0.69 MPa) if web reinforcement minimums are increased is stated to apply to both reinforced beams *and* prestressed beams. While this may be valid, there is no direct experimental justification.

An extensive program of shear research on prestressed concrete beams is described in el-Zanaty *et al.*[49] Results confirmed that the basis of the 1989 ACI Code approach for calculating the 'concrete contribution' V_c, by which the web shear-cracking load and the flexure-shear cracking load are calculated separately, and V_c taken as the lower of the two values, is sound in the sense that both types of shear cracking were observed in 34 prestressed beam tests, and the basic behavior was different in each mode. It was found that the ACI Code equations conservatively predicted the shear cracking load for both flexure-shear and web-shear type cracks for beams without stirrups. It was further disclosed from test results for beam with stirrups that the shear strength predicted by ACI Code equations was on the safe side, although the degree of safety was variable. An overall conclusion was that several important parameters, namely concrete compressive strength, shear-span ratio, shear reinforcement index, and degree of prestress are not properly considered in the 1989 ACI Code equations. The basic flaw in the '$V_c + V_s$' approach is once again evident.

7.9   Slabs

Very little research has been done to date on structural slabs made using high-strength concrete, and in fact concrete for slabs in practice has seldom been over about 6000 psi (41 MPa). Probably this reflects the difficulty that would likely be experienced with deflections for the very thin structural slabs that could be designed using higher strength concrete, although prestressing offers the possibility of controlling deflections using the load-balancing method.

Meanwhile, there seems to be no reason that flexural behavior determined from beam testing could not be applied with safety to slab design. Similarly, shear design methods for beams could probably be applied safely to one-way and two-way edge supported slabs. While stirrups normally would not be used, shear stress in such elements would usually be far below that value permitted on the concrete.

Shear in beamless column-supported slabs, or flat plates, has long been a concern however. Present ACI Code design equations, again largely empirical and based on testing with concrete strengths in the range below 6000 psi (41 MPa), require re-evaluation before they are applied above that level. An experimental investigation[56] studying punching shear in concrete flat plates having concrete with compressive strength of about 10,000 psi (69 MPa) concluded that ACI Code equations, which relate slab shear strength to $\sqrt{f_c'}$, overestimate the benefit of increasing concrete strength on shear capacity. It was suggested that the British Codes CP 110 and BS 8110, which base strength on $\sqrt[3]{f_c'}$, may give a better representation. Clearly, more research is needed over the full range of parameters.

7.10 Eccentrically loaded columns

Compressive stress distribution

As pointed out in Section 7.2, columns in practice must be designed to resist not only axial loads but bending moments as well, either as a result of continuous frame action or from eccentric application of the load. In discussing beams in Section 7.3, it was pointed out that the shape of the compressive stress distribution in high-strength concrete beams is different from that in beams using low-strength concrete, because of the difference in shape of the stress-strain curve in compression. For reinforced concrete beams, which typically are under-reinforced, with strength controlled by yielding of the tensile reinforcement, the actual shape of the concrete compressive stress block used to calculate flexural strength is of little importance as long as the internal lever arm between compressive and tensile resultant is nearly correct. Over-reinforced beams are not permitted in present practice. It follows that present procedures for predicting flexural strength of beams are satisfactory, regardless of whether low- or high-strength concrete is used.

However, in the case of bending combined with axial load, i.e. eccentrically loaded columns, members for which failure would be in flexural compression cannot be avoided. For members with low eccentricity of loading, failure will be initiated when the concrete reaches its compressive strain limit, while the steel on the far side of the column may be well below its tensile yield, may be nearly unstressed, or loaded in compression. It is apparent that the shape of the compressive stress-strain curve used in the column analysis for such cases could assume much greater importance.

Short-column strength

An analytical study was made[57] to predict the failure load of eccentrically loaded columns using concrete strengths to 12,000 psi (83 MPa), with steel ratios from 1 to 4%, and with square or rectangular cross sections, the latter studied for both strong-axis and weak-axis bending. Three different concrete stress distributions were studied: (a) a continuous function based on actual stress-strain curves from uniaxial compression tests, (b) the ACI Code equivalent rectangular stress block, and (c) a trapezoidal stress distribution proposed in Pastor *et al.*[36] to provide a closer approximation to the actual stress distribution in bending than does the rectangular block.

For each of the 54 columns studied, the complete column nominal strength and design strength curves were constructed, as the results from using the three alternative compressive stress distributions were compared.

The continuous stress block can be assumed to be the most accurate predictor of column strength. As expected, the strength predicted by the ACI equivalent rectangular stress block showed the greater difference when compared with the continuous function for all cases, and on the unconservative side, while the trapezoidal stress block gave values between the continuous representation and the rectangular block. The difference between the rectangular and continuous function results were as large as 12% when 12,000 psi (83 MPa) concrete was used, but only up to 6% with 4000 psi (28 MPa) concrete. The difference between the trapezoidal and continuous function results were 3% or less for the high-strength concrete and 5% or less for low-strength concrete. Maximum differences were generally found when the eccentricity ratio e/h was in the range from 0.1 to 0.5.

The overall conclusion from the study was that either the continuous stress block or the trapezoidal approximation might be used in practice for columns with eccentricities that would produce failure in the compression failure range with significant bending, but with the rectangular stress block as much as 12% in error on the unconservative side. For low eccentricities, there was little difference in the predictions of the three representations, and for large eccentricities that would produce failure in the tension range there was almost no difference.

Slenderness effects

Most columns in present practice are short columns, in the sense that their strength is governed solely by material properties and the geometry of the cross section. However, with the increasing use of higher strength steel as well as concrete, and with improved methods for dimensioning of compression members, it is now possible to design much smaller cross sections than in the past. This clearly leads to columns with higher slenderness, and rational and reliable design procedures have become increasingly important.

The slender column design procedures of the 1989 ACI Code, while considerably more complicated than earlier methods, have been generally accepted by the profession. They appear to be suitable for use with high-strength concrete columns as well, although certain adjustments to the materials parameters may be appropriate. The modified value for elastic modulus given by Equation (7.10) is recommended in substitution for the 1989 ACI Code value, which may be as much as 15% too high for high-strength concrete. In addition, some modification of the factor β_d used in the slender column analysis is appropriate, to reflect the much lower creep coefficient typical of high-strength concrete. No more specific recommendation is possible at this time.

Beam-column connections

Current USA practice regarding joint design follows the recommendations of ACI Committee 352.[58] While not a part of the 1989 ACI Code, these recommendations provide a basis for the safe design, using ordinary concretes, of beam-column joints for both common construction and for buildings subject to seismic forces. According to the committee recommendations, joints are classified as Type 1 joints, connecting members in an ordinary structure designed on the basis of strength, according to the main body of the ACI Code, and Type 2 joints connecting members designed to have sustained strength under deformation reversals into the inelastic range, such as members in a structure designed for earthquake motions.

There appear to be no special considerations introduced with the use of high-strength concrete in the design of Type 1 joints, other than a requirement for careful detailing for clearance of bars in the smaller members that are typical. However, for Type 2 joints, in which ductile behavior must be maintained through a number of load reversals, some modifications to certain of the present Committee 352 recommendations have been suggested.[59]

Present recommendations limit the joint shear stress to $\gamma \sqrt{f_c'}$, where the coefficient γ is a function of the type of joint and loading condition, and the value of f_c' used is not to exceed 6000 psi (41 MPa). New test results show that with high-strength concrete, this equation becomes seriously unconservative, and a modified equation is proposed by Ehsani and Alameddine.[59] The γ factor remains unchanged, differentiating between corner, exterior and interior joints.

Further attention has been directed to the joint confinement requirements of the Committee 352 report, because present design recommendations result in excessive congestion of reinforcement for higher-strength concrete frames. Test results reported in Ehsani and Alemeddine[59] indicate that much lower amounts of steel may be adequate for high-strength concrete.

7.11 Summary and conclusions

A brief review has been presented of the special characteristics of high-strength concrete, and how they influence the behavior of structural concrete members of various types. Because of the generally empirical nature of many of the important design equations incorporated in present-day codes and used in practice, it is imperative that these equations be reviewed, and predicted performance compared against the results obtained in comprehensive test programs.

Important experimental research on high-strength concrete has been done over the past 15 years, both on the material itself and on structural members making use of it. This research has in some cases confirmed that present codified procedures are suitable. In other cases serious problems have been disclosed, indicating that design equations should be modified or even discarded in favor of new approaches. Code changes have been slow, and not all the necessary changes have yet been incorporated. For the design engineer proposing to exploit the many advantages of the new material, it is essential to seek out the best current experimental information. To this end, the extensive reference list that follows may be of particular value.

References

1   Carrasquillo, R.L., Nilson, A.H. and Slate, F.O. (1980) Microcracking and engineering properties of high-strength concrete. *Research Report* No. 80-1, Department of Structural Engineering, Cornell University, Ithaca, Feb. 1980, 254 pp.
2   Carrasquillo, R.L., Nilson, A.H. and Slate, F.O. (1981) Properties of high-strength concrete subject to short-term loads. *ACI Journal*, Proceedings **78**, No. 3, May–June, 171–8.
3   Carrasquillo, R.L., Slate, F.O. and Nilson, A.H. (1981) Microcracking and behavior of high-strength concrete subject to short-term loading. *ACI Journal*, Proceedings **78**, No. 3, May–June, 179–86.
4   Ngab, A.S., Slate, F.O. and Nilson, A.H. (1980) Behavior of high-strength concrete under sustained compressive stress. *Research Report* No. 80-2, Department of Structural Engineering, Cornell University, Ithaca, Feb, 201 pp.
5   Ngab, A.S., Nilson, A.H. and Slate, F.O. (1981) Shrinkage and creep of high-strength concrete. *ACI Journal*, Proceedings **78**, No. 4, July–Aug, 255–61.
6   Ngab, A.S., Slate, F.O. and Nilson, A.H. (1981) Microcracking and time-dependent strains in high-strength concrete. *ACI Journal*, Proceedings **78**, No. 4, July–Aug, 262–8.
7   Smadi, M.M., Slate, F.O. and Nilson, A.H. (1982) Time-dependent behavior

of high-strength concrete under high sustained compressive stresses. *Research Report* No. 82-16, Department of Structural Engineering, Cornell University, Ithaca, Nov.

8   Smadi, M.M., Slate, F.O. and Nilson, A.H. (1987) Shrinkage and creep of high-, medium- and low-strength concretes, including overloads. *ACI Materials Journal*, Proceedings **84**, No. 3, May–June, 224–34.

9   Smadi, M.M., Slate, F.O. and Nilson, A.H. (1985) High , medium- and low-strength concrete subject to sustained overloads – strains, strengths and failure mechanisms. *ACI Journal*, Proceedings **82**, No. 5, Sept–Oct, 657–64.

10   (1977) High-strength concrete in Chicago high-rise buildings. *Task Force Report* No. 5, Chicago Committee on High-Rise Buildings, 63 pp.

11   Kaar, P.H., Hanson, N.W. and Capell, H.T. (1978) Stress-strain characteristics of high-strength concrete, Douglas McHenry International Symposium on Concrete and Concrete Structures, *Special Publication SP-55*, American Concrete Institute, Detroit, 161–185. Also, Research and Development Bulletin No. RD051.01D, Portland Cement Association.

12   Perenchio, W.F. and Klieger, P. (1978) Some physical properties of high-strength concrete. *Research and Development Bulletin* No. RD056.01T, Portland Cement Association, Skokie, 7 pp.

13   Shah, S.P. (1979) *Proceedings national science foundation workshop on high-strength concrete*. University of Illinois at Chicago Circle, Dec, 226 pp.

14   Wang, P.T., Shah, S.P. and Naaman, A.E. (1978) Stress-strain curves of normal and lightweight concrete in compression. *ACI Journal*, Proceedings **75**, No. 11, Nov, 603–11.

15   (1987) Research needs for high-strength concrete, Report by ACI Committee 363, *ACI Materials Journal*, Proceedings **84**, No. 6, Nov–Dec, 559–61.

16   (1987) *Proceedings symposium on utilization of high-strength concrete*, Stavanger, Norway, June 15–18. Tapir Publishers, N-7034 Trondheim-NTH, Norway, 688 pp.

17   Nilson, A.H. (1985) Design implications of current research on High-Strength Concrete, high-strength concrete. *Special Publication SP-87*. American Concrete Institute, Detroit, 85–118.

18   Ahmad, S.H. and Shah, S.P. (1982) Stress-strain curves of concrete confined by spiral reinforcement. *ACI Journal*, Proceedings **79**, No. 6, Nov–Dec, 484–90.

19   Martinez, S., Nilson, A.H. and Slate, F.O. (1982) Spirally-reinforced high-strength concrete columns. *Research Report* No. 82-10, Department of Structural Engineering, Cornell University, Ithaca, Aug.

20   Martinez, S., Nilson, A.H. and Slate, F.O. (1984) Spirally-reinforced high-strength concrete columns. *ACI Journal*, Proceedings **81**, No. 5, Sept–Oct, 431–42.

21   Fafitis, A. and Shah, S.P. (1985) Lateral reinforcement for high-strength concrete columns, High-Strength Concrete, *Special Publication SP-87*. American Concrete Institute, Detroit, 213–32.

22   Yong, Y.K., Nour, M.G. and Nawy, E.G. (1988) Behavior of laterally confined high-strength concrete under axial loads. *Journal of Structural Engineering*, ASCE, **114**, No. 2, Feb, 332–51.

23   Iyenger, K.T., Sundara, R., Desayi, P. and Reddy, K.N. (1970) Stress-strain characteristics of concrete confined in steel binders. *Magazine of Concrete Research*, London, **22**, No. 72, Sept, 173–84.

24   Ahmad, S.H. (1981) 'Properties of Confined Concrete Subject to Static and Dynamic Loads', *Ph.D. Thesis,* University of Illinois at Chicago Circle, Mar.

25   Slate, F.O., Nilson, A.H. and Martinez, S. (1986) Mechanical properties of high-strength lightweight concrete. *ACI Journal*, Proceedings **83**, No. 4, July–Aug, 606–13.

26 Vallenas, J., Bertero, V.V. and Popov, E.P. (1977) Concrete confined by rectangular hoops and subjected to axial loads. *Research Report* No. UCB/EERC-77/13. Earthquake Engineering Research Center, University of California, Berkeley.

27 Sheikh, S.A. and Uzumeri, S.M. (1980) Strength and ductility of tied concrete columns. *Journal of Structural Division*, ASCE **106**, No. ST5, May, 1079–1102.

28 Russell, H.G. and Corley, W.G. (1978) Time-dependent behavior of columns in water tower place, Douglas McHenry International Symposium on Concrete and Concrete Structures, *Special Publication SP-55*. American Concrete Institute, Detroit, 347–73. Also, Research and Development Bulletin No. RD052.01B, Portland Cement Association.

29 Bennett, E.W. and Muir, S.E.S. (1967) Some fatigue tests on high-strength concrete in uniaxial compression. *Magazine of Concrete Research*, London **19**, No. 59, June, 113–17.

30 Bresler, B. and Bertero, V.V. (1975) Influence of high strain rate on cyclic loading behavior of unconfined and confined concrete in compression, *Proceedings* Second Canadian Conference on Earthquake Engineering. Department of Civil Engineering, McMaster University, Hamilton, June, 15-1–15-38.

31 Bertero, V.V., Bresler, B. and Liao, H. (1969) Stiffness degradation of reinforced concrete members subject to cyclic flexural moments. *Report No.* EERC-69/12, University of California, Berkeley, Dec.

32 Nilson, A.H. and Winter, G. (1991) *Design of concrete structures*, 11th ed. McGraw-Hill Inc., New York, 904 pp.

33 Leslie, K.E., Rajagopalan, K.S. and Everhard, N.J. (1976) Flexural behavior of high-strength concrete beams. *ACI Journal*, Proceedings **73**, No. 9, Sept, 517–21.

34 Nedderman, H. (1973) 'Flexural Stress Distribution in Very High Strength Concrete', *M.Sc. Thesis*, University of Texas at Arlington, Dec. 182 pp.

35 Zia, P. (1977) Structural design with high-strength concrete, *Research Report* No. PZIA-77-01. Civil Engineering Department, North Carolina State University, Raleigh, Mar, 65 pp.

36 Pastor, J.A., Nilson, A.H. and Slate, F.P. (1984) Behavior of high-strength concrete beams, *Research Report* No. 84-3. Department of Structural Engineering, Cornell University, Ithaca, Feb.

37 Shin, S., Ghosh, S.K. and Moreno, J. Flexural ductility of ultra-high-strength concrete members, accepted for publication in *ACI Structural Journal*.

38 Ahmad, S.H. and Barker, R. (1991) Flexural behavior of reinforced high-strength lightweight concrete beams. *ACI Structural Journal*, Proceedings **88**, No. 1, Jan–Feb, 69–77.

39 Wang, P.T., Shah, S.P. and Naaman, A.E. (1978) High-strength concrete in ultimate strength design, *Proceedings* ASCE **104**, No. ST11, Nov, 1761–73.

40 (1977) Discussion of 'Flexural behavior of high-strength concrete beams', by Leslie, K.E., Rajagopalan, K.S. and Everhard, N.J. *ACI Journal*, Proceedings **74**, No. 3, Mar, 104–45.

41 Ahmad, S.H. and Batts, J. (1991) Flexural behavior of doubly-reinforced high-strength lightweight concrete beams with web reinforcement. *ACI Structural Journal*, Proceedings **88**, No. 3, May–June, 351–8.

42 *Private communications* with ACI Committee 318, Subcommittee D on Flexure and Axial Loads.

43 (1984) State-of-the-art report on high strength concrete, Reported by ACI Committee 363. *ACI Journal*, **81**, No. 4, July–Aug, 364–411.

44 Leubkeman, C.H., Nilson, A.H. and Slate, F.O. (1985) Sustained load deflection of high-strength concrete beams, *Research Report* No. 85-2. Department of Structural Engineering, Cornell University, Ithaca, Feb. 164 pp.

45   Paulson, K.A., Nilson, A.H. and Hover, K.C. (1989) Immediate and long-term deflection of high-strength concrete beams, *Research Report* No. 89-3. Department of Structural Engineering, Cornell University, Ithaca, 230 pp.

46   Paulson, K.A., Nilson, A.H. and Hover, K.C. (1991) Long-term deflection of high-strength concrete beams. *ACI Materials Journal*, Proceedings **88**, No. 2, Mar–Apr, 197–206.

47   Mphonde, A.G. and Frantz, G.C. (1984) Shear tests of high and low strength concrete beams without stirrups. *ACI Journal*, Proceedings **81**, No. 4, July–Aug, 350–7.

48   Mphonde, A.G. and Frantz, G.C. (1985) Shear tests of high and low strength concrete beams with stirrups, High Strength Concrete, *Special Publication SP-87*. American Concrete Institute, Detroit, 179–96.

49   El-Zanaty, A.H., Nilson, A.H. and Slate, F.O. (1986) Shear capacity of prestressed concrete beams using high-strength concrete. *ACI Journal*, Proceedings **83**, No. 3, May–June, 359–68.

50   El-Zanaty, A.H., Nilson, A.H. and Slate, F.O. (1986) Shear capacity of reinforced concrete beams using high-strength concrete. *ACI Journal*, Proceedings **83**, No. 2, Mar–Apr, 290–6.

51   Ahmad, S.H., Khaloo, A.R. and Poveda, A. (1986) Shear capacity of reinforced high-strength concrete beams. *ACI Journal*, Proceedings **83**, No. 2, Mar–Apr, 297–305.

52   Ahmad, S.H. and Lue, D.M. (1987) Flexure-shear interaction of reinforced high-strength concrete beams. *ACI Structural Journal*, Proceedings **84**, No. 4, July–Aug, 330–41.

53   Schlaich, J., Shafer, K. and Jennewein, M. (1987) Toward a consistent design of structural concrete. *J. Prestressed Concrete Institute*, **32**, No. 3, May–June, 74–150.

54   Roller, J.J. and Russell, H.G. (1990) Shear strength of high-strength concrete beams. *ACI Structural Journal*, Proceedings **87**, No. 2, Mar–Apr, 191–8.

55   Treece, R.A. and Jirsa, J.O. (1989) Bond strength of epoxy-coated reinforcing bars. *ACI Materials Journal*, **86** (2), March–April, 167–74.

56   Marzouk, H. and Hussein, A. (1991) Experimental investigations on the behavior of high-strength concrete slabs. *ACI Structural Journal*, **88**, No. 6, Nov–Dec, 701–13.

57   Garcia, D.T. and Nilson, A.H. (1990) A comparative study of eccentrically-loaded high-strength concrete columns, *Research Report* No. 90-2. Department of Structural Engineering, Cornell University, Ithaca, Jan.

58   (1985) 'Recommendations for Design of Beam-Column Joints in Monolithic Reinforced Concrete Structures', Reported by ACI Committee 352. *ACI Structural Journal*, **82**, No. 3, May–June, 266–83.

59   Ehsani, M.R. and Alameddine, F. (1991) Design recommendations for Type 2 high-strength reinforced concrete connections. *ACI Structural Journal*, Proceedings **88**, No. 3, May–June, 277–91.

8 Ductility and seismic behavior

S K Ghosh and Murat Saatcioglu

8.1 Introduction

It has been shown for quite some time that concrete becomes less
deformable or more brittle as its compressive strength increases. Figure 8.1
shows a high-strength concrete cylinder being tested in compression. The
failure is obviously explosive, indicating that the material is brittle. The
same fact is depicted in a different way by Fig. 8.2[1] which shows the

Fig. 8.1   Testing of a high-strength concrete cylinder

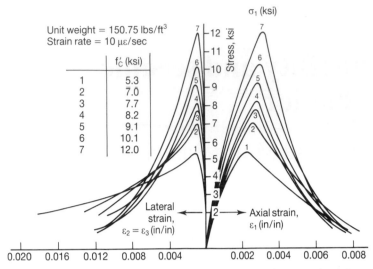

Fig. 8.2 Complete axial stress versus axial strain curves for normal weight concretes of different strengths[1]

axial stress-strain curves and axial-lateral strains curves in compression of normal weight concretes having different strength levels. Low-strength concrete obviously can develop only a modest stress level, but it can sustain that stress over a significant range of strains. High-strength concrete attains a much higher stress level, but then cannot sustain it over any meaningful range of strains. The load-carrying capacity drops precipitously beyond the peak of the stress-strain relationship.

Figure 8.3[1] shows the stress-strain curves of lightweight concretes having different compressive strengths. These curves were obtained by Ahmad in an investigation on mechanical properties of high strength lightweight

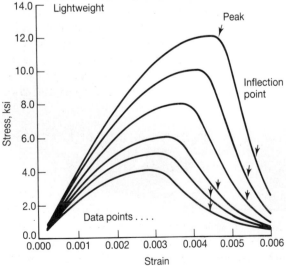

Fig. 8.3 Stress-strain curves of lightweight concretes of different strengths[1]

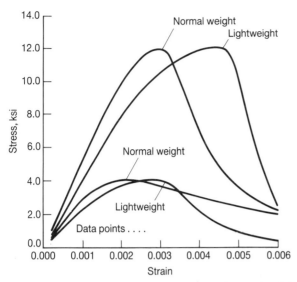

Fig. 8.4   Comparison of stress-strain curves of normal weight and lightweight concretes[1]

aggregate concrete which was conducted at North Carolina State University. In Fig. 8.4,[1] a selected comparison is made between the stress-strain curves of normal weight and lightweight concretes having essentially the same compressive strengths of about 4000 and 12,000 psi (27.6 and 82.8 MPa). It can be seen that for similar strengths, lightweight concrete exhibits a steeper drop of the descending part of the stress-strain curve than normal weight concrete. In other words, lightweight concrete is a more brittle material than normal weight concrete of the same strength.

The lack of deformability of high-strength concrete does not necessarily result in a lack of deformability of high-strength concrete members that combine this relatively brittle material with reinforcing steel. This interesting and important aspect is discussed in detail in this chapter. The application of high-strength concrete in regions of high seismicity is discussed at the end of the chapter. Such application depends, of course, on adequate inelastic deformability of high-strength concrete structural members under reversed cyclic loading of the type induced by earthquake excitation.

8.2   Deformability of high-strength concrete beams

Normal weight concrete beams under monotonic loading

Perhaps the earliest investigation on the deformability of high-strength concrete beams was carried out by Leslie, Rajagopalan and Everard.[2] They tested 12 under-reinforced rectangular beams with f_c' ranging between 9300 and 11,800 psi (64 and 81 MPa). The specimen details are shown in Fig. 8.5 and Table 8.1. It was observed that as the reinforcement ratio ρ increased, the 'maximum ultimate deflection' decreased, and the

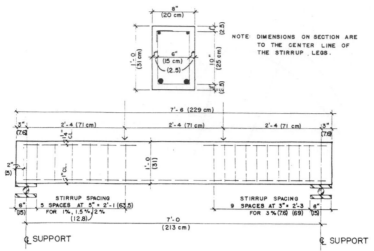

Fig. 8.5 Details of specimens tested by Leslie *et al.*[2]

ductility index μ (the ratio of maximum ultimate deflection to the deflection at the end of the initial linear portion of the load-deflection curve) decreased drastically. It is evident from Table 8.1 that the ductility index drops to quite low values for $\rho/\rho_b = 0.69$, whereas a ratio of up to 0.75 is allowed by the ACI Code.[3] However, this observation probably requires some qualification. Table 8.2 from Leslie *et al.*[2] lists the average ductility indices for increasing values of ρ. Also listed in the table are average values of ultimate deflection and yield deflection. First, neither of these terms was clearly defined in Leslie *et al.*[2] Secondly, as can be seen, the ductility index decreased with increasing ρ not so much because the ultimate deflection decreased as because the yield deflection increased. Thus, without knowing precisely how the yield deflection was determined, it is difficult to tell if this investigation was indicative of any lack of inelastic deformability on the part of high-strength concrete beams containing moderately high amounts of tension reinforcement (above 1.5%). It should be noted from Table 8.1 that μ values varied widely for the beams having the same tension reinforcement ratio, and that there was no correlation between this variation and the variation in concrete strength.

A comprehensive investigation on the deformability of high-strength concrete beams was carried out by Pastor *et al.*[4] Two series of tests, A and B, were conducted.

Series A consisted of four beams of high-strength concrete (HSC), one of medium strength concrete (MSC) and one of low-strength concrete (LSC). The scope was limited to singly reinforced, unconfined rectangular members subject to short-term 1/3 point loading. Beam Series A details are given in Fig. 8.6 and Table 8.3. Concrete compressive strength, f_c' (at the time of testing), and the tensile steel ratio, ρ, were the experimental variables.

The scope of Series B was limited to high-strength concrete rectangular

Table 8.1   Specimen details and ductility indices[2]

Specimen[a]	f_c', psi (MPa)	b, in. (mm)	d, in. (mm)	A_s[b]	ρ	ρ_b	ρ/ρ_b	μ
7.5–1	9310 (64.1)	8.25 (210)	10.63 (270)	2#6	0.01	0.0045	0.30	5.9
8.0–1	10,660 (73.5)	8.25 (210)	10.63 (270)	2#6	0.01	0.046	0.26	8.0
9.0–1	10,620 (73.2)	8.25 (210)	10.63 (270)	2#6	0.01	0.045	0.27	4.3
7.5–1.5	9720 (67.0)	8.00 (203)	10.56 (268)	2#7	0.014	0.051	0.44	4.5
8.0–1.5	11,400 (78.6)	8.13 (207)	10.56 (268)	2#7	0.014	0.051	0.31	2.5
9.0–1.5	11,630 (80.2)	8.50 (216)	10.56 (268)	2#7	0.013	0.051	0.30	4.2
7.5–2	10,850 (748)	8.50 (216)	10.50 (267)	2#8	0.018	0.040	0.69	3.2
8.0–2	10,610 (73.1)	7.88 (200)	10.50 (267)	2#8	0.019	0.040	0.63	2.4
9.0–2	11,780 (81.2)	8.13 (207)	10.50 (267)	2#8	0.019	0.039	0.56	2.7
7.5–3	11,650 (80.3)	8.38 (213)	10.50 (267)	3#8	0.027	0.039	0.82	1.9
8.0–3	11,730 (80.9)	8.25 (210)	10.50 (267)	3#8	0.027	0.039	0.71	2.1
9.0–3	11,210 (77.3)	8.25 (210)	10.50 (267)	3#8	0.027	0.039	0.77	1.5

[a] The first number indicates cement content in sacks/cu.yd. The second number indicates the nominal percentage of longitudinal reinforcement.
[b] The yield strengths for No. 6, No. 7 and No. 8 bars were 60.22 ksi (415 MPa), 55.83 ksi (385 MPa) and 66.88 ksi (461 MPa), respectively.

members, confined and doubly reinforced, under short-term 1/3 point loading. No low-strength concrete or medium strength concrete beams were tested. Beams in this series were modeled after beam A-4 of Series A to establish a basis for comparison between the behavior of unconfined singly reinforced, and confined doubly reinforced beams. Beam Series B details are given in Fig. 8.7 and Table 8.4. Compressive steel ratio, ρ', and lateral reinforcement ratio, ρ_s, were the controlled variables. Concrete strength was kept around 8500 psi (58.7 MPa), except for B-2a which had

Table 8.2   Comparison of ductility indices[2]

Reinforcement ratio ρ	Deflection at the end of initial strength portion, in. (mm)	Ultimate deflection, in. (mm)	Ductility index
0.01	0.28 (71)	1.70 (432)	6.0
0.014	0.32 (81)	1.18 (300)	3.7
0.019	0.36 (91)	1.05 (267)	2.9
0.027	0.55 (140)	1.00 (254)	1.8

Table 8.3   Beam Series A details[4]

Beam	Compressive strength at test, psi	Age at test, days	Test region dimensions			Tensile steel			ρ	ρ/ρ_b
			h, in.	b, in.	d, in.	Rebars	A_s, in.^{2}	f_y, ksi		
A-1	3700	95	12.07	7.34	10.69	2#6	0.88	69.0	0.011	0.52
A-2	6500	113	12.01	7.22	10.63	3#6	1.32	69.0	0.017	0.53
A-3a	9284	7	12.03	7.13	10.65	2#6	0.88	66.0	0.012	0.26
A-4	8535	122	12.00	7.38	10.56	3#7	1.80	71.0	0.023	0.87
A-5	9264	122	12.00	6.56	10.50	3#8	2.37	71.0	0.034	0.87
A-6a	8755	137	12.00	6.94	9.75	6#7	3.60	59.5	0.053	1.10

1 in. = 25.4 mm
1000 psi = 6.895 MPa

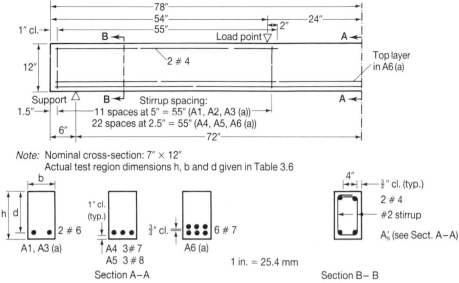

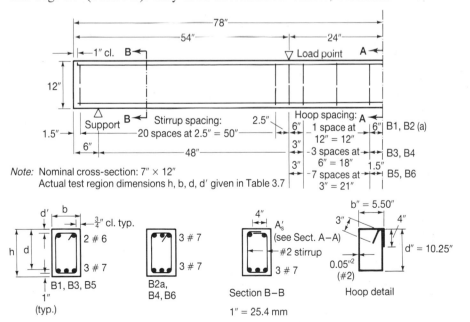

Fig. 8.6   Details of Series A beams tested by Pastor *et al.*[4]

$f_c' = 9284$ psi (64.1 MPa), while the amount of tensile steel was the same for all beams and consisted of three No. 7 (22 mm diameter) deformed bars.

To facilitate the evaluation of the moment-curvature and load-deflection data, experimental curves are represented as shown in Fig. 8.8 (Series A) and Fig. 8.9 (Series B). Key load-deformation values, identified in each

Fig. 8.7   Details of Series B beams tested by Pastor *et al.*[4]

Table 8.4 Beam Series B details[4]

Beam	Compressive strength at test, psi	Age at test, days	Test region dimensions				Tensile steel				Compressive steel			Tie steel[a]	
			h, in.	b, in.	d, in.	d', in.	A_s, in.²	f_y, ksi	ρ	ρ/ρ_b	A_s', in.²	f_y, psi	ρ'	A_s'', in.²	ρ_s
B-1	8534	186	12.13	6.94	10.69	1.76		62	0.024	0.43	2#6 (0.88)	69	0.012	No. 2 at 12 in. (0.05)	0.0023
B-2a	9284	7	12.25	7.19	10.81	1.70		60	0.023	0.31	3#7 (1.80)	60	0.023	No. 2 at 12 in. (0.05)	0.0023
B-3	8578	189	12.25	6.88	10.81	1.88	3#7 (1.80)	62	0.024	0.46	2#6 (0.88)	69	0.012	No. 2 at 6 in. (0.05)	0.0047
B-4	8478	186	12.19	6.69	10.75	1.88		62	0.025	0.36	3#7 (1.80)	62	0.025	No. 2 at 6 in. (0.05)	0.0047
B-5	8516	194	12.06	7.00	10.63	1.69		62	0.024	0.43	2#6 (0.88)	69	0.012	No. 2 at 3 in. (0.05)	0.0093
B-6	8468	190	12.06	6.88	10.63	1.75		62	0.025	0.36	3#7 (1.80)	62	0.025	No. 2 at 3 in. (0.05)	0.0093

(a) Yield strength = 51 ksi
1 in. = 25.4 mm
1000 psi = 6.895 MPa

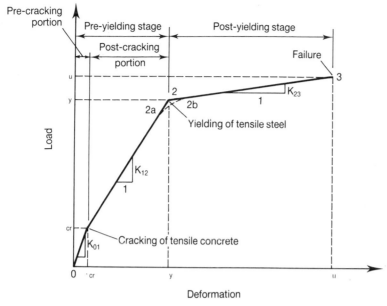

Fig. 8.8   Idealization of beam Series A load-deformation curves[4]

experimental figure, are listed in Table 8.5 (moment-curvature data) and Table 8.6 (load-deflection data). Included in these tables are the corresponding values of the curvature (μ_c) and displacement (μ_d) ductility indices.

Curvature and displacement ductility indices of Series A beams, as listed

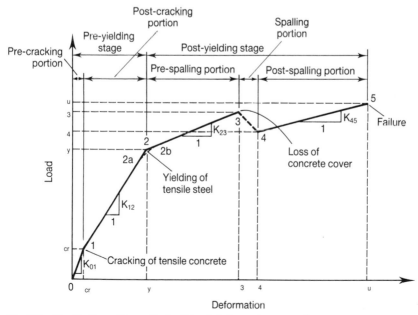

Fig. 8.9   Idealization of beam Series B load-deformation curves[4]

Table 8.5 Moment-curvature data and curvature ductility indices for Series A and B beams[4(a)]

Beam	Moments M (ft – kips)					Curvatures $\phi \times 10^{-6}$ (in.$^{-1}$)					Slopes K $= 10^{-3}$ (kip-in.^{2})				Ductility index
	M_{cr}	M_y	M_3	M_4	M_u	ϕ_{cr}	ϕ_y	ϕ_3	ϕ_4	ϕ_u	K_{01}	K_{12}	K_{23}	K_{45}	μ_c
A-1	8.0	43.0			50.0	50	350			1409	1920	1400	79		4.03
A-2	10.2	68.5			75.0	20	388			1293[b]	6120	1901	86[c]		3.33[c]
A-3a	7.2	45.6	$M_3 = M_u$ Not Applicable		56.0	17	373	$\phi_3 = \phi_4$ Not Applicable		3394	5082	1294	41	Not Applicable	9.10
A-4	12.2	95.0			100.0	30	390			1060	4880	2760	90		2.72
A-5	12.6	112.2			119.0	25	424			673	6048	2995	328		1.59
A-6a	15.0	124.0			142.0	35	427			630	5143	3337	1064		1.48
B-1	15.0	85.0	96.0	84.0	84.0	45	396	1181	1446	1446	4000	2393	168	–	3.65
B-2a	13.0	88.1	106.0	108.0	116.0	25	513	2558	4396	7148	6240	1847	105	35	13.93
B-3	15.0	85.0	100.0	94.0	97.2	30	259	1182	1225	2732	6000	3668	197	25	10.55
B-4[d]	7.6	85.0	103.0	97.0	119.0	17	357	1450	1700	4448	5365	2732	198	96	12.46
B-5[d]	10.0	85.0	97.0	94.0	101.0	30	410	1179	1981	2871	4000	2368	187	94	7.00
B-6[d]	14.4	85.0	97.0	97.0	110.0	50	439	1431	2130	3463	3456	2178	145	117	7.84

[a] See Figs. 8.8 and 8.9 for definition of symbols
[b] Extrapolated
[c] From extrapolated values
[d] Failed prematurely

1 in. = 25.4 mm

1 kip = 4.44822 kN

Table 8.6 Load-deflection data and displacement ductility indices for Series A and B Beams[4(a)]

Beams	Loads P (kips)					Deflections Δ (in.)					Slopes K (kips/in.)				Ductility index
	P_{cr}	P_y	P_3	P_4	P_u	Δ_{cr}	Δ_y	Δ_3	Δ_4	Δ_u	K_{01}	K_{12}	K_{23}	K_{45}	μ_d
A-1	4.0	23.5	$P_3 = P_u$	Not applicable	25.0	0.12	0.95	$\Delta_3 = \Delta_u$	Not applicable	3.50	33.3	23.5	0.6	Not applicable	3.68
A-2	5.1	34.2			37.5	0.06	0.94			2.50[b]	85.0	33.1	2.1[c]		2.66[c]
A-3a	3.6	24.0			28.0	0.05	0.71			3.63	72.0	31.0	1.4		5.11
A-4	6.1	47.5			50.0	0.08	1.05			1.88	76.3	42.7	3.0		1.79
A-5	6.3	56.1			59.5	0.08	1.05			1.21	78.8	51.3	21.3		1.15
A-6a	7.5	64.0			71.0	0.08	0.95			1.20	93.8	64.9	28.0		1.26
B-1	7.5	42.5	48.0	42.0	42.0	0.10	0.83	1.75	1.95	1.95	75.0	48.0	6.0	–	2.35
B-2a	6.5	43.8	53.0	54.0	58.0	0.08	0.79	3.00	4.55	5.63	81.3	52.5	4.2	3.7	7.13
B-3	7.5	43.5	50.0	47.0	48.6	0.11	0.89	2.48	2.60	4.04	68.2	46.2	4.1	1.1	4.54
B-4[d]	3.8	42.5	51.5	48.5	59.5	0.04	0.84	2.63	3.15	7.10	95.0	48.4	5.0	2.8	8.45
B-5[d]	5.0	44.2	48.5	47.0	50.5	0.07	0.79	1.35	2.95	4.36	71.4	54.4	3.7	2.5	5.52
B-6[d]	7.2	42.5	48.5	48.5	55.0	0.09	0.85	2.16	3.13	5.38	80.0	45.9	4.6	2.9	6.26

(a) See Figs. 8.8 and 8.9 for definition of symbols
(b) Extrapolated
(c) From extrapolated values
(d) Failed prematurely

1 in. = 25.4 mm
1 kip = 4.448822 kN

in Tables 8.8 and 8.6, are plotted as functions of the tensile steel ratio ρ in Fig. 8.10. Results show an overall reduction in both μ_c and μ_d with increasing ρ. The loss of ductility with increasing ρ is associated mainly with a decrease in the ultimate deformation of the member. In turn, ultimate deformations are inversely proportional to the neutral axis depth at failure.

Figure 8.11 plots the inverse of the Series A neutral axis depths at failure as a function of ρ. Since c values reflect the influence of fundamental material properties such as f_c' and f_y, the behavior shown in Fig. 8.11 provides a basic explanation to that shown in Fig. 8.10. For high-strength concrete beams, therefore, loss of ductility with increasing ρ can be traced back to the fact that c increases with increasing values of ρ.

Consider the upper half of Fig. 8.10. Points corresponding to the low-strength and medium strength concrete beams lie below the curve defined by the high-strength concrete data. When all other variables are held constant, μ_c clearly increases with f_c'.

Tognon et al.[5] arrived at the same conclusion from tests on singly reinforced model beams of every high-strength concrete. Reinforced concrete beams (4 in. × 8 in. × 6.5 ft or 100 mm × 200 mm × 2 m) were manufactured using very high-strength concrete (VHSC, f_c' = 23,484 psi or 162 MPa) and an ordinary concrete (LSC, f_c' = 5797 psi or 40 MPa) as a reference. The reinforcement consisted of three or six longitudinal steel

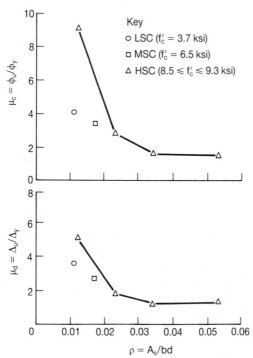

Fig. 8.10   Beam Series A curvature and displacement ductility indices versus tensile steel ratio[4]

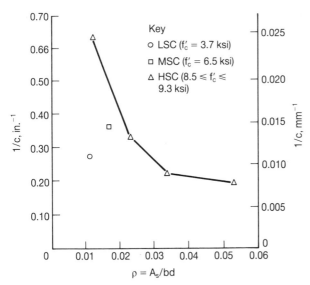

Fig. 8.11   Inverse values of beam Series A neutral axis depths at failure versus tensile steel ratio[4]

bars in the tensile zone with a concrete cover of approximately 0.4 in. or 10 mm. The amount of transverse reinforcement provided in the part of the beam subjected to variable moment was sufficient to prevent shear cracking.

To emphasize the different behavior of the two concretes with low, medium and high percentages of reinforcement, reinforced concrete beams were prepared with 0.87, 1.97 and 4.61% tension steel. 8.88% steel was also used for the very high-strength concrete. Balanced steel percentages were 3.85 and 13.49 for the ordinary and very high-strength concretes, respectively.

Ductility was represented by the ratio of ultimate curvature to yield curvature. The yield curvature was defined as corresponding to a tensile steel strain of 0.2%, and the ultimate curvature to a tensile steel strain of 0.1% or a compressive concrete strain of 0.35%.

Figure 8.12 shows that the beams made with VHSC are more ductile than those made with LSC when the reinforcement is about or over 1%. Moreover, the ductility of the beams made with LSC quickly decreases as the reinforcement ratio increases, whereas that of the beams made with VHSC declines gently up to a reinforcement ratio of about 5%.

It is worth noting that the curvature ductility of all heavily reinforced sections ($\rho > \rho_b$) theoretically approaches unity regardless of concrete strength, f_c'. This suggests that the increase in μ_c with f_c' tends to decrease with ρ. There seems to be, therefore, a limiting tensile steel ratio beyond which μ_c values are practically the same regardless of f_c'.

Results shown in the lower half of Fig. 8.10 pertaining to μ_d exhibit the same general characteristics previously described for μ_c. Differences

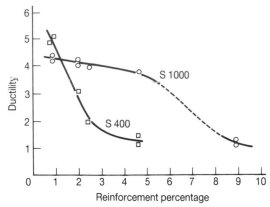

Fig. 8.12   Curvature ductility versus reinforcement percentage for LSC and VHSC beams[5]

between μ_d values due to differences in f_c' are, however, much smaller, mainly because the expected reduction from μ_c to μ_d is greater for the lightly to moderately reinforced HSC beams than for the LSC and MSC members. This suggests that the hypothetical ρ limit beyond which μ_d is no longer influenced by f_c' is comparatively smaller than that for μ_c. In any case, it seems reasonable to assume that μ_d increases with f_c', although much less significantly than μ_c.

The above conclusion emphasizes the difference between material, sectional and member ductility. The ductility of concrete as a material depends on the post-peak deformation of its stress-strain response, a property influenced primarily by f_c'. The ductility of a singly reinforced concrete section, on the other hand, depends on the post-yield deformation of its momentum-curvature response. Similarly, the ductility of a singly reinforced beam as a structural unit depends on the post-yield deformation of its load-deflection response. The two latter characteristics are associated with the position of the neutral axis at failure, a property influenced not only by f_c', but also by ρ and f_y.

As shown in Fig. 8.11, neutral axis depths at failure for the LSC and MSC beams are greater than the corresponding HSC values. It follows, therefore, that the decrease in c with f_c' more than compensates for the loss of material ductility in the HSC range, and helps explain the observed increase in sectional and member ductility with f_c' for constant ρ and f_y.

Consider Fig. 8.13[4] where μ_c and μ_d are plotted versus ρ/ρ_b. It is interesting to note that differences between LSC, MSC and HSC μ_c values shown in Fig. 8.13 are greatly reduced compared to those observed in the upper half of Fig. 8.10. This can be explained in terms of the inverse proportionality that exists between ρ/ρ_b and f_c'. In effect, everything else being equal, ρ_b will be higher for a beam with higher f_c'. Therefore, a beam of low f_c' with the same ρ/ρ_b ratio as that of a higher strength member necessarily has less area of steel. Less tensile ateel area implies a shallower neutral axis depth at failure. Consequently, the difference in ultimate

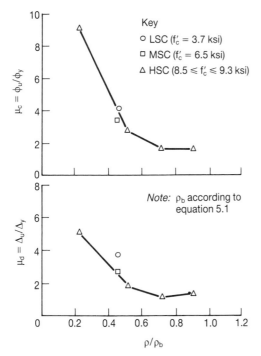

Fig. 8.13   Beam Series A curvature and displacement ductility indices versus tensile steel ratio expressed as a fraction of balanced steel ratio[4]

curvatures and therefore in curvature ductility indices due to differences in f_c' decreases, which is precisely what is illustrated in the upper half of Fig. 8.13.

Results shown in the lower half of Fig. 8.13 for $\rho/\rho_b = 0.45$ indicate a general tendency for μ_d to decrease with increasing f_c', although the difference between the values for MSC and HSC is insignificant. Insofar as the behavior suggested by the LSC and the corresponding HSC values, the observed decrease in μ_d with f_c' agrees with results reported by Leslie *et al.*[2] For reasons previously mentioned while discussing the relationship between μ_d and f_c', differences between LSC and HSC μ_d values should decrease with increasing values of ρ/ρ_b.

Inaccuracies in linear voltage differential transducer (LVDT) measurements introduced by the relatively large rotations of doubly reinforced beams at loads approaching ultimate resulted in erratic ϕ_u values which in turn resulted in unreliable μ_c values that exhibited large scatter. Consequently, curvature ductility indices for Series B beams, listed in Table 8.5, were not included in further analyses. On the other hand, data from related Cornell tests by Fajardo[4] were included to compensate for the loss of Series B data due to the premature failures of B-4, B-5, and B-6. Each of these beams had failed due to rupture of the tensil steel at the first stirrup weld outside the central test region (stirrups were spot welded to the longitudinal rebars). Although in all cases the tensile steel had already yielded when failure occurred and in some cases measured strains were well in the strain hardening portion of the steel stress-strain curve, this type of failure was considered premature. Full details of the Fajardo beams are given in Table 8.7.

Consider Fig. 8.14 where displacement ductility indices (μ_d) of nine doubly reinforced confined HSC beams and one singly reinforced unconfined HSC member (beam A-4) are plotted versus the quantity $\rho''f_y''/f_c'$. The symbol f_c'' is the yield strength of the lateral steel, f_c' is the compressive strength of the concrete, and ρ'' is the combined volumetric ratio of compressive and lateral reinforcement, defined as follows:

$$\rho'' = \rho_s + A_s'/b''d'' \qquad (8.1)$$

where ρ_s = volumetric ratio of lateral reinforcement
$\quad\quad\quad = 2(b'' + d'')A_s''/b''d''s$
$\quad A_s'$ = area of longitudinal compressive steel
$\quad A_s''$ = area of lateral tie steel
$\quad b''$ = width (outside-to-outside of tie steel vertical legs) of the d''confined core
$\quad s$ = depth (outside-to-outside of tie steel horizontal legs) of the confined core
$\quad\quad = $ spacing (center-to-center) of tie steel

The best-fit linear regression curves shown interpolating the plotted data in Fig. 8.14 intersect at a $\rho''f_y''/f_c'$ value of about 0.11. They define two distinctly different modes of behavior that are briefly discussed below.

The relatively flat slope of the line to the left indicates that beams with

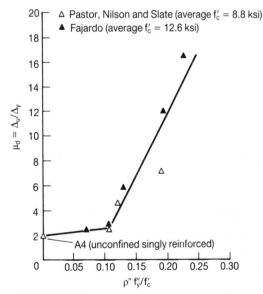

Fig. 8.14   Influence of compression and transverse reinforcement on the displacement ductility of doubly reinforced confined HSC beams[4]

Table 8.7 Properties of beams tested by Fajardo[4]

Beam	Age, days	f'_c, psi	f_y, ksi	$b \times h$, in.	L, ft	d, in.	d', in.	A_s, in.²	ρ, %	A'_s, in.²	ρ', %	ρ_b, %	ρ/ρ_b	Confining hoops
B-7	121	12,800	65.0	7×12	12.0	10.50	1.44	2.35 (3#8)	3.2	1.20 (2#7)	1.6	7.7	0.42	#3 at 12 in.
B-8	134	12,650	65.0	7×12	12.0	10.50	1.50	2.35 (3#8)	3.2	2.35 (3#8)	3.2	9.2	0.35	#3 at 12 in.
B-9	128	12,800	65.0	7×12	12.0	10.50	1.44	2.35 (3#8)	3.2	1.20 (2#7)	1.6	7.7	0.42	#3 at 6 in.
B-10	133	12,000	65.0	7×12	12.0	10.50	1.50	2.35 (3#8)	3.2	2.35 (3#8)	3.2	8.9	0.36	#3 at 6 in.
B-11	132	12,650	65.0	7×12	12.0	10.50	1.25	2.35 (3#8)	3.2	0.40 (2#4)	0.50	6.6	0.48	#3 at 6 in.

1 in. = 25.4 mm
1000 psi = 6.895 MPa

$\rho'' f_y''/f_c' < 0.11$ exhibited a load-deformation behavior with virtually no post-spalling response. These beams failed almost immediately after losing their cover concrete. Since significant inelastic dilatancy of the compression zone concrete occurs only after loss of the cover concrete, the presence of longitudinal compression and lateral tie steel has little or no influence on the ductility of these members. Consequently, the post-yielding load-deformation behavior and the displacement ductility indices were essentially the same as those of the corresponding singly reinforced unconfined member (A-4).

On the other hand, the relatively steep slope of the line to the right suggests that the post-spalling lateral expansion of the compressed concrete was restrained by the compression zone reinforcement. For beams with $\rho'' f_y''/f_c' > 0.11$, therefore, ductility was influenced significantly by the presence of longitudinal compressive and lateral tie steel. It should be mentioned that testing of the two beams that gave the highest μ_d values (B-8 and B-10) was halted when excessive deflections could not be accommodated by the test setup.

The relative efficiency of the longitudinal compressive and the lateral tie steel in increasing μ_d is examined in Fig. 8.15. Figure 8.15(a) plots μ_d versus $\rho_s f_y''/f_c'$ for two values of ρ'/ρ, while Fig. 8.15(b) plots μ_d versus $\rho' f_y''/f_c'$ for three values of ρ_s. In both cases straight lines are used to represent the variation between data points.

Consider first Fig. 8.15(a). For a relatively low $\rho_s f_y''/f_c'$ value of 0.0175, increasing A_s' by a factor of two augments μ_d by a factor of almost four.

Consider now Fig. 8.15(b). For a significant range of $\rho' f_y''/f_c'$ values, increasing ρ_s 2.5 times results in a relatively small increase in μ_d. When ρ_s is increased by a factor of five, the effect of μ_d is, as expected, more pronounced. Nevertheless, the increase in μ_d is still smaller than the increase obtained by doubling the area of longitudinal compressive steel.

Results in Fig. 8.15 suggest that lateral ties are not as efficient in

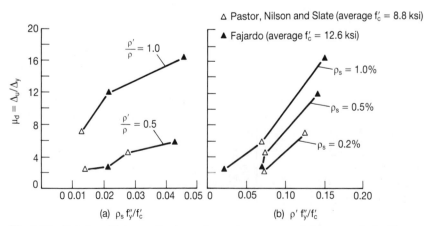

Fig. 8.15   Relative efficiency of compression and transverse reinforcement in doubly reinforced confined HSC beams[4]

increasing the post-yield deformations of beams as they are, for example, in increasing the post-peak deformations of concentrically loaded members. This is mainly because lateral deformations in beams tend to be large at the extreme compression fiber, and practically non-existent at the neutral axis location. Consequently, lateral confining stresses are unevenly and therefore inefficiently distributed across the depth of the compression zone.

On the other hand, the presence of properly restrained cxompression steel allows the member to behave (after loss of cover) much like a two-flanged steel beam (particularly when $A_s' = A_s$), the core concrete acting in this case as the connecting web. Consequently, significant inelastic deformations and large μ_d can be obtained. For such behavior to occur, however, the compression steel must remain stable in the strain hardening range. The stability of the compression bars is influenced strongly by the lateral restraint provided by the transverse ties. Hence, it may be reasonable to think that the primary role of the transverse steel in increasing beam ductility is not as a provider of concrete confinement, but rather as a lateral support mechanism for the longitudinal steel. Reducing the spacing of the ties did not, therefore, increase the confinement of the concrete core as much as it reduced the unsupported length of the compression bars.

Summarizing, the addition of longitudinal compression steel and lateral tie steel increases the displacement ductility of singly reinforced HSC beams (the former more efficiently than the latter), provided that the reinforcement index $\rho''f_y''/f_c'$ is greater than 0.11. Beams with $\rho''f_y''/f_c' < 0.11$ exhibit μ_d values very similar to that of the corresponding singly reinforced member. On the other hand, for indices >0.11, a practically linear $\mu_d - \rho''f_y''/f_c'$ relationship develops in which μ_d increases noticeably for relatively small values of $\rho''f_y''/f_c'$.

Swartz et al.[6] tested four beams ($f_c' = 11,500$ psi or 79 MPa for the first two, and 12,300 psi or 85 MPa for the other two) with longitudinal reinforcing varying from $0.5\rho_b$ to $1.5\rho_b$ based on an assumed triangular stress block (ρ/ρ_b ratios would be somewhat lower, based on the ACI rectangular stress block). The amount of shear reinforcement also varied from zero to 100% based on ACI 318-83.[3] These designs are shown in Fig. 8.16. Load versus midspan displacement is plotted for each beam in Fig. 8.17. The beams with little or no shear reinforcement and also the over-reinforced beam exhibited very little ductility. No detail other than that shown on Fig. 8.17 was available.

The flexural ductility of ultra-high-strength concrete members (concrete strength ranging up to 15 ksi or 103.4 MPa) under monotonic loading was investigated by Shin.[7-9] All specimens were 6 in. (150 mm) × 12 in. (300 mm) in cross-section, and 10 ft (3 m) long (Fig. 8.18). Three sets of twelve specimens each were manufactured, using concrete compressive strengths of 4 ksi (27.6 MPa), 15 ksi (103.4 MPa) and 12 ksi (82.7 MPa) for sets A, B and C, respectively. There were six groups of two identical

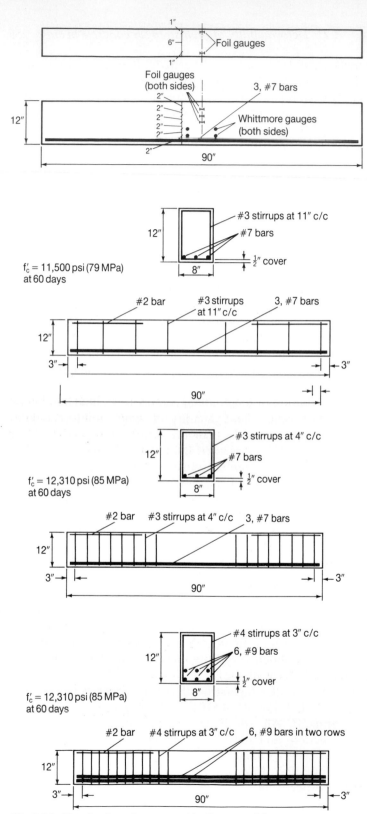

Fig. 8.16 Details of specimens tested by Swartz *et al.*[6]

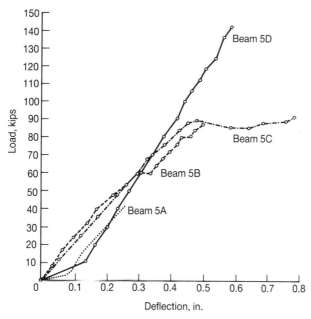

Fig. 8.17   Load versus midspan displacement of beams tested by Swartz *et al.*[6]

specimens in each set, two groups containing four No. 3 (10 mm diameter) bars, two more groups having four No. 5 (16 mm diameter) bars, and the last two groups being reinforced with four No. 9 (29 mm diameter) bars at the four corners. The difference between two groups of specimens with the same concrete strength and longitudinal reinforcement was in the spacing

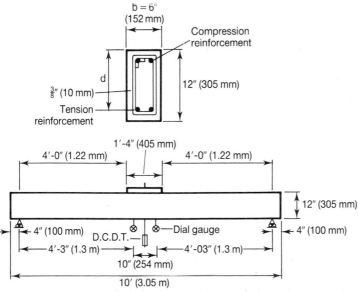

Fig. 8.18   Specimens and setup – Shin *et al.*[7–9]

of No. 3 (10 mm diameter) ties which was either 3 in. (75 mm) or 6 in. (150 mm).

The specimens were reinforced as if they were columns. They were cast horizontally under field conditions, and were also cured under field conditions. Each specimen was tested under two-point loading, which subjected a considerable portion of the specimen to pure flexure. The longitudinal reinforcement was divided equally into tension and compression reinforcement areas. While the tension reinforcement always yielded at advanced load stages, the compression reinforcement quite often developed only small stresses all the way up to beam failure.

Member ductility was first defined as:

$$\mu_0 = \Delta_0/\Delta_y \tag{8.2}$$

where Δ_0 is member deflection corresponding to the maximum load on the member, and Δ_y is member deflection at first yielding of the tension reinforcement.

In view of the fact that the beams continued to sustain substantial loads well beyond the peak of the load-deflection diagram (the load-carrying capacity of heavily reinforced beams temporarily dropped off immediately following the peak, but then picking up again), a second definition of ductility was also considered:

$$\mu_f = \Delta_f/\Delta_y \tag{8.3}$$

where Δ_f is the 'final' deflection corresponding to 80% of the maximum load along the descending branch of the load-deflection curve. Since the concept of ductility is related to the ability to sustain inelastic deformations without a substantial decrease in the load-carrying capacity, the definition of ductility as given by Equation (8.2) was felt to be logical and practical.

The ratio ρ/ρ_b turned out to be the most dominant factor influencing the magnitudes of the ductility indices. For a doubly reinforced section in which the compression reinforcement does not yield at the ultimate stage (defined by the extreme compression fiber concrete strain attaining a value of 0.3%), ρ_b is given by the following equation:

$$\rho_b = \rho_b + \rho'(f_b'/f_y) \tag{8.4}$$

where

$$\rho_b = k_1 k_3 (f_c'/f_y)[E_s \varepsilon_u/(E_s \varepsilon_u + f_y)] \tag{8.5}$$

is the balanced reinforcement ratio for the corresponding singly reinforced section, and

where

$$f_b' = E_s \varepsilon_u - (d'/d)(E_s \varepsilon_u + f_y) \leqslant f_y \tag{8.6}$$

is the stress in the compression reinforcement at balanced strain conditions.

Plots of μ_0 versus ρ/ρ_b and μ_f versus ρ/ρ_b are presented in Figs. 8.19 and 8.20, respectively. The figures clearly show that for the same concrete strength, the ductility indices decrease rather drastically as the ratio ρ/ρ_b increases. However, even at the large ρ/ρ_b ratio of 0.8, μ_f, which probably

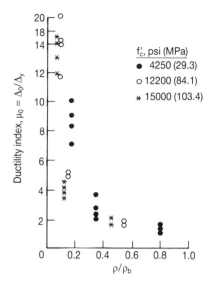

Fig. 8.19   Flexural ductility, as defined by Equation (8.2), under monotonic loading[7–9]

is a more practical measure of ductility for beams tested than μ_0, has rather substantial values.

Figures 8.19 and 8.20 generally show that the same amounts of longitudinal and confinement reinforcement, the ductility indices rise sharply as the concrete strength increases from 4 ksi (27.6 MPa) to 12 ksi (82.7 MPa) (nominal values), but then decrease somewhat as f_c' increases further from 12 ksi (82.7 MPa) to 15 ksi (103.4 MPa).

The confinement reinforcement spacing, within the range studied, did

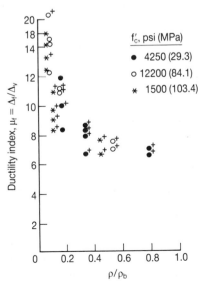

Fig. 8.20   Flexural ductility, as defined by Equation (8.3), under monotonic loading[7–9]

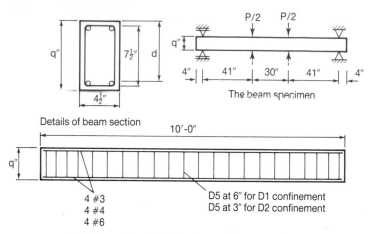

Fig. 8.21   Specimens and setup – Kamara *et al.*[9,10]

not have an appreciable effect on member ductility, for reasons discussed earlier and in Shin *et al.*[8]

Normal weight concrete beams under reversed cyclic loading

Shin[8] and Kamara[9,10] investigated the flexural ductility of ultra-high-strength concrete members (concrete strength ranging up to 15 ksi or 103.4 MPa) under fully reversed cyclic loading.

All specimens (Fig. 8.21) were 4.5 in. (112.5 mm) × 9 in. (225 mm) in cross-section, and 10 ft (3 m) long. Four sets of six specimens each were manufactured, using concrete compressive strengths of 5 ksi (34.5 MPa), 11 ksi (75.9 MPa), and 15 ksi (103.4 MPa) for sets A, B, and C, respectively. Set D was a duplicate of set C. In each set there were two beams reinforced with four No. 3 (10 mm diameter) bars, two beams reinforced with four No. 4 (13 mm diameter) bars and two beams reinforced with four No. 6 (19 mm diameter) bars. The difference between two specimens with the same concrete strength and longitudinal reinforcement was in the spacing of No. 2 (6 mm diameter) ties which was either 3 in. (75 mm) or 6 in. (150 mm).

The specimens, reinforced like columns, were cast horizontally under field conditions, and were also cured under field conditions. The specimens were tested under two-point loading which subjected a considerable portion of the specimens to pure flexure. The applied cyclic loading followed the displacement controlled schedule shown in Fig. 8.22.

The measured hysteretic load-deflection curves for the twenty-four beams tested have been presented in Kamara.[10] A sample is illustrated in Fig. 8.23. The envelopes of the hysteretic load-deflection curves for the beams were also included in Kamara.[10] It was observed that the envelope of the load-deflection curves for a cyclically loaded beam showed the same

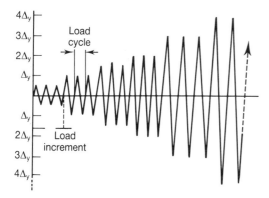

Fig. 8.22   Deformation sequence for reversed cyclic loading tests by Kamara *et al.*[9,10]

features as the load-deflection curve for a comparable monotonically loaded beam.

Member ductility, as defined by Equation (8.3), was investigated. The final deflection Δ_f was considered to be 2.4 in. (61 mm) which was the deflection at the end of cycling. The value of the applied load at that deflection was found to be equal to or slightly larger than the conservative value of the calculated ultimate load. The ductility ratios for downward and upward load cycles are plotted in Figs. 8.24 and 8.25, respectively.

In view of previous research work and building codes, a deflection ductility of 4 appears to represent a reasonably conservative minimum requirement for members subjected to gravity plus wind or moderate seismic loads. This requirement was more than met by all the specimens tested in Kamara's program. Thus it was concluded that, in the absence of axial loads acting simultaneously with flexure, high-strength reinforced concrete members subjected to reversed cyclic loading possess as much

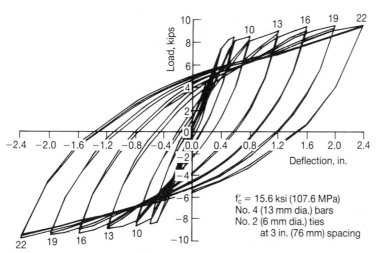

Fig. 8.23   Hysteretic load-deflection curve of specimen tested under reversed cyclic loading[9,10]

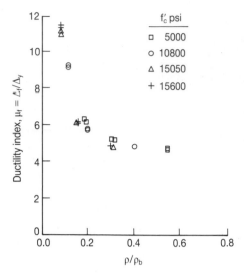

Fig. 8.24   Flexural ductility, as defined in Equation (8.3), under reversed cyclic loading –
upward deflection[9,10]

ductility as is likely to be required of them in practical situations. It should
be remembered, however, that the specimens tested in the course of
Kamara's investigation were under zero axial load.

For the same amounts of longitudinal reinforcement and confinement
reinforcement, the ductility ratios were found to increase with increasing
concrete strength. For the same concrete strength, the ductility ratio
decreased with increasing amounts of longitudinal reinforcement. Within
the range studied in Kamara's work, the spacing between ties appeared to
have virtually no effect on the ductility of the tested specimens.

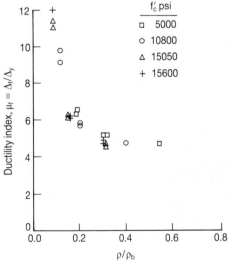

Fig. 8.25   Flexural ductility, as defined in Equation (8.3), under reversed cyclic loading –
downward deflection[9,10]

Lightweight concrete beams under monotonic loading

Ahmad and Barker[11] reported limited experimental data on the flexural behavior of high-strength lightweight concrete beams. Flexural tests were conducted on six singly reinforced beams. Experimental variables were the compressive strength of concrete ($5200 < f_c' < 11,000$ psi or $35.9 < f_c' < 75.9$ MPa) and the reinforcement ratio, ρ/ρ_b ($0.18 < \rho/\rho_b < 0.54$). No compression or lateral reinforcement was used in the beams. A summary of the experimental program is presented in Table 8.8.

Deflection ductility was defined as the ratio of the deflection at ultimate to the deflection at yielding of the tensile steel. Ultimate was defined as the stage beyond which it was felt during testing that a beam would not be able to sustain additional deformation at the same load intensity.

The deflection ductility index μ_d decreased with an increase in tensile steel content ρ (Fig. 8.26 from Ahmad and Barker[11]). The results of Figs. 8.26 and 8.27[11] show that for an approximately equal ρ/ρ_b ratio, μ_d decreases with an increase in f_c'. As discussed earlier in connection with Pastor's investigation,[4] this is because the value of ρ_b increases with greater concrete strengths. Therefore, for a constant value of ρ/ρ_b, a beam with a higher strength concrete contains more steel than one with a lower strength concrete, which in turn decreases μ_d. For the beams with compressive strengths of 5000 and 8000 psi (34.5 and 55.9 MPa), a sharp reduction in μ_d occurred with an increase in ρ/ρ_b. However, for beams of 11,000 psi (75.9 MPa) concrete, the value of μ_d appeared to be less sensitive to changes in ρ/ρ_b. The trend indicated that beyond a certain range of ρ/ρ_b,

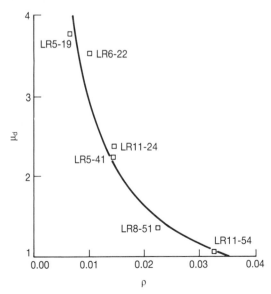

Fig. 8.26 Effect of the reinforcement ratio, ρ, on the displacement ductility of singly reinforced lightweight concrete beams tested by Ahmad and Barker[11]

Table 8.8   Test program[11]

Beam[a]	b, in. (mm)	h, in. (mm)	d, in. (mm)	Age at testing, days	f'_c[b] psi (MPa)	f'_c[c] psi (MPa)	E_c ksi (MPa)	Reinforcing bar detail	A_s, in.² (mm²)	ρ	ρ_g	ρ/ρ_b
LR5-19	6.0 (152.4)	12.0 (304.8)	10.25 (260)	2	5470 (37.7)	5200 (35.9)	3500 (2415)	2#4	0.4 (258)	0.0065	0.0344	0.189
LR5-41	6.0 (152.4)	12.0 (304.8)	9.25 (235)	2	5690 (39.3)	5410 (37.3)	3520 (2429)	4#4	0.8 (5160)	0.0144	0.0354	0.407
LR8-22	6.0 (152.4)	12.0 (304.8)	10.19 (259)	5	8770 (60.5)	8330 (57.5)	3760 (2594)	2#5	0.62 (400)	0.0101	0.0454	0.222
LR8-51	6.0 (152.4)	12.0 (304.8)	9.13 (231)	5	8550 (59.0)	8120 (56.0)	3750 (2588)	4#5	1.24 (800)	0.0226	0.0443	0.511
LR11-24	6.0 (152.4)	12.0 (304.8)	10.13 (257)	49	11560 (79.8)	10980 (75.8)	4410 (3043)	2#6	0.88 (568)	0.0145	0.0599	0.242
LR11-54	6.0 (152.4)	12.0 (304.8)	9.00 (229)	49	11590 (80.0)	11010 (76.0)	4770 (3291)	4#6	1.76 (1135)	0.0326	0.0600	0.543

(a) Beam nomenclature: for Beam LR5-19, '5' indicates the approximate concrete compressive strength in ksi and '19' indicates reinforcement ratio ρ/ρ_b

(b) Based on 4 × 8-in. cylinder strength

(c) Using equivalent 6 × 12-in. cylinder strength, assumed to be 95 per cent of 4 × 8-in. cylinder strength

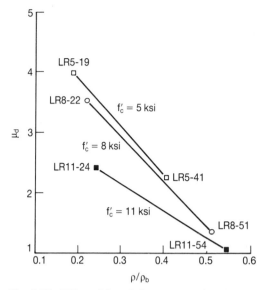

Fig. 8.27   Effect of the reinforcement ratio, ρ/ρ_b, on the displacement ductility of singly reinforced lightweight concrete beams tested by Ahmad and Barker[11]

deflection ductility μ_d becomes essentially independent of concrete strength, as suggested earlier by Pastor and Nilson.[4] Figure 8.28 shows the effect of concrete strength on μ_d for beams with different values of ρ/ρ_b. The comparison of deflection ductilities for lightweight and normal weight high-strength concrete beams (Fig. 8.29 from Ahmad and Barker[11])

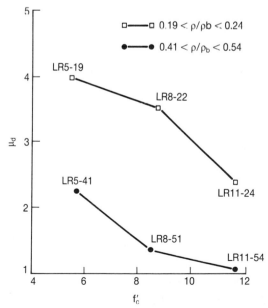

Fig. 8.28   Effect of the concrete strength, $f_c' f_g$ on the displacement ductility of singly reinforced lightweight concrete beams tested by Ahmad and Barker

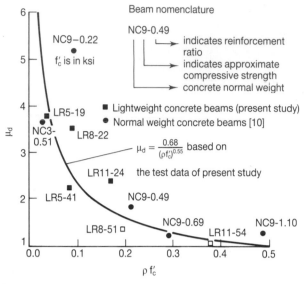

Fig. 8.29 Comparison of ductilities of lightweight and normal weight high-strength singly reinforced lightweight concrete beams tested by Ahmad and Barker[11]

indicates that ρ has a similar influence on μ_d for beams made of both types of concrete. The μ_d values obtained in Ahmad's study[11] were lower than those reported by Pastor and Nilson[4] for reinforced normal weight high-strength concrete beams.

Ahmad and Batts[12] developed limited experimental data on the flexural behavior of doubly reinforced high-strength lightweight concrete beams with web reinforcement. Flexural tests were conducted on six doubly reinforced beams. Experimental variables were the compressive strength of concrete ($6700 < f_c' < 11,060$ psi or $46.2 < f_c' < 76.3$ MPa) and the reinforcement ratio, ρ/ρ_b ($0.16 < \rho/\rho_b < 0.47$). All the beams had compression and web reinforcement. The compression reinforcement was kept to approximately half of the tension reinforcement, and web reinforcement was provided by No. 2 (6 mm diameter) smooth bars placed as stirrups at a spacing equal to half the depth of the section. A summary of the experimental program is presented in Table 8.9.

Deflection ductility was defined as in Ahmad and Barker.[11] The index μ_d decreased with an increase in the tensile steel content ρ (Fig. 8.30 from Ahmad and Batts[12]). The results in Figs. 8.31 and 8.32[12] show that for an approximately equal ρ/ρ_b ratio, μ_d decreases with an increase in f_c', for reasons discussed earlier. For beams with concrete strengths of 8000 and 11,000 psi (56 and 77 MPa), the ductility decreased relatively less with an increase in ρ/ρ_b than for beams with 5000 psi (35 MPa) concrete. Test results (Fig. 8.33 from Ahmad and Batts[12]) showed that the ductility decreases with increasing values of the product $\rho f_c'$. However, for larger values of the product $\rho f_c'$, the ductility essentially becomes constant.

A comparison of the results presented in Figs. 8.26–8.29 and those given

Table 8.9   Test program[12]

Beam	Compressive strength at test f_c'[(a)], psi	f_c'[(b)], psi	Test region dimensions				Tensile steel			Web steel			
			h, in.	b, in.	d, in.	d', in.	A_s, in.²	ρ	ρ/ρ_b, in.²	A_s', in.²	ρ'	A_s'', in.²	ρ_s
LJ-6-16	6700	6380	12	6	10.00	1.90	0.40	0.0067	0.16	0.22	0.0037	0.05	0.012
LJ-7-31	7720	7330	12	6	9.00	2.00	0.80	0.0148	0.31	0.40	0.0074	0.05	0.017
LJ-8-21	8080	7680	12	6	9.94	2.00	0.62	0.0104	0.21	0.40	0.0067	0.05	0.017
LJ-8-44	8360	7940	12	6	8.88	2.10	1.24	0.0233	0.44	0.62	0.0116	0.05	0.023
LJ-11-22	11740	11150	12	6	9.88	2.00	0.88	0.0148	0.22	0.40	0.0067	0.05	0.017
LJ-11-47	11060	10510	12	6	8.75	2.10	1.76	0.0335	0.47	0.88	0.0168	0.05	0.030

(a) Based on 4 × 8-in. cylinder strength

(b) Using equivalent 6 × 12-in. cylinder strength, assumed to be 95 per cent of 4 × 8-in. cylinder strength

The tension and compression reinforcement consisted of ASTM A 615 Grade 60 deformed bar

The stirrups used were No. 2 smooth bars with yield strength of 60 ksi

1 in. = 25.4 mm

1000 psi = 6.895 MPa

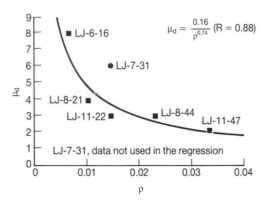

Fig. 8.30 Effect of reinforcement ratio, ρ, on the displacement ductility of doubly reinforced ($\rho' \approx 0.5\rho$) lightweight concrete beams tested by Ahmad and Batts[12]

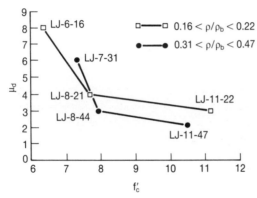

Fig. 8.31 Effect of concrete strength on the displacement ductility of doubly reinforced ($\rho' \approx 0.5\rho$) lightweight concrete beams tested by Ahmad and Batts[12]

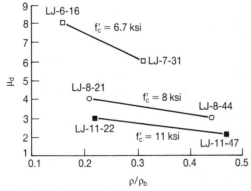

Fig. 8.32 Effect of the reinforcement ratio, ρ/ρ_b, on the displacement ductility of doubly reinforced ($\rho' \approx 0.5\rho$) lightweight concrete beams tested by Ahmad and Batts[12]

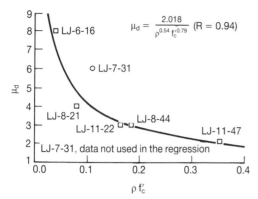

Fig. 8.33   Displacement ductility versus $\rho f_c'$ for doubly reinforced ($\rho' \approx 0.5\rho$) lightweight concrete beams tested by Ahmad and Batts[12]

in Figs. 8.30–8.33 clearly shows that the addition of compression reinforcement equal to approximately half the amount of tension reinforcement had a distinctly beneficial effect on deflection ductility.

Lightweight concrete beams under reversed cyclic loading

Ghosh *et al.*[13] conducted an experimental investigation aimed at gathering information on the flexural properties, including ductility, of high-strength lightweight concrete members (concrete with a dry unit weight of approximately 120 pcf and with compressive strength approaching 9 ksi at 56 days) under reversed cyclic loading.

Two sets of six specimens each were manufactured using lightweight aggregate concrete having compressive strengths of 5 ksi (34.5 MPa) at 28 days and 9 ksi (62 MPa) at 56 days. The test variables were the concrete strength, the amount of longitudinal reinforcement, and the spacing of ties. The test results, including hysteretic load-deflection curves, for the specimens representing columns under zero axial load are reported in Ghosh *et al.*[13]

The specimen dimensions, test procedure and loading history were identical to those used earlier by Kamara.[9,10] All specimens of each series were cast at the same time.

Except for one specimen that failed in shear, the moderate – as well as high-strength lightweight – concrete specimens exhibited stable hysteretic behavior all the way up to the limiting stroke of the testing machine. Flexure-dominated behavior could be ensured by supplying design shear strength in excess of the shear corresponding to the probable flexural strength.

The maximum deflection that could be imposed on the lightweight concrete beams of this investigation and on the normal weight concrete beams tested under reversed cyclic loading in the previous investigation by Kamara[9,10] was limited by the maximum stroke of the testing machine, so that full potential values of the ductility index μ_f (Equation (8.3)) could not be measured. The ductility indices of the lightweight concrete beams were lower than those of the corresponding normal weight concrete beams. The reason is that, because of the lower modulus of elasticity of lightweight concrete, the neutral axis depth at yield in a lightweight concrete beam was larger than that in a corresponding normal weight concrete beam. This made the yield curvature, and consequently the yield deflection, Δ_y, larger in the lightweight beam. In terms of the drop in maximum load carrying capacity at the maximum deflection that could be imposed, there was no significant difference between lightweight and normal weight concrete beams having the same longitudinal reinforcement and comparable concrete strengths, if the shear-dominated lightweight beam were excluded from consideration.

The ratio ρ/ρ_b once again turned out to be the most dominant factor influencing the magnitudes of the ductility indices. Plots of μ_0 versus ρ/ρ_b and μ_f versus ρ/ρ_b are presented in Figs. 8.34 and 8.35, respectively. The figures clearly show that for the same concrete strength, the ductility indices decrease rather drastically as the ratio ρ/ρ_b increases. However, all specimens, with the exception of the one that failed in shear, developed rather substantial values of μ_f, which is probably a more practical measure of ductility for the specimens tested than μ_0. It needs to be pointed out again that the values of μ_f in Fig. 8.35 are the largest values that could be measured, and not the largest values that could be attained.

Figures 8.34 and 8.35 generally show that the same amount of longitudinal reinforcement, ductility increases with increasing concrete strength.

The confinement reinforcement spacing, within the range studied, did not have an appreciable effect on member ductility, as is to be expected in view of earlier discussions.

Conclusions

The following conclusions can be drawn with respect to the deformability of high-strength concrete beams:

1 Although high-strength concrete is a less deformable material than lower strength concrete, the curvature ductility, μ_c, of a singly reinforced concrete section increases with f_c' for the same value of the reinforcement ratio, $\rho = A_s/bd$. This is because the neutral axis depth c decreases with increasing concrete strength, and the decrease in c with f_c' more than compensates for the loss of material ductility in the HSC range. For the same f_c', μ_c decreases with increasing ρ, because the neutral axis depth, c increases with increasing values of ρ.

2 The curvature ductility of all heavily reinforced sections $(\rho > \rho_b)$ theoretically approaches unity regardless of concrete strength, f_c'. Thus, the increase in μ_c with f_c' tends to decrease with ρ. There

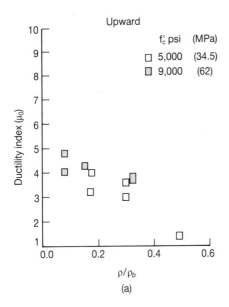

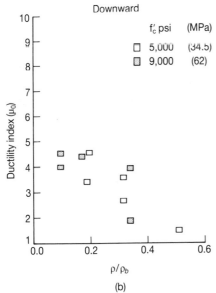

Fig. 8.34   Ductility index μ_0 versus ρ/ρ_b for doubly reinforced ($\rho = \rho'$) lightweight concrete beams[13]

appears to be a limiting tensile steel ratio beyond which μ_c values are practically the same regardless of f_c'.

3   The curvature ductility, μ_c, decreases with increasing values of the reinforcement ratio, ρ/ρ_b. At the same ρ/ρ_b, the differences between LSC, MSC and HSC curvature ductility (μ_c) values are greatly reduced compared to those observed for a constant value of ρ. This is

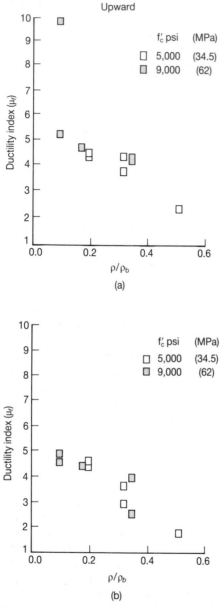

Fig. 8.35   Ductility index μ_f versus ρ/ρ_b for doubly reinforced ($\rho = \rho'$) lightweight concrete beams[13]

because the value of ρ_b increases with greater concrete strengths. For a constant value of ρ/ρ_b, a beam with a higher strength concrete contains more steel than one with lower strength concrete.

4   The member deflection ductility, μ_d, decreases with increasing values of ρ. For the same value of ρ, μ_d increases with increasing

concrete strength, f_c'. However, the magnitude of the increase is less than in the case of μ_c.

5   The deflection ductility, μ_d, also decreases with increasing values of ρ/ρ_b. For the same ρ/ρ_b, depending on that value of ρ/ρ_b, μ_d may decrease with increasing concrete strength. The differences between LSC and HSC μ_d values decrease with increasing values of ρ/ρ_b.

6   The ρ limit beyond which μ_d is no longer influenced by f_c' is comparatively smaller than that for μ_c.

7   The addition of longitudinal compression steel and lateral tie steel increases the displacement ductility of singly reinforced HSC beams (the former more efficiently than the latter), provided that the reinforcement index $\rho''f_y''/f_c'$ is greater than a certain critical value.

8   For low values of the above reinforcement index, beams fail amost immediately after losing their concrete cover. Since significant inelastic dilatancy of the compression zone concrete occurs only after loss of the cover concrete, the presence of longitudinal compression and lateral tie steel has little influence on the ductility of these members.

9   For moderate to larger values of the index $\rho''f_y''/f_c'$, displacement ductility is influenced significantly by the presence of longitudinal compressive and lateral tie steel, because the post-spalling lateral expenasion of the compressed concrete is restrained by the compression zone reinforcement.

10   Lateral ties are not as efficient in increasing the post-yield deformations of beams as they are in increasing the post-peak deformations of concentrically loaded members. This is because lateral confining stresses are unevenly and inefficiently distributed across the beam compression zone. The primary role of the transverse steel in increasing beam ductility is not as a provider of concrete confinement, but rather as a lateral support mechanism for the longitudinal steel.

11   The above conclusions drawn from tests on normal weight concrete beams under monotonic loading also apply, from all indications, to lightweight concrete beams under monotonic loading, except that, all other variables being the same, ductility is likely to be lower for the lightweight beam than for the corresponding normal weight member.

12   In the absence of axial loads acting simultaneously with flexure, high-strength reinforced concrete members subjected to reversed cyclic loading possess as much ductility as is likely to be required of them in practical situations. Under the same loading, the ductility indices of lightweight concrete beams are lower than those of the corresponding normal weight concrete beams. The reason is that, because of the lower modulus of elasticity of lightweight concrete, the neutral axis depth at yield is a lightweight concrete beam is larger than that in a corresponding normal weight concrete beam. This makes the yield curvature, and consequently the yield deflection larger in the lightweight beam.

8.3 Deformability of high-strength concrete columns

Deformability of high-strength concrete columns plays a major role in providing overall strength and stability to earthquake resistant structures. High deformability requirements in the first-story columns of multistory buildings can only be achieved through confinement of the core concrete. Inelastic deformability of high-strength concrete columns and associated confinement requirements are assessed in this section by reviewing test data from the available literature.[14] The data included the results of column tests conducted either under monotonically increasing concentric compression[15-19] or under lateral load reversals.[20-24] Concrete strength in the range of 4800 to 16,800 psi (33 to 116 MPa) has been considered. The tests include both normal weight and lightweight concrete columns with circular, square and rectangular cross-sections.

Deformability of concrete is evaluated in terms of a strain ductility ratio. Stress-strain relationships of confined high-strength concrete, obtained from column tests under concentric compression, are used for this purpose. The stress-strain relationship of the confined core is obtained by subtracting the contributions of longitudinal reinforcement and cover concrete from the recorded column capacity. The strain ductility ratio is defined as the ratio of concrete strain at 85% of the peak stress on the descending branch, to the strain corresponding to the peak stress. This is illustrated in Fig. 8.36.

Column deformability is evaluated in terms of a displacement ductility ratio. The displacement ductility ratio is defined as the ratio of maximum displacement recorded prior to exceeding 20% strength decay under cyclic loading, to the yield displacement. Cyclic loading consists of at least two cycles at each of the incrementally increasing deformation levels, where each increment is less than twice the yield displacement. Although approximately, these restrictions on imposed deformations reflect effects of loading history, and eliminate variations in column ductility resulting from distinctly different loading histories. Figure 8.37 depicts the definition of the displacement ductility ratio. The following notation is used in this section:

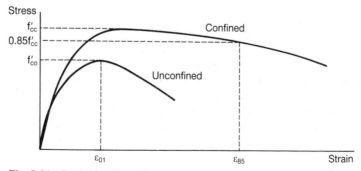

Fig. 8.36 Strain ductility ratio

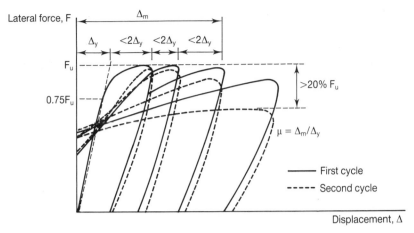

Fig. 8.37 Displacement ductility ratio

A_g : Gross area of concrete column section
b : Column cross-sectional dimension
D : Diameter of circular column section
f_c' : Concrete cylinder strength
f_{co}' : Unconfined concrete strength in column, determined by testing a specimen of the same size and shape as the column
f_{yt}' : Yield strength of confinement reinforcement
h : Column cross-sectional dimension
LWC : Lightweight concrete
M : Maximum bending moment sustained by test specimen
NWC : Normal weight concrete
P : Axial compression load
s : Tie spacing
s_1 : Spacing of legs of column ties in the horizontal plane
v : Maximum shear stress sustained by test specimen
V : Maximum shear force sustained by test specimen
ε_0 : Strain corresponding to unconfined strength of concrete
ε_{85} : Strain corresponding to 85% of confined strength on the descending branch of concrete stress-strain relationship
μ : Displacement ductility ratio
ρ_s : Volumetric ratio of confinement steel

Tables 8.10 and 8.11 provide a summary of the test data considered, for columns tested under concentric and lateral cyclic loadings, respectively. The tables also include strain and displacement ductility ratios. Figure 8.38 illustrates cross-sectional shapes and reinforcement arrangements considered in the experimental programs. The results indicate that deformability of confined high-strength concrete, and columns made with such concrete, are affected by *three* groups of parameters: (1) those related to applied loading, (2) those related to concrete confinement, and (3) those

Table 8.10   Columns tested under concentric compression

Column no.	Test label	Section type	Section dimensions	Ref.	f_{co}' (psi)	ρ_s (%)	f_{yt} (ksi)	s (in.)	$\varepsilon_{85}/\varepsilon_0$
1	III-3	1	D = 3 in.	[8.15]	5500	1.6	60	1.00	2.8
2	III-4				5500	3.2	60	0.50	7.2
3	V-2				9500	1.6	60	1.00	2.3
4	V-3				9500	3.2	60	0.50	4.5
5	PH5-5B	2	b = 5.8 in.	[8.16]	6200	2.1	198	2.00	4.4
6	PH5-9A		h = 5.8 in.		9000	2.1	198	2.00	2.5
7	PS5-5C				5900	2.1	198	2.00	5.1
8	PS5-7A				8100	2.1	198	2.00	3.5
9	PS5-9A				10,000	2.1	198	2.00	1.8
10	PD5-9A				11,000	4.2	198	2.00	7.6
11	NC164-1	1	D = 4 in.	[8.17]	7280	2.2	55	0.20	3.4
12	NC164-2				7280	2.2	55	0.20	3.2
13	NC164-3				7280	2.2	55	0.20	3.2
14	NC166-1				7281	7.5	60	0.25	12.6
15	NC166-2				7281	7.5	60	0.25	15.2
16	NC166-3				7281	7.5	60	0.25	11.9
17	NC168-1				9958	7.7	60	0.25	4.1
18	NC168-2				9958	7.7	60	0.25	4.3
19	NC168-3				9958	7.7	60	0.25	4.0
20	NC82-1				6191	2.3	60	0.36	4.9
21	NC82-2				6191	2.3	60	0.36	4.4
22	NC83-1				8447	3.4	60	0.25	3.3
23	NC83-2				8447	3.4	60	0.25	2.7
24	NC242-1	1	D = 5 in.	[8.17]	6108	2.2	60	0.65	3.3
25	NC242-2				6108	2.2	60	0.65	3.0
26	NC243-1				8255	3.1	60	0.65	2.0
27	NC243-2				8255	3.1	60	0.65	1.8
28	LC167-1	1	D = 4 in.	[8.17]	7990	2.7	55	0.16	2.1
29	LC167-2				7990	2.7	55	0.16	2.0
30	LC167				7990	2.7	55	0.16	1.0

31	ND95-3		D = 6 in.	[8.18]	12,660	3.2	90	1.00	5.2
32	ND115-3				14,445	3.2	90	1.00	3.5
33	LWA75-3	1			10,210	3.2	90	1.00	2.9
34	ND95-10-1-3	3	b = 6 in.	[8.18]	13,230	3.2	90	1.00	3.0
35	LWA75-10-1-3		h = 6 in.		11,880	3.2	90	1.00	1.4
36	3	1	D = 3 in.	[8.19]	8200	2.4	85	1.25	5.6
37	9				8200	2.4	85	1.25	3.6

Notes: Section type and reinforcement arrangement are shown in Fig. 8.38. Columns No. 28, 29, 30, 33, and 35 have lightweight concrete,
Column No. 37 was subjected to $5\sqrt{f'_c}$ psi shear stress.
1 in. = 25.4 mm; 1000 psi = 6.895 MPa

Table 8.11 Columns tested under lateral load reversals

Column no.	Test label	Section type	Section dimensions	Ref.	f_c' (psi)	ρ_s (%)	f_{yt} (ksi)	s (in.)	$P/A_g F_c'$ (%)	M/Vh	μ
38	600-02 77-1.2-1	4	b = 8 in. h = 6 in.	[8.20]	8500	3.00	50	20	2.75	2.5	10.8
39	600-02 79-1.2-10	4			8660	3.00	52	20	2.32	2.5	7.2
40	AL-2	5	b = 8 in. h = 8 in.	[8.21]	12,430	4.37	48	1.34	63	2.5	2.0
41	AH-2	5			12,430	4.37	115	1.34	63	2.5	7.3
42	BL-2	5			16,800	4.37	48	1.34	42	2.5	3.3
43	BH-2	5			16,800	4.37	115	1.34	42	2.5	8.0
44	1	5	b = 10 in. h = 10 in.	[8.22]	9670	2.00	121	1.77	31	2.0	6.7
45	2	5			12,260	2.57	121	1.38	28	2.0	6.3
46	3	5			5020	1.64	121	2.17	60	2.0	5.0
47	4	5			9670	2.60	46	1.97	57	2.0	2.5
48	5	5			9670	2.25	121	1.57	57	2.0	3.5
49	6	6			9670	2.08	198	1.77	57	2.0	5.0
50	7	5			12,260	2.57	121	1.38	51	2.0	5.0
51	N-1	7	b = 10 in. h = 10 in.	[8.23]	13,200	1.41	193	1.57	35	2.0	5.0
52	N-3	7			13,200	1.41	193	1.57	52	2.0	2.6
53	B1	5	b = 10 in. h = 10 in.	[8.24]	14,500	1.28	112	2.36	35	2.0	2.0
54	B2	5			14,500	1.92	112	1.57	35	2.0	3.3
55	B3	5			14,500	1.55	50	2.36	35	2.0	1.7
56	B4	5			14,500	1.28	163	2.36	35	2.0	3.3
57	B5	8			14,500	1.28	112	1.18	35	2.0	2.0
58	B6	8			14,500	1.26	124	2.36	35	2.0	2.0

Note: Section type and reinforcement arrangement are shown in Fig. 8.38.
1 in. = 25.4 mm, 1000 psi = 6.895 MPa

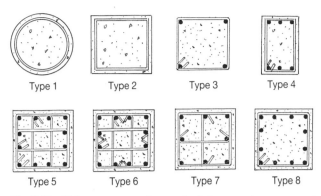

Type 1   Type 2   Type 3   Type 4

Type 5   Type 6   Type 7   Type 8

Fig. 8.38 Column cross-sections and reinforcement arrangements considered in tests

related to concrete type. These groups of parameters are discussed in the following sections.

Parameters related to applied loading

Effect of axial load

Axial compression reduces column deformability. Table 8.12 and Fig. 8.39 provide comparisons of columns tested under different levels of constant axial compression and incremently increasing lateral load reversals. The results indicate that column deformability, as indicated by displacement ductility ratio, decreases with increasing axial compression.

The effect of axial tension on column deformability was investigated by Watanabe *et al.*[25] A specimen subjected to a constant level of axial tension showed reduced capacity but increased deformability, under lateral load reversals.

Table 8.12 Effect of axial compression on column deformability

Column no.	f_c' (psi)	ρ_s (%)	f_{yt} (ksi)	s/h	s_1/h	$v/\sqrt{f_c'}$ (psi)	$P/(A_g f_c')$ (%)	μ
45	12,260	2.57	121	0.14	0.25	8.4	28	6.3
50	12,260	2.57	121	0.14	0.25	8.4	51	5.0
44	9670	2.00	121	0.18	0.25	9.0	31	6.7
48	9670	2.25	121	0.16	0.25	9.6	57	3.5
40	12,430	4.37	48	0.17	0.26	8.9	63	2.0
42	16,800	4.37	48	0.17	0.26	8.9	42	3.3
41	12,430	4.37	115	0.17	0.26	9.0	63	7.3
43	16,800	4.37	115	0.17	0.26	8.9	42	8.0
51	13,200	1.41	193	0.16	0.38	9.4	35	5.0
52	13,200	1.41	193	0.16	0.38	10.0	52	2.6

1 in. = 25.4 mm; 1000 psi = 6.895 MPa

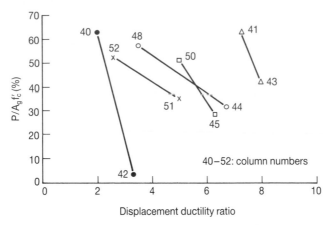

Fig. 8.39   Effect of axial compression on column deformability

Effect of shear stress

Data on the effect of shear on deformability of high-strength concrete is scarce. Limited tests reported by Abdel-Fattah and Ahmad,[19] conducted on concrete cylinders, indicate that the shear effect on high-strength concrete is less pronounced than that on normal-strength concrete. Table 8.13 shows that the strain ductility ratio decreases with imposed shear.

Table 8.13   Effect of shear on concrete deformability

Column no.	f_{co}' (psi)	ρ_s (%)	f_{yt} (ksi)	s/h	s_1/h	$v/\sqrt{f_c'}$ (psi)	$\varepsilon_{85}/\varepsilon_0$
36	8200	2.4	85	0.42	N/A	0	5.6
37	8200	2.4	85	0.42	N/A	5.1	3.6

1 in. = 25.4 mm; 1000 psi = 6.895 MPa

Effect of loading history

Two identical columns were tested by Chung *et al.*[20] under successively increasing lateral deformation cycles to investigate the effect of loading history. One of the columns was subjected to one cycle at each deformation level, while the other was subjected to ten cycles at each level. Table 8.14 indicates that the deformability of high-strength concrete columns decreases with increasing numbers of inelastic deformation cycles.

Table 8.14   Effect of load history on column deformability

Column no.	f_c' (psi)	ρ_s (%)	f_{yt} (ksi)	s/h	s_1/h	$v/\sqrt{f_c'}$ (psi)	$P/(A_g f_c')$ (%)	No. of cycles	μ
38	8500	3.00	50	0.35	0.66	5.9	20	1	10.8
39	8600	3.00	50	0.29	0.66	5.8	20	10	7.2

1 in. = 25.4 mm; 1000 psi = 6.895 MPa

Effect of rate of loading

Experimental data on high-strength concrete columns subjected to high strain rates is limited. Small-scale columns tested by Bjerkeli *et al.*[18] under strain rates of 3.33×10^{-5} per sec and 0.33×10^{-6} per sec indicate no significant effect of loading rate on the deformability of confined high-strength concrete.

Parameters related to concrete confinement

Volumetric ratio of confinement reinforcement

Volumetric ratio of confinement steel is one of the main parameters that affects concrete confinement. An increase in the volumetric ratio directly translated into a corresponding increase in confinement pressure, and resulting improvements in the deformability of concrete. This is shown in Table 8.15 and Fig. 8.40. While 7280 psi (50 MPa) concrete with 2.2% confinement steel shows a strain ductility ratio of approximately 3, the same concrete confined with 7.5% of a similar grade steel shows an increased ductility ratio of about 12. A comparison made with the deformability of normal-strength concrete[26] indicates that higher volumetric ratio of confinement steel is required for high-strength concrete to attain deformabilities usually expected of normal-strength concrete.

Strength of confinement reinforcement

Confinement pressure is a passive pressure that is activated by lateral expansion of concrete under axial compression. This pressure is dependent on the ability of concrete to expand laterally prior to failure. It is also limited by the strength of the lateral steel. If the transverse strain of high-strength concrete is not high enough to strain the confinement steel to its capacity, then the full capacity of steel is not utilized. It is speculated that high-strength concrete, being a brittle material, may not develop

Table 8.15   Effect of volumetric ratio of confinement steel on concrete deformability

Column no.	f_c' (psi)	ρ_s (%)	f_{yt} (ksi)	s/h	s_1/h	$\varepsilon_{85}/\varepsilon_0$
9	10000	2.1	198	0.34	N/A	1.8
10	10000	4.2	198	0.34	N/A	7.6
11	7280	2.2	55	0.05	N/A	3.4
14	7281	7.5	60	0.06	N/A	12.6
12	7280	2.2	55	0.05	N/A	3.2
15	7281	7.5	60	0.06	N/A	15.2
13	7280	2.2	55	0.05	N/A	3.2
16	7281	7.5	60	0.06	N/A	11.9
23	8447	3.4	60	0.06	N/A	2.7
16	7281	7.5	60	0.06	N/A	11.9

1 in. = 25.4 mm; 1000 psi = 6.895 MPa

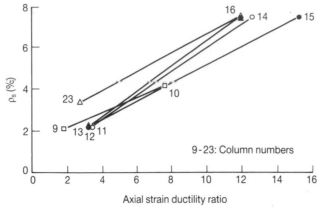

Fig. 8.40   Effect of volumetric ratio of confinement steel on concrete deformability

transverse strains high enough to strain the steel to its yield level. Research by Ahmad and Shah[15] has shown that spiral reinforcement is less effective for high-strength concrete columns, and that the stress in the spiral steel at maximum column load is often significantly less than yield stress, as assumed commonly. These results are consistent with Martinez *et al.*[17] who did not observe any yielding in the confinement reinforcement of high-strength concrete. Therefore, the same researchers recommended an upper limit for strength of confinement reinforcement. The use of high-strength steel for confinement of high-strength concrete is often questioned by researchers.

Experimental research by Muguruma *et al.*,[21] however, indicates that very high ductilities can be achieved in high-strength concrete when it is confined by high-strength steel. Table 8.16 and Fig. 8.41 show the improvements achieved in column deformability with the use of high-strength steel as confinement reinforcement. Columns with 12,500 to 16,800 psi (86 to 116 MPa) concretes, confined with 4.4% volumetric ratio of steel, show approximately 250% increase in displacement ductility ratios when the steel yield strength is increased by 140% from 48 to 115 ksi (328 to

Table 8.16   Effect on transverse steel strength on column deformability

Column no.	f_c' (psi)	ρ_s (%)	s/h	s_1/h	$v/\sqrt{f_c'}$ (psi)	$P/(A_g f_c')$ (%)	f_{yt} (ksi)	μ
40	12430	4.37	0.17	0.26	8.9	63	48	2.0
41	12430	4.37	0.17	0.26	9.0	63	115	7.3
42	16800	4.37	0.17	0.26	8.9	42	48	3.3
43	16800	4.37	0.17	0.26	8.9	42	115	8.0
47	9670	2.60	0.20	0.25	9.9	57	46	2.5
48	9670	2.25	0.16	0.25	9.6	57	121	3.5
49	9670	2.08	0.18	0.25	9.5	57	198	5.0
55	14500	1.55	0.24	0.25	9.5	35	50	1.7
53	14500	1.28	0.24	0.25	8.7	35	112	2.0
56	14500	1.28	0.24	0.25	8.7	35	163	3.3

1 in. = 25.4 mm; 1000 psi = 6.895 MPa

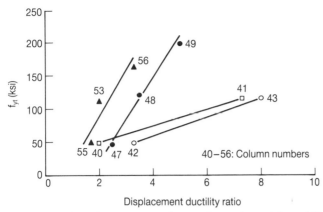

Fig. 8.41 Effect of transverse steel strength on column deformability

792 MPa). These columns develop displacement ductility ratios in excess of 7 when confined with high-strength steel. The results clearly show the beneficial effect of high-strength steel in confining high-strength concrete.

Further examination of the test data indicates that the improvement associated with the use of high-strength steel as confinement reinforcement is approximately the same as that obtained by increasing the volumetric ratio of confinement steel (ρ_s). This is to be expected if the confinement steel develops its strength prior to concrete strength decay, since in this case the tensile force in the confinement steel is directly related to the amount as well as the yield strength of steel (f_{yt}). It is also observed that higher confinement pressure (i.e. higher $\rho_s f_{yt}$) is required for higher strength concretes to maintain the same level of deformability. Therefore the test data are re-evaluated based on the non-dimensional ratio $\rho_s f_{yt}/f_c'$, as shown in Table 8.17. Deformability of columns with normal-strength

Table 8.17 Effect of $\rho_s f_{yt}/f_c'$ on column deformability

Column no.	f_c' (psi)	ρ_s (%)	f_{yt} (ksi)	s/h	s_1/h	$v/\sqrt{f_c'}$ (psi)	$P/(A_g f_c')$ (%)	$\rho_s f_{yt}/f_c'$	μ
C9	4800	1.49	57	0.25	0.66	2.1	26	0.18	1.9
58	14500	1.26	124	0.24	0.73	9.4	35	0.11	2.0
C14	3300	0.95	34	0.17	0.72	3.8	0	0.10	2.0
57	14500	1.28	112	0.12	0.75	9.0	35	0.10	2.0
C45	5100	1.60	41	0.17	0.26	8.9	20	0.13	3.8
56	14500	1.28	163	0.24	0.25	8.7	35	0.14	3.3
C44	5260	1.60	41	0.17	0.26	8.4	32	0.13	2.8
54	14500	1.92	112	0.16	0.25	9.1	35	0.15	3.3
44	9670	2.00	121	0.18	0.25	9.0	31	0.25	6.7
45	12260	2.57	121	0.14	0.25	8.4	28	0.25	6.3

Note: Columns C9, C14, C44 and C45 are normal strength concrete columns, previously investigated by Saatcioglu[25]
1 in. = 25.4 mm; 1000 psi = 6.895 MPa

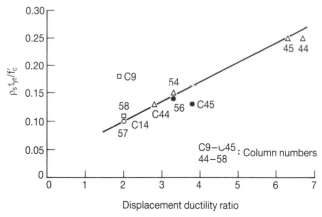

Fig. 8.42   Effect of $\rho_s f_{yt}/f_c'$ ratio on column deformability

concrete, previously evaluated by Saatcioglu,[26] is also included in Table 8.17 to extend the lower end of the strength range, and to relate deformability of high-strength concrete to that of normal-strength concrete for which extensive research data and knowledge already exist. Table 8.17 and Fig. 8.42 show the variation of column displacement ductility ratio with the $\rho_s f_{yt} f_c'$ ratio. Examination of the test data also indicates that higher strength concretes require proprtionately higher values of the product $\rho_s f_{yt}$. Higher values of this product can be achieved either by the use of higher volumetric ratio and/or higher strength of confinement steel. Figure 8.43 shows column pairs with constant $\rho_s f_{yt}/f_c'$ ratios, where the columns in a pair have distinctly different concrete strengths, and yet develop similar displacement ductility ratios. This implies that within the range of parameters considered in the test programs evaluated here, deformabilities usually expected of normal-strength concrete columns can be attained in high-strength concrete columns by maintaining the $\rho_s f_{yt}/f_c'$ ratio approximately constant.

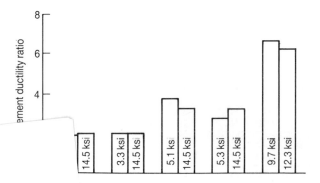

…mparisons of displacement ductility ratios of columns with different concrete
…constant $\rho_s f_{yt}/fc'$ ratios

Table 8.18 Effect of tie spacing on concrete deformability

Column no.	f_{co}' (psi)	ρ_s (%)	f_{yt} (ksi)	s_1/h	s/h	$\varepsilon_{85}/\varepsilon_0$
26	8255	3.1	60	N/A	0.17	2.0
22	8447	3.4	60	N/A	0.06	3.3
27	8255	3.1	60	N/A	0.17	1.8
23	8447	3.4	60	N/A	0.06	2.7
24	6108	2.2	60	N/A	0.13	3.3
20	6191	2.3	60	N/A	0.09	4.9
25	6108	2.2	60	N/A	0.13	3.0
21	6191	2.3	60	N/A	0.09	4.4

1 in. = 25.4 mm; 1000 psi = 6.895 MPa

Spacing of the transverse reinforcement

The spacing of transverse reinforcement is one of the important parameters that affects the distribution of confinement pressure. Closer spacing
of transverse reinforcement increases uniformly of lateral pressure along
the column height and improves the effectiveness of confinement reinforcement. However, this improvement may not result in significantly
different column deformabilities unless the other confinement parameters
are favorable. For example, the volumetric ratio and/or yield strength steel
must be sufficiently high for the spacing to make a difference in column
deformability. Table 8.18. and Fig. 8.44 show the effect of tie spacing on
the strain ductility ratio of high-strength concrete.

Arrangement of transverse reinforcement

The arrangement of transverse reinforcement is another parameter that
affects the distribution of confinement pressure. The tension force that
develops in a transverse leg of a hoop or a crosstie is distributed on the side

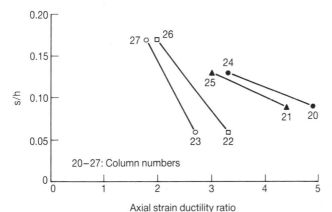

Fig. 8.44 Effect of tie spacing on concrete deformability

of the core concrete. If the transverse legs are closely spaced in the horizontal plane, the resulting confinement pressure becomes more uniform in this plane, improving effectiveness of the confinement reinforcement. This has been clearly shown for normal-strength concrete columns.[27] Tests on high-strength concrete columns with different reinforcement arrangements are scarce in the literature. Although the effect of transverse reinforcement arrangement is expected to be the same for high-strength concrete columns, the available data are not sufficient to illustrate this point clearly.

Section geometry and size

Circular spirals are more effective in confining concrete than rectilinear ties. The superiority of circular spirals comes from their geometric shape which produces uniform pressure around the circumference of the core. Rectilinear ties produce non-uniform pressure which peaks at loacations of transverse legs of tie steel. It was shown earlier for normal-strength concrete that full effectiveness of circular geometry can be achieved by rectilinear reinforcement if the tie arrangement includes close spacing of transverse legs in both longitudinal and transverse directions.[27] Table 8.19 provides a comparison of strain ductility ratios for high-strength concretes, confined by either circular spirals or square perimeter hoops. The results indicate that circular spirals produce higher deformabilities than square perimeter hoops.

Experimental data on the effect of size on the deformability of high-strength concrete are not sufficient. Most high-strength concrete columns tested in the past were small-scale specimens. However, tests conducted by Martinez[17] indicate that small specimens produce lower strength than specimens of larger sizes.

Parameters related to concrete type

Concrete strength

Strength and deformability of concrete are known to be inversely prop-

Table 8.19   Effect of section geometry

Column no.	f_{co}' (psi)	ρ_s (%)	f_{yt} (ksi)	s/h	s_1/h	Cross section	$\varepsilon_{85}/\varepsilon_0$
31	12660	3.2	90	0.17	N/A	Circular	5.2
34	13230	3.2	90	0.17	0.89	Square	3.0
33	10210	3.2	90	0.17	N/A	Circular	2.9
35	11880	3.2	90	0.17	0.89	Square	1.4

Note: Columns 33 and 35 are lightweight concrete columns
1 in. = 25.4 mm; 1000 psi = 6.895 MPa

Table 8.20 Effect of strength on concrete deformability

Column no.	f_{c0}' (psi)	ρ_s (%)	f_{yt} (ksi)	s/h	s_1/h	$\varepsilon_{85}/\varepsilon_0$
31	12660	3.2	90	0.17	N/A	5.2
32	14445	3.2	90	0.17	N/A	3.5
1	5500	1.6	60	0.33	N/A	2.8
3	9500	1.6	60	0.33	N/A	2.3
2	5500	3.2	60	0.17	N/A	7.2
4	9500	3.2	60	0.17	N/A	4.5
7	5900	2.1	198	0.34	N/A	5.1
8	8100	2.1	198	0.34	N/A	3.1
9	10000	2.1	198	0.34	N/A	1.8
5	6200	2.1	198	0.34	N/A	4.4
6	9000	2.1	198	0.34	N/A	2.5
14	7281	7.5	60	0.06	N/A	12.6
17	9958	7.7	60	0.06	N/A	4.1
15	7281	7.5	60	0.06	N/A	15.2
18	9958	7.7	60	0.06	N/A	4.3
16	7281	7.5	60	0.06	N/A	11.9
19	9958	7.7	60	0.06	N/A	4.0

1 in. = 25.4 mm; 1000 psi = 6.895 MPa

ortional. It is therefore reasonable to expect the deformability of unconfined high-strength concrete to decrease with increasing strength.
Table 8.20 includes test data obtained from concentrically tested confined
high-strength concrete columns. The data are arranged such that the strain
ductility of confined concrete with different unconfined concrete strengths
can be compared. The results show the expected trend, i.e. concrete
ductility decreases with increasing strength. This is also shown in Fig. 8.45.
A comparison between 5700 and 10,800 psi (40 MPa and 74 MPa) concretes indicates a 40% reduction in strain ductility ratio in the higher

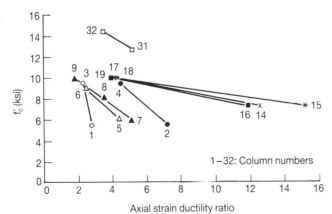

Fig. 8.45 Effect of strength on concrete deformability

Table 8.21   Effect of concrete strength on column deformability

Column no.	f_c' (psi)	ρ_s (%)	f_{yt} (ksi)	s/h	s_1/h	$v/\sqrt{f_c'}$ (psi)	$P/(A_g f_c')$ (%)	μ
46	5020	1.64	121	0.22	0.25	8.2	60	5.0
48	9670	2.25	121	0.16	0.25	9.6	57	3.5

1 in. = 25.4 mm; 1000 psi = 6.895 MPa

strength concrete. Similar observations can be made with regard to high-strength concrete columns tested under reversed cyclic loading. Table 8.21 and Fig. 8.46 illustrate the effect of unconfined concrete strength on the displacement ductility ratio. In all cases, ductility decreases with increasing concrete strength. This shows that the confinement steel requirements for high-strength columns must be more stringent than those for normal-strength concrete columns.

Unit weight of aggregate

The deformability of high-strength concrete decreases with the unit weight of the aggregate. Lightweight aggregate concrete shows a significantly lower ductility than normal weight concrete of the same strength. Martinez[17] reported that the confinement efficiency of circular spirals is approximately 60%, less for lightweight concrete than for normal weight concrete. Table 8.22 shows comparisons of strain ductility ratios for normal weight and lightweight confined high-strength concretes, indicating a significant drop in deformability with the use of lightweight aggregate in concrete.

Conclusions

The following conclusions can be made with respect to the deformability of high-strength concrete columns:

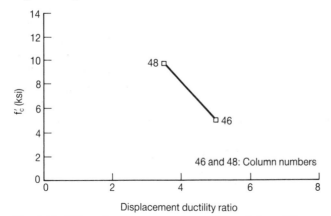

Fig. 8.46   Effect of concrete strength on column deformability

Table 8.22 Effect of unit weight of aggregate on concrete deformability

Column no.	f_{co}' (psi)	ρ_s (%)	f_{yt} (ksi)	s/h	s_1/h	Concrete type	$\varepsilon_{85}/\varepsilon_0$
31	12660	3.2	90	0.17	N/A	NWC	5.2
33	10210	3.2	90	0.17	N/A	LWC	2.9
34	13230	3.2	90	0.17	0.89	NWC	3.0
35	11880	3.2	90	0.17	0.89	LWC	1.4
11	7280	2.2	55	0.05	N/A	NWC	3.4
28	7990	2.7	55	0.04	N/A	LWC	2.1
12	7280	2.2	55	0.05	N/A	NWC	3.2
29	7990	2.7	55	0.04	N/A	LWC	2.0
13	7280	2.2	55	0.05	N/A	NWC	3.2
30	7990	2.7	55	0.04	N/A	LWC	1.9

1 in. = 25.4 mm; 1000 psi = 6.895 MPa

1   High-strength concrete columns are brittle members unless confined by proper reinforcement. For the same reinforcement arrangement, concrete strength and column deformability are inversely proportional.

2   High-strength concrete columns can be confined to develop deformabilities usually expected from seismic resistant elements. The parameters of confinement are essentially the same as those for normal strength concrete. However, the confinement requirements for high-strength concrete columns are more stringent than those for normal strength concrete columns.

3   The major parameters of confinement for high-strength concrete are the volumetric ratio, yield strength and spacing of confinement reinforcement. Experimental evidence points out that, with volumetric ratios and/or steel yield strengths higher than those used for normal strength concrete, it is possible to attain deformabilities usually achieved in normal strength concrete.

4   The use of high strength steel as confinement reinforcement appears to be an effective way of increasing confinement pressure to the level needed for improved deformability of high-strength concrete.

5   There is a strong indication that the extra confinement pressure required in high-strength concrete, relative to that required for normal strength concrete, is proportional to the difference in concrete strengths. Within the range of column tests investigated, columns with distinctly different concrete strengths show approximately the same displacement ductility ratio, if the ratio $\rho_s f_{yt}/f_c'$ is kept constant.

6   Axial compression reduces column deformability under reversed cyclic loading.

7   Lightweight high-strength concrete columns show significantly less deformability than those observed in corresponding normal weight concrete columns.

8   More research is needed to establish ductility and confinement requirements for seismic resistant high-strength concrete columns.

8.4   Deformability of high-strength concrete beam-column joints

ACI-ASCE 352 (1985)[28] recommends that for joints which are part of the
primary system for resisting seismic lateral loads, the sum of the nominal
moment-strengths of the column sections above and below the joint
(ΣM_c), calculated using the axial load which gives the minimum column-
moment strength, should not be less than 1.4 times the sum of the nominal
moment strengths of the beam sections at the joint (ΣM_g). It may be noted
that in ACI 318-89,[3] the minimum required flexural strength ratio $\Sigma M_c/$
ΣM_g is 1.2, instead of 1.4.

ACI-ASCE 351 (1985)[28] further recommends that where rectangular
hoop and crosstie reinforcement is used to confine the concrete within a
joint, the center-to-center spacing between layers of transverse reinforcement should not exceed the least of one-quarter of the minimum column
dimension (h_c), six times the diameter of the longitudinal bars to be
restrained, or 6 in. (150 mm). It may again be noted that in ACI 318-89,[3]
the maximum spacing of transverse reinforcement is restricted to the
smaller of $h_c/4$ or 4 in. (100 mm).

The 352 recommendations[28] specify maximum allowable joint shear
stresses in the form of $\gamma\sqrt{f_c'}$, where the joint shear stress factor γ is a
function of the joint type (i.e., interior, exterior or corner). The primary
function of the joint transverse reinforcement is to provide confinement for
the concrete in the joint region. Thus, if the joint shear stresses are in
excess of the maximum allowable limits, the problem cannot be remedied
by providing additional transverse reinforcement.

The above recommendations are based on tests of normal-strength
concrete connections with compressive strength, f_c', ranging between 3500
and 5500 psi (24 and 38 MPa). Several experimental investigations have
recently been conducted to check the validity of the recommendations for
high-strength concrete beam-column joints, and are discussed below.

Ehsani *et al.*[29] conducted a study to investigate the effect of various shear
stress levels on beam-column connections constructed with high-strength
concrete, and to compare the results with those of subassemblies constructed with ordinary-strength concrete.

Four reinforced concrete beam-column subassemblies were constructed
with high-strength concrete. The results were compared with those from a
similar specimen constructed with ordinary-strength concrete and reported
by Ehsani and Wight[30] (Specimen 5 of Ehsani *et al.*[29] is identical to
Specimen 4b of Ehsani and Wight[30]). The configuration of the specimens
qualifies them as corner connections.[28] As shown in Fig. 8.47, a corner
connection is one in which the beam frames into only one side of the

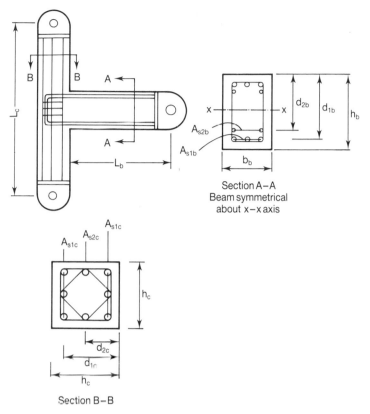

Fig. 8.47   Beam-column subassemblage tested by Ehsani *et al.*[29]

column and no spandrel beams are present. Reinforcing details for the specimens are presented in Table 8.23. Note that the beam and column of each specimen were properly reinforced to eliminate the possibility of shear failure of these elements. All speciments were cast flat rather than vertically as in actual construction.

The primary variable for the high-strength concrete specimens was the joint shear stress. According to ACI 352,[28] the maximum allowable joint shear stress for a corner connection constructed with ordinary-strength concrete is $12\sqrt{f_c'}$ psi ($1.0\sqrt{f_c'}$ MPa). As shown in Table 8.24, the joint shear stress for the high-strength specimens varied between $7.52\sqrt{f_c'}$ and $12.84\sqrt{f_c'}$ psi ($0.63\sqrt{f_c'}$ and $1.07\sqrt{f_c'}$ MPa). The joint shear stress in Specimen 5 was $12.55\sqrt{f_c'}$ psi ($1.04\sqrt{f_c'}$ MPa).

For the high-strength concrete specimens, the column-to-beam flexural strength ratio was held fairly constant between 1.7 and 1.9. The concrete compressive strength for these specimens was either 9380 or 9760 psi (64.7 or 67.3 MPa). This minor difference in concrete compressive strength was assumed to have no significant influence on the test results.

The joints of all specimens were reinforced with three layers of No. 4,

Table 8.23   Physical dimensions and properties of specimens[29]

	Specimen number				
	1	2	3	4	5
L_c, in.	136.0	136.0	136.0	136.0	84.0
L_b, in.	62.0	62.0	62.0	62.0	60.0
h_c, in.	13.4	13.4	11.8	11.8	11.8
d_{1c}, in.	11.4	11.4	9.8	9.8	9.6
d_{2c}, in.	6.7	6.7	5.9	5.9	5.9
A_{s1c}	2 No. 7 + 1 No. 6	2 No. 7 + 1 No. 6	2 No. 7 + 1 no. 6	2 No. 8 + 1 No. 7	4 No. 6
A_{s2c}	2 No. 6	2 No. 6	2 No. 6	2 No. 7	2 No. 6
h_b, in.	18.9	18.9	17.3	17.3	17.3
b_b, in.	11.8	11.8	10.2	10.2	10.2
d_{1b}, in.	16.9	16.9	15.4	15.4	15.4
d_{2b}, in.	15.0	15.0	13.4	13.4	13.4
A_{s1b}	2 No. 6 + 1 No. 5	3 No. 6	3 No. 6	3 No. 7	3 No. 7
A_{s2b}	2 No. 5	2 No. 6	2 No. 5	2 No. 5	3 No. 6

1 in. = 25.4 mm

Table 8.24   Design parameters of test specimens[29]

	Specimen number				
	1	2	3	4	5
Column axial load P, kips	30	76	86	73	50
Column balanced axial load, kips	498	513	362	319	267
$M_{n,\text{col}}$ at P, in.-kips	1638	1865	1582	1917	1478
$M_{n,\text{beam}}$ in.-kips	1729	2041	1663	2290	2101
M_R	1.89	1.83	1.90	1.67	1.41
Joint shear force V_j, kips	123	150	133	165	131
Joint shear stress v_j, psi	729	888	1027	1269	1008
f_c', psi	9380	9760	9380	9760	6470
Joint shear stress/$\sqrt{f_c'}$	7.52	8.99	10.61	12.84	12.55
h_b/column bar diameter	21.6	21.6	19.8	17.3	23.0
Required l_{dh}, in.	7.7	7.6	7.7	8.9	7.3
Provided l_{dh}, in.	10.8	10.8	9.2	9.3	8.6

1 in. = 25.4 mm; 1000 psi = 6.895 MPa; $1.0\sqrt{f_c'}$ psi = $0.083\sqrt{f_c'}$ MPa

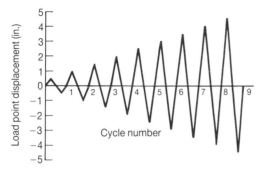

Fig. 8.48 Deformation sequence for reversed cyclic loading tests by Ehsani *et al.*[29]

Grade 60 hoops. As shown in Fig. 8.47, each layer of hoop consisted of a large square tie, enclosing all column longitudinal corner reinforcement, and a smaller square tie supporting the four intermediate longitudinal bars of the column. The area and the spacing of the hoops satisfied all recommended values.[28]

The selection of Specimen 5 for comparison with the high-strength concrete specimens was based on the fact that this specimen very closely satisfies all requirements of ACI 352.[28] The flexural strength ratio for Specimen 5 is 1.4. The joint shear stress for this specimen exceeds the maximum allowable value by only 4%. Other design parameters for this specimen, such as the spacing of the hoops and the development of longitudinal bars within the joint, satisfy or are very close to those recommended in ACI 352.[28] As a result, the performance of Specimen 5 can be interpreted as minimally acceptable by the joint ACI-ASCE Committee 352 criteria. This specimen can thus serve as a benchmark.

An axial load, less than 40% of the balanced axial load, was applied to each column and kept constant throughout the test. Displacement-controlled loads were applied to the free end of the beams, as shown in Fig. 8.48. Specimen 5 was loaded to 1.5 times the observed beam yield displacement during the first cycle of loading. The maximum displacement in each subsequent cycle of loading was increased by one-half of the observed yield displacement.

Plots of the applied shear versus the displacement at the free end of the beam for all specimens were presented in Ehsani *et al.*[29] The hysteresis diagrams indicated that Specimens 1, 2, 3 and 5, with large areas enclosed within each cycle of load-displacement curves, had the capacity to dissipate large amounts of energy. This was not true of Specimen 4, which had relatively smaller areas for each cycle.

Comparison of the loss of strength for the specimens is shown with the aid of Fig. 8.49. This figure shows the plot of percentage yield strength versus the displacement ductility. The yield load and displacement for each specimen were determined from strain gage data based on the yielding of the beam longitudinal reinforcement at the face of the column. The

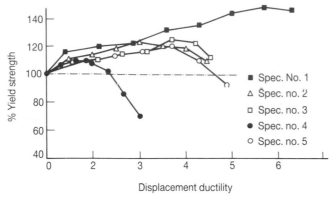

Fig. 8.49   Maximum load carried by each specimen at various displacement levels[29]

average of the maximum positive and negative loads during each cycle was divided by the yield strength to obtain percentage yield strength. The displacement ductility, plotted as the abscissa in Fig. 8.49, is defined as the displacement at the end of each cycle divided by the yield displacement of the specimen. It is clear that Specimens 1, 2, 3 and 5 performed very well by maintaining their yield strength for displacement ductilities of 4 or higher. For Specimen 4, however, the load-carrying capacity was sharply reduced to values below its yield strength after a displacement ductility of 2.5.

Within the limitations of the study by Ehsani *et al.*[29] the following conclusions could be drawn:

1   Properly detailed connections constructed with high-strength concrete exhibit ductile hysteretic response similar to those for ordinary-strength concrete connections.
2   The maximum permissible joint shear stress factor γ should probably be a function of the concrete compressive strength.
3   The mode of failure is determined by a combination of flexural strength ratio and joint shear stresses rather than by the flexural strength ratio alone.
4   High joint shear stresses significantly reduce the energy-absorption capability of subassemblies even in the presence of high flexural strength ratios.

Ehrani and Alameddine[31] investigated the effects of key variables on the behavior of high-strength reinforced concrete beam-column corner connections subjected to inelastic cyclic loading.

Twelve specimens were constructed and tested (Fig. 8.50 and Table 8.26). The configuration of the specimens (Fig. 8.50) qualified them as corner connections.[28] All specimens were cast flat rather than vertically as in actual construction. The variables studied were concrete compressive strength (8.1, 10.7, and 13.6 ksi or 55.8, 73.8, and 93.8 MPa), joint-shear stress (1100 and 1400 psi or 7.6 and 9.7 MPa), and the degree of joint

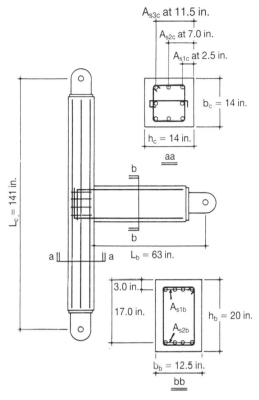

Fig. 8.50 Beam-column subassemblage tested by Ehsani and Alameddine[31]

confinement provided in the form of stirrups and crossties. For all specimens, the flexural strength ratio M_R, defined as the sum of the flexural capacities of the columns to that of the beam, was equal to 1.4, which is the minimum value allowed by the 253 recommendations.[28]

The 352 recommendations[28] limit the joint-shear stress in corner connections to $12\sqrt{f_c'}$ psi ($1.0\sqrt{f_c'}$ MPa). The low joint-shear stress investigated in this study is considered equivalent to $12\sqrt{8000} \approx 1100$ psi (7.6 MPa), while the high joint-shear stress is equivalent to $12\sqrt{8000} \approx 1400$ psi (9.7 MPa).

Table 8.25 Required and provided joint confinement[31]

f_c', psi	s, in., Eq. (21-3) Ref. [8.3]	s, in., Eq. (21-4) Ref. [8.3]	No. of sets of hoops, Eq. (21-3) Ref. [8.3]	No. of sets of hoops, Eq. (21-4) Ref. [8.3]	No. of sets of hoops used
8000	2.2	4.5	6	2	4 & 6
11,000	1.6	3.3	8	4	4 & 6
14,000	1.3	2.6	11	5	4 & 6

1 in. = 2.54 mm; 1000 psi = 6.895 MPa

Table 8.26 Physical dimensions and properties of specimens[31]

Specimen	LLs	LHs	HLs	HHs
L_c, in.	141	141	141	141
L_b, in.	63	63	63	63
h_c, in.	14	14	14	14
d_{1c}, in.	2.5	2.5	2.5	2.5
d_{2c}, in.	7.0	7.0	7.0	7.0
d_{3c}, in.	11.5	11.5	11.5	11.5
A_{s1c}	2 No. 8 & 1 No. 7	2 No. 8 & 1 No. 7	3 No. 8	3 No. 8
A_{s2c}	2 No. 7	2 No. 7	2 No. 8	2 No.8
A_{s3c}	2 No. 8 & 1 No. 7	2 No. 8 & 1 No. 7	3 No. 8	3 No. 8
h_b, in.	20	20	20	20
b_b, in.	12.5	12.5	12.5	12.5
d_{1b}, in.	3.00	3.00	3.00	3.00
d_{2b}, in.	17.00	17.00	17.00	17.00
A_{s1b}	4 No. 8	4 No. 8	4 No. 9	4 No. 9
A_{s2b}	4 No. 8	4 No. 8	4 No. 9	4 No. 9
No. of hoops	4	6	4	6
ρ_s	1.20	1.80	1.22	1.84
$h_b/d_{b,\text{col}}$	20	20	20	20
Required l_{dh}, in. for f_c' = 8000 psi	8.9	8.9	10.0	10.0
Provided l_{dh}, in.	10.5	10.5	10.5	10.5

1 in. = 25.4 mm; 1000 psi = 6.895 MPa

To ensure adequate confinement of the joint, the 352 recommendations[28] as well as ACI 318-89[3] requires that the total cross-sectional area of hoops and crossties A_{sh} be at least equal to that given by Eqs. (21-3) and (21-4) of ACI 318.[3]

The area of transverse reinforcement A_{sh} in each set included a Grade 60, No. 4, closed rectangular hoop with a 135-deg standard hook, in addition to a Grade 60, No. 4 crosstie, as shown in Fig. 8.50. The required and provided joint confinement for all specimens are given in Table 8.25. Satisfying these requirements would have resulted in joints congested with reinforcement and impractical to construct. The amount of joint transverse reinforcement was either low, corresponding to four layers of reinforce-

Table 8.27 Design parameters of test specimens[31]

Specimen	P_{col} psi	$P_{b,col}$ kips	$M_{n,col}$ in.-kips	$M_{pr,beam}$ in.-kips	$M_{n,beam}$ in.-kips	V_j, kips	v_j, psi
LL8	66	531	2119	3719	3027	210.7	1136
LH8	66	531	2119	3719	3027	210.7	1136
HL8	114	530	2546	4448	3637	268.4	1447
HH8	114	530	2546	4448	3637	268.4	1447
LL11	64	652	2183	3823	3118	209.8	1131
LH11	62	639	2157	3772	3081	210.2	1133
HL11	132	637	2692	4690	3845	266.3	1435
HH11	136	637	2710	4754	3872	266.3	1435
LL14	53	736	2178	3826	3112	209.8	1131
LH14	50	764	2178	3826	3112	209.8	1131
HL14	110	762	2689	4686	3842	266.7	1437
HH14	107	762	2681	4652	3830	267.0	1439

1 kip = 4.45 kN; 1 in.-kip = 0.133 kN-m

ment ($\rho_s \approx 1.2\%$), or high, corresponding to six layers of reinforcement ($\rho_s \approx 1.8\%$). The beam and the columns of the specimens were reinforced to eliminate any possibility of shear failure of these elements.

For all 12 specimens tested, the terminating beam longitudinal bars were hooked within the transverse reinforcement of the joint using 90 deg standard hooks. The provided development length from the critical secton defined by the 352 recommendations[28] was higher than that required by the recommendations.

An axial load, smaller than the balanced column load, was applied to the column (Table 8.27) and kept constant throughout the test. The magnitude of this load was determined from the column interaction diagram to satisfy a flexural strength ratio M_R of 1.4.

Reversed cyclic displacements were applied to the free end of the beam. Each test consisted of nine cycles of loading. The maximum displacement for the first cycle strarted as $\frac{1}{2}$ in. (13 mm) and was incremented by $\frac{1}{2}$ in. (13 mm) during each subsequent cycle. The final cycles of loading, corresponding to story drifts of 7%, are very unlikely to be experienced by any real structure.

With the flexural strength ratio M_R held constant at a minimum value of 1.4, the joint shear stress as well as the joint confinement level were both found to be key factors in achieving adequate strength and ductility of the joint. Specimens with low shear level and high joint confinement were able to develop the ultimate capacities in the beams. These same specimens had the least stiffness degradation and loss of load-carrying capacity at dis-

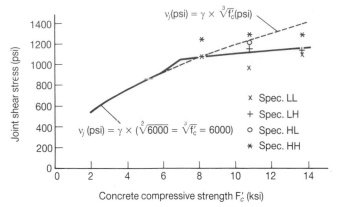

Fig. 8.51   Permissible shear stress in earthquake-resisting high-strength concrete beam-column joint[31]

placements beyond the yield displacement. Other specimens with high joint shear stresses and/or low joint confinement levels suffered greater strength loss and lower ductility.

Based on the test results, the maximum permissible joint shear stress was proposed to be modified as indicated in Fig. 8.51. In addition, modified joint confinement requirements for high-strength concrete were proposed.

Shin *et al.*[32] reported on the testing of one normal strength and ten high-strength concrete half-scale beam-column joint specimens. One of the high-strength specimens was tested under monotonic loading, all others were tested under reversed cyclic loading. According to the classification in the ACI-ASCE recommendations,[28] the configuration of the specimens qualified them as corner connections. The variables investigated were:

1   The concrete compressive strength (f_c' = 4380 and 11,380 psi or 30.2 and 78.5 MPa).
2   Number of confining hoops within the joint core (3, 2, 1 and 0; corresponding spacing = $h_c/4$, $h_c/3$, $h_c/2$ and infinity, respectively).
3   Type of loading (cyclic, monotonic).
4   Column-to-beam flexural strength ratio ($M_R = \Sigma M_c / \Sigma M_g$ = 1.4, 1.6, 1.8 and 2.0) and
5   Number of bent-up bars within the joint core (1 and 2).

Details of the specimens are given in Table 8.28 and Fig. 8.52. Each specimen was cast horizontally with one batch of concrete.

Figure 8.53 schematically shows the test setup. The column of each specimen was kept under a constant axial load equal to 40% of the balanced axial load. The vertical load on the beam was applied at 36 in. (980 mm) from the column face. All reversed cyclic loading tests were run under displacement control. The displacement sequence applied at the end of each beam is illustrated in Fig. 8.22. Δ_y is the vertical beam displacement at the point of loading, corresponding to yielding of the longitudinal beam

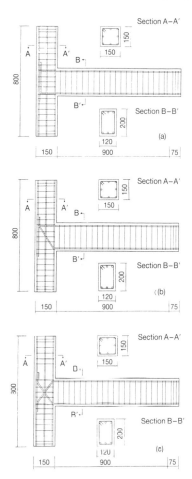

Fig. 8.52 Specimen reinforcing details [32] (2) all specimens except HJC3-R0-B1, HJCS-R0-B2, (b) HJC3-R0-B2, (c) HJCS-R0-B1

reinforcement at the column face. In the case of the specimen resisting monotonic loads, the test was conducted under load control until one-third of the expected maximum load was attained; a switch was made to displacement control thereafter.

In the normal strength concrete specimen NJC3-R0, many hairline flexural cracks formed uniformly along the beam at low displacement stages; failure occurred in the beam region between the column face and a section one-quarter of the beam depth away from the column face. In the corresponding high-strength concrete specimen HJC3-R0, wide cracks were concentrated at the beam-column joint face. Visually and otherwise, damage to the high-strength specimen was more severe than damage to the normal-strength specimen. Figure 8.54 graphically shows the effects of the other test variables on the failure mode.

A comparison between the hysteresis loops for NJC3-R0 ($f_c' =$

Table 8.28   Description of test specimens and selected test results[32]

Specimens	f'_c psi (MPa)	Loading type[a]	Number of hoops in core	Number of M_R ($\Sigma M_c/\Sigma M_g$)	Number of bent-up bars in core	Column Section in. (mm)	Column Reinforcement	Beams Section in. (mm)	Beams Reinforcement	P_{max} kips (kN)	Shear stress constant[c] Theoretical	Shear stress constant[c] Experimental
NJC3-R0	4380 (30.2)	C	3	1.4	0	6 × 6 (150 × 150)	8 No. 3 (10 mm dia.)	4⅘ × 8 (120 × 120)	4 No. 4 (13 mm dia.) 2 No. 3 (10 mm dia.)	8.82 (39.23)	11.83	10.72
HJM3-R0	11,380 (78.5)	M	3	1.4	0	–	–	–	6 No. 4 (13 mm dia.)	13.78 (61.29)	8.21	6.35
HJC3-R0	–	C	3	1.4	0	–	–	–	–	10.03 (44.62)	8 21	7.86
HJC2-R0	–	C	2	1.4	0	–	–	–	–	12.68 (56.39)	8 21	6.80
HJC1-R0	–	C	1	1.4	0	–	–	–	–	10.36 (46.09)	8 21	7.73
HJC0-R0	–	C	0	1.4	0	–	–	–	–	9.56 (42.66)	8.21	8.05
HJC3-R1	–	C	3	1.6	0	–	–	–	4 No. 4 (13 mm dia.) 2 No. 3 (13 mm dia.)	10.77 (48.05)	7.24	5.78
HJC3-R2[b]	–	C	3	1.8	0	–	–	–	4 No. 3 (10 mm dia.) 2 No. 4 (13 mm dia.)	–		
HJC3-R3	–	C	3	2.0	0	–	–	–	6 No. 3 (10 mm dia.)	6.81 (30.40)	≥53	3.80

HJC3-R0-B1	–	C	3	1.4	2	–	–	–	6 No. 4 (13 mm dia.)	–	7.72 (34.32)	8.21	8.79
HJC3-R0-B2	–	C	3	1.4	1	–	–	–		–	10.03 (44.62)	8.21	7.86

H J C 3 - R 0 B 1

- High-strength concrete
- Beam-column joint
- Cyclic loads
- Number of transverse hoops in joint core
- Two tent-up bars in joint core
- Bent-up bars
- $M_r = 1.4$
- Flexural strength ratio

(a) C—Cyclic load, M—Monotonic load
(b) HJC3-R2 failed prematurely
(c) $V_j/A_j \sqrt{f'_c}$

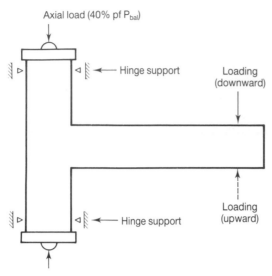

Fig. 8.53   Beam-column joint test setup[32]

430.2 MPa) and HJC3-R0 ($f_c' = 78.5$ MPa) showed distinctly pinched hysteresis loops for HJC3-R0, probably because failure of HCJ3-R0 was largely concentrated at the beam-column joint face. The amount of transverse confinement reinforcement within the joint core had little effect on hysteretic behavior of the specimens. However, at advanced deformation stages, the load resisting capacity decreased less for specimens with more confinement reinforcement within the joint. Specimen HJC3-R0-B1, with two bent-up bars within the joint core, exhibited distinctly pinched hysteresis loops.

The hysteresis loops were less pinched for HJC2-R0-B2, with one bent-up bar within the joint core. However, the hysteresis loops were more pinched for HJC3-R0-B2 than for HJC3-R0 with no bent-up bars. In HJC3-R0, damage was spread over the beam and the beam-column joint core; in HJC3-R0-B1, damage was concentrated between the beam-column joint face and the nearest layer of longitudinal column reinforcement. Energy dissipation capacity, as defined in Fig. 8.55, is plotted in Fig. 8.56 for specimens with different confinement reinforcements within the joint core, different bent-up bars within the joint core, different beam-colum flexural strength ratios, respectively. In all these figures, the energy dissipation capacities of high-strength concrete specimens should be compared to that of the normal-strength concrete specimen, NJC3-R0. Figure 8.56(a) shows that only HJC2-R0 exhibited higher energy dissipation capacity than NJC3-R0. It should be noted, however, that the column of HJC2-R0 was severely damaged through load cycles 10–15. Figure 8.56(b) shows that with the addition of bent-up bars within the joint core, the energy dissipation capacities of high-strength concrete specimens approach that of the comparable normal-strength concrete specimen.

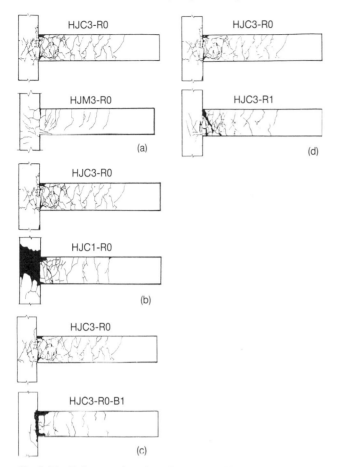

Fig. 8.54   Failure modes of specimens with (a) different loading patterns, (b) different confinement reinforcement, (c) different bent-up bars, and (d) different flexural strength ratios[32]

Figure 8.56(c) shows a definite correlation between energy dissipation capacity and column-beam flexural strength ratio, and indicates that a ratio of at least 1.6 was needed for the energy dissipation capacity of a high-strength concrete beam-column joint specimen to match that of a comparable normal-strength concrete specimen.

It should be noted that the column-to-beam flexural strength ratio, M_R, was varied in this investigation by varying the amount of flexural reinforcement in the beams. As can be seen from Table 8.28, this meant that as M_R increased, the intensity of shear stress within the joint decreased. Thus, what appears to be the effect of increasing M_R on energy dissipation capacity may very well have been the effect of a decreasing shear stress level within the joint.

The following observations could be made from the investigation by Shin et al.[32]

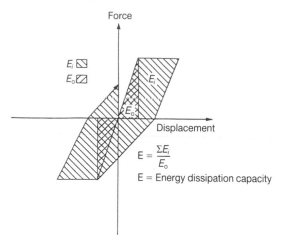

Fig. 8.55 Definition of energy dissipation capacity, as plotted in Fig. 8.56[32]

1 In high-strength concrete beam-column joints subjected to reversed cyclic loading, combined bending and shear was the final failure mode. Under monotonic loading, the final failure was dominated by bending.

2 In high-strength concrete beam-column joints designed with wider transverse reinforcement spacing than recommended by ACI-ASCE 352,[28] failure occurred in the beam-column joint core, and extended throughout the panel zone, into the upper and lower columns.

3 The behavior of high-strength concrete beam-column joint specimens with bent-up bars within the joint core was shear dominated; the hysteretic load-displacement loops were severely pinched due to stress concentration at the beam-column joint face. Thus it appeared desirable to avoid the use of bent-up bars.

4 Energy dissipation capacity increased with increasing values of M_R, beam-to-column flexural strength ratios of joints. This may have been the effect of simultaneous decreases in joint shear stress levels.

Conclusions

The following general conclusions can be drawn with respect to the deformability of high-strength concrete beam-column joints.

1 Properly detailed connections constructed with high-strength concrete exhibit ductile hysteretic response similar to those for ordinary-strength connections.

2 The maximum permissible joint shear stress factor γ should probably be a function of the concrete compressive strength.

3 For a given amount of transverse reinforcement within the joint, the mode of failure is determined by a combination of flexural strength ratio and joint shear stress rather than by the flexural strength ratio alone. High joint shear stresses significantly reduce the energy-

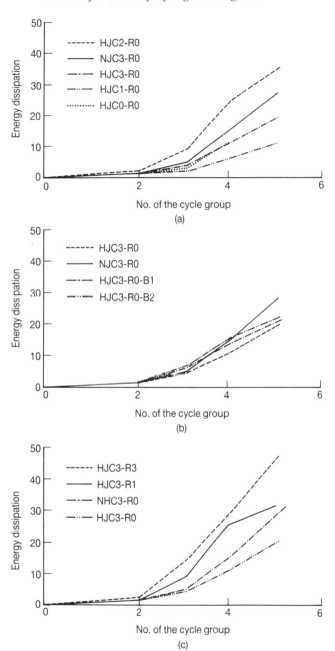

Fig. 8.56   Energy dissipation capacity, as affected by (a) confinement reinforcement within joints, (b) different bent-up bars within joints, and (c) different column-beam flexural strength ratios[32]

absorption capability of subassemblies even in the presence of high flexural strength ratios.

4   For a given column-to-beam flexural strength ratio, the joint shear

stress as well as the joint confinement level are key factors in achieving adequate strength and ductility of the joint. Specimens with low joint shear stresses and high joint confinement are able to develop the flexural strengths of the beams. These same specimens have the least stiffness degradation and loss of strength at post-yield displacements. Specimens with high joint shear stresses and/or low joint confinement levels suffer greater strength loss and lower ductility.

8.5   Application of high-strength concrete in regions of high seismicity

The application of high-strength concrete in highly seismic regions has lagged behind its application in regions of low seismicity. One of the primary reasons has been a concern with the inelastic deformability of high-strength concrete structural members under reversed cyclic loading of the type induced by earthquake excitation. This section discusses the current state of application of high-strength concrete (with specified compression strength in excess of 6000 psi or 40 MPa) in buildings across the USA, including major west coast cities.

As late as the early 1950s, the tallest concrete buildings were in the 20-story height range. By 1975, the 74-story high Water Tower Place, till recently the tallest concrete building in the world, had already been constructed. This virtual revolution within a very short time span was made possible by a number of factors, the most important amongst which were the availability of: new, improved construction methods; bigger cranes; high-strength materials; innovative structural systems; and high-storage, high-speed computer hardware plus the corresponding software that gave the structural engineer unprecedented analytical capabilities. It is futile to speculate which of the factors was more or less important than the others; all of them contributed to the dramatic growth in height of concrete buildings.

Figure 8.57 shows a series of nine concrete buildings, each of which, with the exception of Two Prudential Plaza, was the tallest concrete building in the world at the time of its completion. It is clear that the growth in the height of concrete buildings has gone hand-in-hand with the availability of higher and higher strength concretes.

Almost incredibly, seven of the nine record-setting buildings are located in Chicago, a city that in many ways has pioneered the evolution of high-strength concrete technology. However, very recently there has been an impressive spread in the availability of ultra-high-strength concrete (with specified compression strength in excess of 10,000 psi or 70 MPa). Figure 8.58 shows that 12,000 psi (80 MPa) of higher-strength concrete has been used in the last three or four years in Atlanta, Cleveland, Minneapolis, New York, and most significantly, Seattle which is in Uniform Building Code[33] Seismic Zone 3.

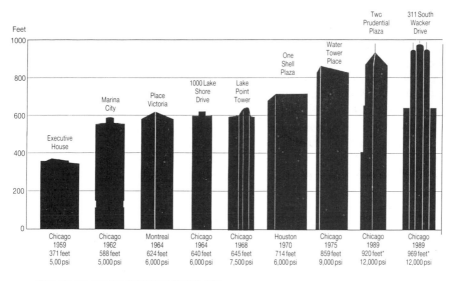

Fig. 8.57 High-strength concrete in high-rise construction

In fact, the highest concrete strength ever used in a building has been 19,000 psi (130 MPa) in the composite columns of Seattle's 62-story, 759-ft high Two Union Square (Skilling Ward Magnusson Barkshire Inc , Structural Engineers). The strength was obtained by use of: what may be a

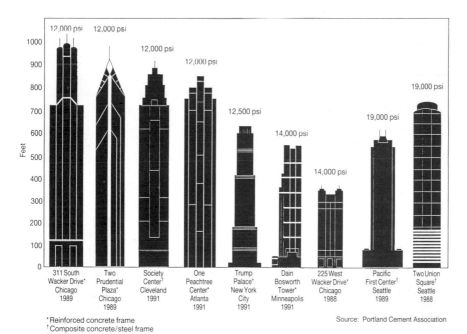

Fig. 8.58 Ultra-high-strength concrete shapes new skylines

record low water cementitious ratio of 0.22 (this is the single most important factor in increasing strength and reducing shrinkage and creep); the strongest of available cements; a superplasticizer which reduces the need for water, and provides the necessary workability; a very high cement content; a very strong, small (3/8 in. or 10 mm), round glacial aggregate available locally; silica fume (increasing strength by about 25%); a design strength obtained at 56 rather than the usual 28 days; and an extraordinarily thorough quality assurance program. The 19,000 psi (130 MPa) strength was the byproduct of the design requirement for an extremely high modulus of elasticity of 7.2 million psi (49,650 MPa). The stiffness was desired in order to meet the occupant-comfort criterion for the completed building. The same concrete strength was later used by Skilling Ward in the composite columns of the shorter 44-story Pacific First Center (Fig. 8.59).

Concrete with strengths of 95,000 psi (65 MPa) concrete has been used by Skilling Ward at 600 California in San Francisco, and at 1300 Clay in Oakland, both composite buildings. The Watry Design Group has used 8000 psi (55 MPa) concrete in several all-concrete Bay Area buildings, including the 19-story Fillmore Building (Fig. 8.60).

The spread in the use of high-strength concrete in Southern California has been hampered by the City of Los Angeles Code provision restricting the strength of concrete to a maximum of 6000 psi (40 MPa). Even then,

Fig. 8.59 44-story Pacific First Center in Seattle using 19,000 psi (130 MPa) concrete [photo courtesy of Skilling Ward Magnusson Barkshire Inc., structural engineers]

Fig. 8.60 19-story Fillmore Building in San Francisco using 8000 psi (55 MPa) concrete [photo courtesy of Watry Design Group, structural engineers]

concrete strength in excess of 6000 psi has been used in the Great American Plaza office-hotel-garage complex in San Diego, a 14-story residential building at 5th and Ash in San Diego, and in the 22-story Pacific Regent (senior citizen housing) at LaJolla.

The three biggest advantages of high-strength concrete that make its use attractive in high-rise buildings are that it provides more

- strength/unit cost
- strength/unit weight
- stiffness/unit cost

than most other building materials including normal weight concrete.

Commercially available 14,000 psi (96 MPa) concrete costs significantly less in dollars per cu. yd. (m^3) than $3\frac{1}{2}$ times the price of 4000 psi (27 MPa) concrete. In fact, the unit price goes up relatively little as concrete strength increases from 4000 to 10,000 psi (27 to 70 MPa). Thus, high-strength concrete gives the user more strength per dollar.

The unit weight of concrete goes up only insignificantly as concrete strength increases from moderate to very high levels. Thus, more strength per unit weight is obtained, which can be a significant advantage for construction in high seismic regions, since earthquake induced forces are directly proportional to mass.

The modulus of elasticity of concrete remains proportional to the square root of the compressive strength of concrete at the age of loading even for high-strength concrete.[34] The user thus obtains a higher stiffness per unit weight and unit cost. Indeed, it is quite common for a structural engineer to consider and specify high-strength concrete for its stiffness, rather than for its strength.

It is important to mention that the specific creep (ultimate creep strain per unit of sustained stress) of concrete decreases significantly as the concrete strength increases. The most recent verification of this is available in Bjerkeli *et al.*[18] This is indeed a fortunate coincidence without which the application of high strength concrete in highrise buildings would have been seriously hampered. Because of the lower specific creep, high-strength concrete columns with their high stress levels suffer no more total shortening than normal-strength concrete columns with their lower strength levels. Otherwise, the problem of differential shortening of vertical elements within highrise buildings would have been aggravated by the use of high-strength concrete in columns.

The use of high-strength concrete in taller buildings (and, of course in other applications not discussed here) is bound to increase across the world in the ensuing years, because of the very significant advantages discussed here.

8.6   Summary

Information available in the literature on the deformability of high-strength concrete beams, columns and beam-joints is reviewed in three separate sections of this chapter. At the end of each of those sections, relevant practical conclusions are drawn. The last section is devoted to the application of high-strength concrete in regions of high seismicity which has been on the increase in recent years. The latter trend should continue into the future.

References

1   Khaloo, A.R. and Ahmad, S.H. (1988) Behaviour of normal and high strength concrete under combined compression-shear loading. *ACI Materials Journal* **85**, No. 6, November–December, 551–9.

2   Leslie, K.E., Rajagopalan, K.S. and Everard, N.J. (1976) Flexural behavior of high-strength concrete beams. *ACI Journal*, Proceedings **73**, No. 9, September, 517–21.

3   ACI Committee 318 (1989) *Building code requirements for reinforced concrete and commentary*, ACI 318-83, ACI 318-89 and ACI 318 R-89. American Concrete Institute, Detroit, MI, 1983.

4   Pastor, J.A., Nilson, A.H. and Slate, F.O. (1984) Behavior of high-strength concrete beams, *Research Report 84-3*. Department of Structural Engineering, Cornell University, Ithaca, New York, February, 311 pp.

5   Tognon, G., Ursella, P. and Coppetti, G. (1980) Design and properties of

concretes with strength over 1500 kgf/cm^2. *ACI Journal*, Proceedings **77**, No. 3, May–June, 171–8.

6 Swartz, S.E., Nikaeen, A., Narayan Babu, H.D., Periyakaruppan, N. and Refai, T.M.E. (1985) Structural bending properties of higher strength concrete, *Special Publication 87*. American Concrete Institute, Detroit, MI, 147–78.

7 Shin, S.-W. (1986) 'Flexural Behavior including Ductility of Ultra-High-Strength Concrete Members', *Ph.D. Thesis*, University of Illinois at Chicago, Chicago, IL, 232 pp.

8 Shin, S.-W., Ghosh, S.K. and Moreno, J. (1989) Flexural ductility of ultra-high-strength concrete members. *ACI Journal*, Proceedings **86**, No. 4, July–August, 394–400.

9 Shin, S.-W., Kamara, M. and Ghosh, S.K. (1990) Flexural ductility, strength prediction and hysteretic behavior of ultra-high-strength concrete members, *Special Publication 121*. American Concrete Institute, Detroit, MI, 239–64.

10 Kamara, M.E. (1988) 'Flexural Behavior including Ductility of Ultra-High-Strength Concrete Members Subjected to Reversed Cyclic Loading', *Ph.D. Thesis*, University of Illinois at Chicago, Chicago, IL, 258 pp.

11 Ahmad, S.H. and Barker, R. (1991) Flexural behavior of reinforced high-strength lightweight concrete beams. *ACI Structural Journal*, **88**, No. 1, January–February, 69–77.

12 Ahmad, S.H. and Batts, J. (1991) Flexural behavior of doubly reinforced high-strength lightweight concrete beams with web reinforcement. *ACI Structural Journal*, **88**, No. 3, May–June, 351–8.

13 Ghosh, S.K., Narielwala, D.P., Shin, S.W. and Moreno, J. (1991) Flexural behavior including ductility of high-strength lightweight concrete members under reversed cyclic loading, Presented at the International Symposium on Performance of Structural Lightweight Concrete, ACI Fall Convention, Dallas, TX, November (to be published in an *ACI Special Publication*).

14 Razvi, S.R. and Saatcioglu, M. (1992) Confinement and deformability of high-strength concrete columns, Submitted for publication, *ACI Structural Journal*.

15 Ahmad, S.H. and Shah, S.P. (1982) Stress-strain curves of concrete confined by spiral reinforcement. *ACI Journal*, **79**, No. 6, November–December, 484–90.

16 Muguruma, H., Watanabe, F., Iwashimizu, T. and Mitsueda, R. (1983) Ductility improvement of high strength concrete by lateral confinement. *Transactions of the Japan Concrete Institute*, **5**, 403–10.

17 Martinez, S., Nilson, A.H. and Slate, F.O. (1984) Spirally reinforced high strength concrete columns. *ACI Journal*, **81**, No. 5, September–October, 431–42.

18 Bjerkeli, L., Tomaszewicz, A. and Jensen, J.J. (1990) Deformation properties and ductility of high strength concrete, *Special Publication 121*. American Concrete Institute, Detroit, MI, 215–38.

19 Abdel-Fattah, H. and Ahmad, S.H. (1989) Behavior of hoop confined high strength concrete under axial and shear loads. *ACI Structural Journal*, **86**, November–December, 652–9.

20 Chung, H., Hayashi, S. and Kokusho, S. (1990) Reinforced high strength columns subjected to axial forces, bending moments, and shear forces. *Transactions of the Japan Concrete Institute*, **2**, 335–42.

21 Muguruma, H., Watanabe, F. and Komuro, T. (1990) Ductility improvement of high strength concrete columns with lateral confinement, *Special Publication 121*. American Concrete Institute, Detroit, MI, 47–60.

22 Sugano, S., Nagashima, T., Kimura, H., Tamura, A. and Ichikawa, A. (1990) Experimental studies on seismic behavior of reinforced concrete members of

high strength concrete, *Special Publication 121*. American Concrete Institute, Detroit, MI, 61–87.

23   Kabeyasawa, T., Li, K.N. and Huang, K. (1990) Experimental study on strength and deformability of ultrahigh strength concrete columns. *Transactions of the Japan Concrete Institute*, **12**, 315–22.

24   Sakai, Y., Hibi, J., Otani, S. and Aoyama, H. (1990) Experimental study on flexural behavior of reinforced concrete columns using high-strength concrete. *Transactions of the Japan Concrete Institute*, **12**, 323–30.

25   Watanabe, F., Muguruma, H., Matsutani, T. and Sanda, D. (1987) Utilization of high strength concrete for reinforced concrete high-rise buildings in seismic area, *Proceedings of the Symposium* on Utilization of High Strength Concrete, Tapir, Publisher, Stavangar, Norway, 655–66.

26   Saatcioglu, M. (1991) Deformability of reinforced concrete columns, *Special Publication 127*. American Concrete Institute, Detroit, MI, 421–52.

27   Saatcioglu, M. and Razvi, S.R. (1992) Strength and ductility of confined concrete. *ASCE Journal of Structural Engineering*, **118**, No. 6, 1590–607.

28   (1985) Monolithic reinforced concrete structures, *ACI 352 R-85*. American Concrete Institute, Detroit, 19 pp.

29   Ehsani, M.R., Moussa, A.E. and Vallenilla, C.R. (1987) Comparison of inelastic behavior of reinforced ordinary- and high-strength concrete frames. *ACI Structural Journal*, **84**, No. 2, March–April, 161–9.

30   Ehsani, M.R. and Wight, J.K. (1985) Exterior reinforced concrete beam-to-column connections subjected to earthquake-type loading. *ACI Journal*, Proceedings **82**, No. 4, July–August, 492–9.

31   Ehsani, M.R. and Alameddine, F. (1991) Design recommendations for Type 2 high-strength reinforced concrete connections. *ACI Structural Journal*, **88**, No. 3, May–June, 277–91.

32   Shin, S.-W., Lee, L.-S. and Ghosh, S.K. (1992) High-strength concrete beam-column joints, Presented at the *10th World Conference on Earthquake Engineering*, Madrid, Spain, July.

33   (1991) *Uniform building code*. International Conference of Building Officials, Whittier, CA.

34   Russell, H.G. (1990) Shortening of high-strength concrete members, *Special Publication 121*. American Concrete Institute, Detroit, MI, 1–20.

9 Structural design considerations and applications

Henry G Russell

9.1 Introduction

The successful application of high-strength concrete requires the complete cooperation of the owner, architect, structural engineer, contractor, concrete supplier and testing laboratory. In locations where applications of high-strength concrete have been successfully accomplished, the construction team has worked together for their mutual benefit and for the benefit of the engineering community. In the case of buildings, the owner must be willing to allow the building to become a state-of-the-art structure. The architect must be willing to design a structure that will utilize the benefits of high-strength concrete. The structural engineer must have the knowledge and ability to adapt structural design concepts based on lower strength concretes for use with higher strength materials. At the same time, the designer must work within the boundaries of the codes and specifications. The contractor must be willing to work with different materials and must accept the need for a higher degree of quality control. The concrete supplier must be able to supply concrete of the specified strength. Finally, the quality control testing laboratory must have the capability to prepare the test specimens and to test them appropriately. This requires that all members of the team work together from the inception of the project until its completion. Where this approach has been adopted, the development and applications of high-strength concrete have been successful; higher strength materials have been utilized and taller buildings have been built.

As the development of high-strength concrete has continued, the definition of high-strength concrete in North America has changed. In the 1950s, a compressive strength of 5000 psi (34 MPa) was considered high strength. In the 1960s, commercial usage of 6000 and 7500 psi (41 and

52 MPa) concrete was achieved. In the early 1970s, 9000 psi (62 MPa) concrete was being used. In the 1980s, design strengths of 14,000 psi (97 MPa) were used for commercial applications in buildings. Strengths as high as 19,000 psi (131 MPa) have been used, although their commercial application has been limited to one geographic location. The primary applications of these higher strength concretes have been in the columns of high-rise buildings. However, there is an increasing interest in the use of higher strength concretes in long-span bridges and offshore structures. The following sections of this chapter describe structural design considerations, construction considerations and quality control aspects and some specific applications of high-strength concrete. Although the specific applications are predominantly in North America, there have been applications in other countries.[1-4]

9.2   Structural design considerations

Many of the design provisions that exist in the codes and standards of North America and Europe are based on experimental results obtained with conventional-strength concretes. As higher-strength concretes have become available, the applicability of the design provisions for higher-strength concrete has been questioned. This section highlights four areas where designers should give special consideration.[5]

Flexural strength

The ACI Building Code[6] allows the use of rectangular, trapezoidal, parabolic or other stress distribution in design provided that the predicted strength is in substantial agreement with the results of comprehensive tests. However, in most situations, it is convenient to utilize an equivalent rectangular compressive stress distribution. Based on published data,[7] it appears that, for under-reinforced beams, the present ACI methods can be used for concretes with compressive strengths up to 15,000 psi (103 MPa.) Additional information is required for concretes with compressive strengths in excess of 15,000 psi (103 MPa).

Shear strength

Recent tests[8] have indicated the need to modify the shear strength design provisions of the ACI Building Code.[6] Test results have indicated that, with the utilization of higher-strength concretes in reinforced concrete beams, the specified minimum amount of web reinforcement must be increased as the concrete compressive strength increases. This has been found necessary for concretes with compressive strengths in excess of

10,000 psi (69 MPa). This increase in the minimum amount of web reinforcement is needed to control the extent of shear cracking in the beams and to provide ductile behavior. The 1989 version of the ACI Building Code[6] contains provisions to achieve this.

Development length

Design provisions in the United States permit the use of shorter development lengths or anchorage lengths for reinforcing bars as the concrete compressive strength increases. However, due to a lack of experimental data, the ACI Building Code[6] was modified to limit the design provisions for development length to concrete with a compressive strength of less than 10,000 psi (69 MPa). Consequently, although a concrete with a compressive strength in excess of 10,000 psi (69 MPa) may be used, the design must be based on a concrete compressive strength of 10,000 psi (69 MPa). This limitation removes one of the advantages of utilizing higher-strength concretes. Currently, there does not seem to be any reason why the limitation should apply. However, data are needed to substantiate removal of the limitation.

Long-term deformations

The use of high-strength concretes in high-rise buildings requires that special attention be paid to the long-term length changes that occur in high-strength concrete members.[9] Long-term deformations result from creep and shinkage. In addition, instantaneous deformations occur whenever load is added to the building.

As with lower-strength concretes, creep deformations in a member depend on the creep properties of the concrete at age of loading, stress level in the concrete, size of member and amount of reinforcement. The creep per unit stress of high-strength concretes is less than the creep per unit stress of lower-strength concretes. This means that, at the same stress level, creep deformations will be less for higher-strength concretes. Alternatively, for concretes loaded to the same ratio of stress to strength, the creep deformations will be about the same irrespective of concrete strength.

Shrinkage of most high-strength concrete is about the same as that of lower-strength concretes and is more dependent on factors other than the strength level. Some admixtures are said to reduce shrinkage. However, the reduced shrinkage may be the result of lower water content in the mix rather than the use of a specific admixture.

Instantaneous deformations also constitute a major source of shortening in the lower story columns of high-rise buildings. Instantaneous deformations are primarily a function of the modulus of elasticity at the age of loading and can be calculated from the following equation:

$$\varepsilon = \frac{P}{A_c E_c + A_s E_s} \tag{9.1}$$

where A_c = area of concrete
A_s = area of steel reinforcement
E_c = modulus of elasticity of concrete at age of loading
E_s = modulus of elasticity of steel reinforcement
P = applied axial load.

The ACI Building Code[6] contains the following equation for calculation of the modulus of elasticity:

$$E_c = w_c^{1.5} 33(f_c')^{1/2} \tag{9.2}$$

where w_c = unit weight of concrete in lb/cu ft
f_c' = compressive strength of concrete as measured on 6×12-in. (152×305-mm) cylinders.

Equation (9.2) was developed by Pauw[10] on the basis of concretes with compressive strengths up to about 5500 psi (38 MPa). As additional data have become available on higher-strength concretes, various investigators have made comparisons between the equation and the data for higher-strength concretes. Based on published data, Martinez *et al.*[11] recommended a modified equation for the calculation of modulus of elasticity. This revised equation predicted a lower modulus of elasticity for higher-strength concretes when compared with Equation (9.2). However, more recent data published by Cook[12] indicate that Equation (9.2) underestimates the modulus of elasticity for the higher-strength concretes. It should be noted that the higher-strength concretes reported by Martinez *et al.* were obtained using a smaller aggregate size. However, Cook was able to produce higher-strength concretes with larger-size aggregates.

It should be recognized that Equation (9.2) was based on a statistical analysis of the available data. As such, there is considerable scatter in the relationship between modulus of elasticity and concrete compressive strength. Consequently, it is recommended that, when accurate calculations are required for high-strength concretes, the modulus of elasticity should be measured as part of the concrete mix design preparation. Alternatively, the engineer must assume that there is going to be some deviation from the values predicted by Equation (9.2) or any other selected equation.

The value of modulus of elasticity used in Equation (9.1) should be the modulus of elasticity of the concrete in the column. However, the modulus of elasticity is measured on smaller specimens. The question, therefore, arises whether the modulus of elasticity as measured on 6×12-in. (152×305-mm) concrete cylinders or other plain concrete specimens is applicable to large structural members. Currently, there are no published data on this topic related to high-strength concrete. However, Hester[4] and Cook[12] have extracted cores from large concrete members and measured their modulus of elasticity. For the strength levels used in his program,

Cook showed that the modulus of elasticity as measured on cores varied as the measured compressive strength varied. If the core strength was low, the modulus of elasticity was also low. Hester[4] also observed the same phenomena. However, in Hester's work, both strength and modulus of elasticity of cores were considerably lower than corresponding values measured on cylinders.

Despite the variations that exist in high-strength concrete and the lack of information about some of the properties, good correlation has been obtained between calculated and measured deformations on real structures.[13,14] These correlations have been conducted for structural members with concrete compressive strengths up to 10,000 psi (69 MPa). There is currently a need to extend these types of correlations to members with higher concrete compressive strengths.

In the design of 311 South Wacker Drive, special consideration was given to vertical shortening of different structural elements. In the design process, column loads were first calculated using a conventional analysis. The column loads were then used in the calculation of vertical shortening of column stacks. The calculated differential shortening was then utilized in subsequent analysis of the forces in the structural frame. If the differential movements were too large, concrete compressive strengths and percentages of reinforcement for the columns were revised to minimize the differential. This iterative analysis process was repeated several times. The calculated differential movements were then utilized to specify formwork chamber for the floor slabs. During construction differential movements were monitored. Very close agreement with design values was obtained.

9.3   Construction considerations

Most of the construction procedures used for high-strength concrete are similar to those for conventional concrete. However, high-strength concretes are less tolerant of errors than lower-strength concretes. Some special considerations are noted below.[5]

Vertical load transmission

The use of high-strength concrete in the columns of high-rise buildings, together with the use of lower-strength concrete in beams and slabs, gives rise to construction problems at the slab-column and beam-column intersections. The ACI Building Code[6] addressed this specific situation. The Code requires that when the specific compressive strength of the concrete in the column is more than 40% greater than the concrete strength specified for the floor system, transmission of column load through the floor system shall be provided by one of three alternatives.

In the first alternative, concrete of the strength specified for the column must be placed in the floor at the column location. The top surface of the concrete column must extend 2 ft (600 mm) into the slab from the face of the column. The second alternative requires that the design strength of the column through the floor system be based on the lower concrete strength. If necessary, additional vertical reinforcement through the intersection shall be provided. In the third alternative, when the column is supported laterally on all four sides by beams, a combination of strengths based on the concrete column strength and the slab concrete strength may be utilized.

In most instances, the designer and contractor will select the first alternative known as 'puddling'. Application of this procedure requires the placing of two different concrete mixes in the floor system. The lower-strength concrete must be placed while the higher-strength concrete is still plastic. Both concretes must be adequately vibrated to ensure that they are well integrated. This requires careful coordination of the concrete deliveries and the possible use of retarders. In some cases, additional inspection services will be required when this procedure is used. It is important that the higher-strength concrete in the floor of the region of the column be placed before the lower-strength concrete in the remainder of the floor. This procedure prevents accidental placing of the lower-strength concrete in the column area. The designer has responsibility to indicate on the drawings where the high- and low-strength concretes are to be placed. In some instances, designers have elected not to use this approach because of the difficulty of maintaining the required coordination. However, inspection of this procedure after the concrete has been placed and hardened is relatively easy. High-strength concrete contains higher cement contents and mineral admixtures. Consequently, the high-strength concretes tend to have a darker color than the lower-strength concretes. Therefore, location of the high-strength concrete can be observed visually. As higher and higher concrete strengths are being used in columns with no corresponding increase in the strength of concrete used in floor slabs, this design and construction provision is becoming more important.

At 311 South Wacker Drive, the contractor elected to use 9000 psi (62 MPa) concrete in some floor slabs to avoid puddling 12,000 psi (83 MPa) concrete. At the same time, the contractor was able to reduce the floor thickness by post-tensioning. This, in turn, reduced the dead load on the structure.

Heat of hydration

Heat development in cement-rich high-strength concretes can result in high internal temperatures. Field measurements on 30-in. (760-mm) thick columns have shown internal temperatures of 150 to 180 °F (66 to 82 °C). Thus, consideration must be given to minimizing thermal gradients as these can result in cracking. The most common solutions to this problem are to keep mixing temperatures as low as possible, use the lowest amount of cement needed to obtain the specified strength level, use mineral admixtures or low-heat-generation cement, and insulate forms as necessary to maintain a more uniform temperature distribution until concrete strengths are sufficient to resist thermally induced tensile stresses. Thermal blankets and expanded polystyrene insulation have been used to insulate the concrete, even in hot climates.

Consolidation and finishing

High-strength concretes are produced with low ratios of water-to-cementitious materials. They also have higher cement contents than conventional concretes and will contain mineral admixtures in the form of fly ash and/or silica fume. These combined factors would normally make high-strength concrete difficult to place. However, the use of high-range water reducers (HRWR) commonly known as superplasticizers has made high-strength concretes extremely workable and easy to consolidate. It is not unusual to have high-strength concretes with slumps as high as 10 in. (250 mm). Care is needed, however, to ensure that the HRWR is added at the optimum time and at the optimum rate.

   With the use of silica fume in concrete, the mixtures have become stickier and difficulties have been encountered in obtaining an adequate surface finish. This is not important for the top surfaces of columns. However, it is very critical when high-strength concrete is used in beams or floor slabs.

Plastic shrinkage cracks

When high-strength concrete is used in floor slabs, plastic shrinkage cracks may result. The primary cause for plastic shrinkage cracks for freshly placed slabs is very rapid loss of moisture from the concrete caused by low humidity, high wind and/or high temperature. When moisture evaporates from the freshly placed concrete surface faster than it is replaced by bleedwater, the surface shrinks. With restraint, tensile stresses develop in the weak concrete, and the concrete cracks. With high-strength concrete and low water-cementitious ratios, the mixing water content is very low. Less water will generally mean less bleeding. If a concrete has the tendency not to bleed, then it will probably exhibit plastic shrinkage cracks if conditions of low humidity, high wind and/or high temperatures exist.

   To eliminate plastic shrinkage cracks, several jobs have used fog nozzles to saturate the air above the slab surface and plastic sheeting to cover the surface between final finishing operations. While it may be desirable to increase bleeding for crack reduction, this may not be feasible due to other strength or durability requirements that the concrete must possess.

Plastic settlement cracks

Generally, plastic settlement cracking occurs after initial placement, vibration and finishing, as the concrete has a tendency to continue to consolidate or settle. During this time, the unhardened concrete may be locally restrained by reinforcing steel or formwork. The local restraint results in voids or cracks adjacent to the restraining element.

With the use of high-slump, high-strength concretes, there is a greater tendency for plastic settlement cracking. In many cases, this has been associated with reinforcing steel in columns. When this situation arises, high slump and low concrete increase the probability of cracking. Furthermore, plastic settlement cracking will increase with a marginal amount of vibration and with flexible forms.

These problems can be avoided by using the lowest practicable slump, preferably less than 4 in. (100 mm). The concrete should be adequately vibrated, the greatest concrete cover used (with the engineer's approval), and proper formwork design shuld be followed. In addition, it has been found that revibration (post-vibration or back-vibration) will reduce plastic settlement cracking and close cracks once they have formed. The concrete should be revibrated as late as possible before initial set and while the internal vibrator will sink under its own weight into the concrete and liquefy it momentarily. External or form vibrators can also be used for revibration to reduce plastic settlement cracking.

9.4   Quality control

Quality control measures for high-strength concrete are eessentially the same as for conventional strength concrete. However, as strengths increase, concrete becomes much less forgiving of improper sampling and testing procedures. In fact, inadequacies in sampling and testing procedures can generally only result in lower apparent concrete strengths. Marginal testing practices that have insignificant effects on results for conventional-strength concretes can give erroneous or erratic results for higher-strength concretes.

Testing procedures

As concrete strengths increase, the use of standard 6×12-in. (152×305-mm) cylinders becomes problematic for many laboratories in North America because test machine capabilities may be insufficient. However, since building code provisions are referenced to 6×12-in. (152×305-mm) cylinders, designers are naturally reluctant to accept test results from smaller specimens. A potential solution to this dilemma is to develop (by testing) correlation curves between larger and smaller specimens for the job concrete mixes. The smaller cylinders can then be used for quality control with direct correlation to equivalent larger specimen strengths. This approach was successfully used on the construction of Two Union Square in Seattle. It should also be noted that, in areas where the use of high-strength concrete is becoming more prevalent, testing laboratories are upgrading to higher-capacity machines.

Tests have shown that concrete specimens made in either cardboard, plastic or tin molds attain lower strengths than specimens made in steel molds. The Canadian Standards Association (CSA) requires that molds other than steel are acceptable, if documentation is available and the cylinders produced from non-steel molds have compressive strengths equivalent to those obtained using steel molds. The most common practice is to use plastic molds for lower-strength concretes. It is reasonable to continue this practice for high-strength concretes. If measured strengths come into question, the CSA correlation approach can be used to verify that the mold material is not causing the problem. Re-use of plastic molds should not be permitted for high-strength concretes.

For high-strength concrete, test strengths are particularly sensitive to specimen end conditions. The American Society for Testing and Materials (ASTM) C 39,[15] *Standard Test Method for Compressive Strengths of Cylindrical Concrete Specimens*, provides guidance on perpendicularity and planeness, and ASTM C 617,[16] *Standard Practice for Capping Cylindrical Concrete Specimens*, covers capping procedures. High-strength capping compounds should be used with a uniform thickness of $\frac{1}{8}$ in. (3 mm). As an alternative to capping, cylinders can be lapped to meet ASTM end-condition requirements, but this procedure is generally more costly. If lapping is used, it should be done a day or two after casting because lapping becomes more difficult as the concrete gains strength.

Compression testing machines should meet the requirements of ASTM C 39,[15] or other applicable standards. Careful attention should be paid to platen smoothness as small deviations can cause increased errors in measured strengths. It is recommended that machines be calibrated every six months rather than annually when used for high-strength concrete testing.

In-place strengths

Several national codes address the use of concrete core data when strengths of concrete quality control specimens do not exceed the specified strengths. For example, the ACI Building Code[6] allows concrete to be accepted if the average strength measured on the cores exceeds 85% of the specified strength and no single strength is less than 75% of the specified strength. The factors of 75 and 85% are based on lower-strength concretes and were determined from comparisons of core strengths with cylinder strengths. Data on concretes with compressive strengths up to 17,000 psi (83 MPa), have indicated that the factors are still valid. However, in at least two projects using concrete with specified strengths in excess of 12,000 psi (83 MPa), the concrete core strengths have been lower than 85% of the cylinder strengths. This has raised the question of the validity of the factors for high-strength concrete. Designers and contractors are cautioned to address this issue before construction begins.

9.5　High rise buildings

The development and availability of higher strength concretes in specific geographic locations and the desire to build taller concrete structures have been synonymous. As a result, growth in the usage of higher strength concretes and increases in the height of the tallest concrete building, have proceeded together. The availability of the higher strength concretes makes it economically feasible to achieve the structures. The increased costs of higher strength concrete materials and increased quality control are more than offset by the ability of the high-strength concrete columns to carry a higher load, to require less reinforcing steel, and to need less formwork. Schmidt and Hoffmann[17] were the first to publish data indicating that the most economical way to design a column was with the highest available strength concrete and the least amount of reinforcing steel. Similar data are shown in Fig. 9.1. This figure takes into account the higher costs of the higher strength concretes, the costs of the reinforcing steel and the cost of the formwork. As the figure illustrates, high-strength concrete with minimum reinforcing steel represents the most economical solution. Although today's unit costs are different from those used by Schmidt and Hoffman, the trend remains the same.

The use of high-strength concretes also permits the use of smaller size

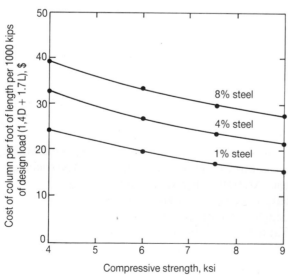

Fig. 9.1　Cost of concrete columns

columns. Since the columns in high-rise buildings are likely to be located in prime retail rentable floor space, minimization of the size of these columns is advantageous for the architect's layout as well as for the owner who wants to maximize rentable floor space. The desire to have smaller and smaller column sizes is facilitated by the availability of the high-strength concretes. In some structures, smaller size columns have also been required in order to minimize their interruption of the packing spaces in the lower stories of high-rise buildings. In the Richmond-Adelaide Centre in Toronto, Canada, the use of high-strength concrete columns enabled the architect to increase the use of the underground parking garage by approximately 30%. In a 15-story parking garage at 900 N. Michigan Avenue, Chicago, high-strength concrete was specified for the columns to reduce the lateral bending stiffness, yet provide sufficient capacity to carry the vertical load. Consequently, when the floor system was post-tensioned, the amount of post-tensioning force being resisted by the columns was minimized.

Although it was originally thought that high-strength concrete would only be used in the columns of very high-rise buildings, more recent applications have indicated that the same benefits can be achieved with shorter buildings.[18] As new applications are realized, it is anticipated that further economical benefits will be achieved.

Table 9.1 lists some of the more significant achievements in the applications of high-strength in high-rise buildings in North America. The table does not include every building that used concrete with a design compressive strength in excess of 6000 psi (41 MPa). However, it serves to illustrate the increase in height and concrete strength with time as well as the range of locations. Some of the structures are described in more detail in the following sections.

Lake Point Tower, Chicago

Lake Point Tower was constructed in Chicago in 1966–1967. Construction of this building represented the first major use of 7500 psi (52 MPa) concrete in a major commercial application. This required a carefully planned research and development program to optimize on all available materials in order to achieve the desired strength. As shown in Fig. 9.2, the building has three wings. Each wing is 65 ft (20 m) wide and extends 117 ft (36 m) from the center of the building. Total height of the structure above ground is 645 ft (197 m). The floors of the building are 8 in. (205 mm) thick flat plates reinforced with Grade 60 deformed bars. They contain lightweight aggregate concrete having a partial replacement of fines with sand to produce a density of 100 lb/ft^3 (1.60 Mg/m^3). The interior columns and core walls have diameters and thicknesses that vary over the height of the structure as listed in Table 9.2. The columns and walls were designed for normal weight concretes having compressive strengths of either 7500, 6000, 5000 or 3500 psi (52, 41, 34 or 24 MPa). Despite the availability of 7500 psi

Table 9.1   Buildings in North America with high-strength concrete

Building	Location	Year[1]	No. of stories	Maximum design strength	
				psi	MPa
One Shell Plaza	Houston	1968	52	6000L[2]	41L[2]
Pacific Park Plaza	Emeryville,	1983	30	6500	45
SE Financial Center	CA	1982	53	7000	48
Petrocanada Building	Miami	1982	34	7250	50
Lake Point Tower	Calgary	1966	70	7500	52
1130 S. Michigan Ave.	Chicago	–	–	7500	52
Texas Commerce Tower	Chicago	1981	75	7500	52
Helmsley Place Hotel	Houston	1978	53	8000	55
Trump Tower	New York	–	68	8000	55
City Center Project	New York	1981	52	8000	55
Larimar Place Condominiums	Minneapolis	1980	31	8000	55
NCNB Corporate Center	Denver	1990	68	8000	55
499 Park Avenue	Charlotte, NC	–	27	8500	59
Royal Bank Plaza	New York	1975	43	8800	61
Richmond-Adelaide Centre	Toronto	1978	33	8800	61
Midcontinental Plaza	Toronto	1972	50	9000	62
Frontier Towers	Chicago	1973	55	9000	62
Water Tower Place	Chicago	1975	76	9000	62
River Plaza	Chicago	1976	56	9000[3]	62[3]
Chicago Mercantile Exchange	Chicago	1982	40	9000[4]	62[4]
Columbia Center	Chicago	1983	76	9500	66
Interfirst Plaza	Seattle	1983	72	10,000	69
Scotia Plaza	Dallas	1987	68	10,150	70
311 S. Wacker Drive	Toronto	1988	70	12,000	83
One Peachtree Center	Chicago	1990	62	12,000	83
Bay Adelaide Center	Atlanta	1991	57	12,300	85
Society Tower	Toronto	1990	63	12,000	83
900 N. Michigan Annex	Cleveland	1986	15	14,000	97
Two Union Square	Chicago	1988	58	14,000[5]	97
255 W. Wacker Drive	Seattle	1988	31	14,000[6]	97
Dain Bosworth/ Niemann Marcus Plaza	Chicago Minneapolis	–	39	14,000	97

1.   Approximate year in which high-strength concrete was cast
2.   L = lightweight concrete
3.   Two experimental columns of 11,000 psi (76 MPa) strength were included
4.   Two experimental columns of 14,000 psi (97 MPa) strength were included
5.   19,000 psi (131 MPa) achieved because of high modulus of elasticity required by the designer
6.   One experimental column of 17,000 psi (117 MPa) strength was included

(52 MPa) concrete at the time Lake Point Tower was built, the use of reinforcement with a yield stress of 75 ksi (520 MPa) was also required.

Construction of the first story of the building began in June 1966 and the 70th story was completed in December 1967. Weather permitting, the building was built at the rate of one floor every three working days. The

Fig. 9.2   Lake Point Tower

building was extensively instrumented for purposes of verifying the time-dependent deformations in the columns.[13] In addition, an extensive program was used to develop basic information about the elastic properties, creep and shrinkage of the various concretes used in the building. Shrinkage of the high-strength concrete was found to be very similar to that of lower strength concretes. However, the specific creep in millionths/psi was found to be significantly lower than measured on the lower strength concretes. Although the maximum specified compressive strength concrete for the lower columns was 7500 psi (52 MPa), concrete strengths at an age of one year exceeded 10,000 psi (69 MPa).

Table 9.2   Properties of interior columns and core walls at Lake Point Tower

Story	Concrete design strength psi	Grade of reinforcement ksi	Column diameter in.	Core wall thickness in.
68–59	3,500	60	30	0, 12
58–44	5,000	60	30	12
43–35	5,000	60	36	12, 14
34–30	6,000	60	36	16
29–17	6,000	60	40	16, 18, 20
16–12	7,500	60	40	22
11–1	7,500	75	40	24, 30

Metric equivalents:
   1000 psi = 1 ksi = 6.895 MPa
   1 in. = 25.4 mm

Fig. 9.3   Water Tower Place

Water Tower Place, Chicago

Water Tower Place shown in Fig. 9.3 is a 76-story reinforced concrete building situated in Chicago, Illinois. When construction was completed in 1976, Water Tower Place was the tallest reinforced concrete building in the world and represented the most significant application of high-strength concrete. The building consists of a 13-story lower portion containing commercial and office space, and a 63-story tower that rises from one quadrant of the base structure. The lower portion of the building is 214×531 ft (65×162 m). The tower structure measures 94×221 ft (29×67 m) in plan.

In the lower portion of the structure, floor loads are carried by reinforced concrete flat slabs with bay sizes of 30×31 ft (9×9 m). Lightweight concrete is used in all floor slabs above ground level. Normal weight concrete is used in the floor slabs of four basements below ground level. In the lower structure, floor loads are carried by flat slabs constructed with lightweight concrete. Horizontal forces on the building are resisted by the tubular design of the tower which consists of two equal tubes formed by the perimeter columns and a transverse wall across the shorter dimension of the rectangular tower.

Since the column framing system in the tower is different from the framing system in the lower portion, transfer between the two framing systems occurs at the 14th floor level. The transfer is achieved through the use of 15-ft deep (4.5-m) by 4-ft (1.2-m) wide reinforced concrete transfer

Table 9.3   Properties of columns and walls at Water Tower Place

Story	Concrete design strength psi	Grade of reinforcement ksi	Column size Interior* in.	Exterior in.	Wall thickness in.
75–73	4000**	60	18×24	25×44	16
73–72	4000**	60	18×24	10×48	16
72–63	4000	60	18×24	10×48	12
63–60	4000	60	18×30	10×48	12
60–55	5000	60	18×30	10×48	12
55–53	6000	60	18×30	10×48	12
53–50	6000	60	18×40	10×48	12
50–45	6000	60	18×40	14×48	12
45–40	6000	75	18×40	14×48	12
40–34	7500	75	18×40	14×48	12
34–33	7500	75	18×40	14×48	16
33–32	7500	75	18×54	16×44	16
32–31	7500	75	18×54	16×48	16
31–25	7500	75	18×54	16×48	14
25–15	9000	75	18×54	16×48	16
15–14	9000		Transfer girder		
14–B2	9000	75	48×48	48×48	12 and 18
B2–B4	9000	75	48×48	48 dia.	12 and 18

* Size of column supporting a full interior bay. Some individual columns change size at levels other than those shown
** 5000 psi concrete used in wall
Metric equivalents:
   1 ksi = 6.89 MPa
   1 in. = 25.4 mm

girders spanning 30 or 31 ft (9 m). These transfer girders contain concrete with a design compressive strength of 9000 psi (62 MPa).

Column sizes, wall thicknesses, and reinforcement details change continuously throughout the height of the structure. The highest specified concrete compressive strength which is used in the lower columns and walls is 9000 psi (62 MPa). The lowest strength concrete used at the top of the building is 4000 psi (28 MPa). Normal weight concrete is used in all columns and walls. Details of the floor levels where major changes in column and wall properties occur are given in Table 9.3. Concrete with a compressive strength of 9000 psi (62 MPa) is used from the fourth basement level through to story 25. Thereafter, 7500 psi (52 MPa) concrete is used through story 40. At the time of its construction, Water Tower Place represented the greatest use of concrete with a compressive strength in excess of 6000 psi (41 MPa). The complete building contains approximately 160,000 cu yd (122,000 cu m) of concrete, of which 90,000 cu yd (69,000 cu m) are lightweight. Water Tower Place, with a height of 859 ft (262 m), was the world's tallest reinforced concrete building until Two Prudential Plaza and 311 South Wacker Drive were completed. More importantly, however, the building was originally conceived as a structural steel frame building. Without the availability of high-strength concrete, Water Tower

Place would not have beeen built as a reinforced concrete structure.

An extensive program of field measurements of column shortening and laboratory programs to determine concrete properties was performed in connection with the construction of the building.[14] Field measurements for a total of 13 years have been reported.[19] Based on the observed creep and shrinkage deformations in laboratory specimens, the time-dependent shortening of the columns was calculated. The effects of loading history, column size, amount of reinforcement and concrete properties were taken into account. The calculated values were compared with those measured in the building and satisfactory agreement was obtained. In constructing the building, the designer had allowed for a shortening of $\frac{3}{8}$ in. (10 mm) per story height in determining joint sizes between the exterior marble panels. Measured shortening was approximately one half of this amount.

Texas Commerce Tower, Houston

The Texas Commerce Tower in United Energy Plaza, Houston, Texas is a 75-story composite steel and concrete building.[20,21] Although a composite structure, the building contains approximately 95,000 cu yd (73,000 cu m) of cast-in-place concrete and 8500 tons of reinforcing steel. Prior to the construction of the Texas Commerce Tower, the maximum concrete strength utilized in building construction in Houston was 6000 psi (42 MPa) lightweight concrete used for One Shell Plaza. However, most of the building construction in Houston utilized normal weight concrete with 5000 psi (35 MPa) as a maximum strength. From the owner's perspective, the use of these lower strength concretes would have resulted in extremely large and unacceptable column sizes. Consequently, extensive work was done to develop a higher strength concrete using limestone aggregate which had to be imported into the Houston area. This enabled the column sizes to be reduced to an acceptable level.

The structural frame of the building uses exterior columns consisting of steel erection columns and cast-in-place concrete columns. At the time of its construction, the Texas Commerce Tower was unique in that all the concrete placed on the project was pumped; the highest concrete placement being approximately 1000 ft (304 m) above street level. The 7500 psi (52 MPa) concrete had an average 28-day concrete compressive strength of 8146 psi (56 MPa) with a 56-day strength of 9005 psi (62 MPa).

In addition to providing the vertical load resisting system for the structure, high-strength concrete also provided increased stiffness for interstory drift. Since the modulus of elasticity of the high-strength concretes is greater than that for lower strength concretes, a structure with high-strength concrete and the same member sizes will have a greater stiffness compared to a similar structure with lower strength concretes. As a result, the maximum deflection of the building under design wind loads will be less with the high-strength concretes. Differential movements between the exterior composite frame and the interior columns of structural steel were also a major design consideration in this structure. The use of high-strength concrete, with a higher modulus of elasticity, reduced the actual shortening of the concrete columns. Consequently, there was less need for compensation between the interior steel frame and the exterior concrete columns. High-strength concrete was also advantageous from the standpoint of speed of construction. Construction of the concrete portion of the building proceeded at the rate of two floors per week, which at least equaled that for the structural steel construction. Concrete compressive strengths of 5200 psi (36 MPa) at three days for the 7500 psi (52 MPa) concrete allowed the columns to be stripped at an early age.

Interfirst Plaza, Dallas

Interfirst Plaza is a 72-story structure which utilizes composite steel and concrete columns.[22,23] The concrete design strength was 10,000 psi (69 MPa) for all columns in the building. High-strength concrete was specified to provide a high stiffness or high modulus of elasticity for the full height of the structure. This was needed because the height to width ratio of the building was 7.24. High-strength concrete was specified to provide maximum stiffness per unit dollar. It has been estimated that high-strength concrete in this structure provided six times as much stiffness per dollar compared to a structural steel framing system. The building has a height of 921 ft (281 m) and, at the time of its construction, was the tallest in Dallas. Vertical load in the building is carried by 16 columns which range in size from 8 × 8 ft (2.4 × 2.4 m) to 10 × 6 ft (3 × 1.8 m) and contain as much as 96 No. 18 bars. Even with 10,000 psi (69 MPa) concrete available, a large amount of reinforcing steel was still needed.

Two Prudential Plaza, Chicago

Two Prudential Plaza, shown in Fig. 9.4, is a 920-ft (280-m) tall building and, prior to the completion of 311 South Wacker Drive, had a short period in which it was the world's tallest reinforced concrete building.[24] The foundation system for the building consists of rock caissons containing 8000 psi (55 MPa) compressive strength concrete. The lower floor shear walls and columns contain 12,000 psi (83 MPa) concrete while the mid-range floors utilize 10,000 and 8000 psi (69 and 55 MPa) compressive strength concrete. The top floor columns contain 9000 psi concrete. Floor slabs and beams utilize 4000 psi (28 MPa) concrete. The cone-shaped portion of the top of the building uses a structural steel framing system.

Initially, the building was designed with a slip formed concrete core and structural steel perimeter columns. A value engineering study by the project team revealed that concrete was a more economical material in terms of both time and money. The concrete building was also more rigid against lateral loading than the structural steel equivalent.

Fig. 9.4   Two Prudential Plaza

311 South Wacker Drive, Chicago

311 South Wacker Drive is currently the tallest reinforced concrete building in the world.[24,25] The 70-story structure rises 969 ft (295 m) above street level. In plan, the tower design is trapezoidal up to the 51st floor where it becomes octagonal to offer additional corner officer space. During design, several alternatives were considered for the structural framing system.

These included reinforced concrete, composite steel frame with concrete shear wall, and steel structure with concrete shear wall core. The selected reinforced concrete framing system uses a continuous shear wall tied to the columns with beams at each floor level to form a diaphragm. Concrete with strengths up to 12,000 psi (83 MPa) are used in the lower columns of the building with strengths of 10,000 and 9000 psi (69 and 62 MPa) being used at the higher levels.

As illustrated in Fig. 9.5, the lower portions of the building consist of a six sided structure 225×135 ft (69×41 m). The building is then stepped back at different levels resulting in an octagonal tower that rises to the 70th level. The building is topped by a 65 ft (20 m) diameter, 75 ft (23 m) high ornamental drum. The vertical load carrying elements of the building consist of reinforced concrete columns and shear walls. The shear walls are located in the central area of the building. Lateral load is resisted by a combination of exterior columns and interior concrete shear walls.

Fig. 9.5   311 South Wacker Drive

311 South Wacker Drive contains approximately 110,000 cu yd (84,000 cu m) of reinforced concrete. A special feature of the construction was the decision to pump virtually all of the concrete.[19,26] Pumping rates as high as 100 cu yd per hour were achieved for the first 36 stories of the building. Another unique feature of the building was the decision to utilize 9000 psi (62 MPa) concrete in some of the floor slabs. This decision was made in order to avoid the more conventional option of puddling high-strength concrete into the floor slabs around the columns. In addition to the utilization of the 9000 psi (62 MPa) concrete, the floors were also post-tensioned to reduce their thickness and self weight of the structure. The mean strengths of the 9000, 10,000 and 12,000 psi (62, 69, 83 MPa) concrete at 90 days were 12,570, 11,240, and 13,900 psi (85.7, 77.5, 95.8 MPa). The high-strength concrete columns were generally stripped at an age of 16 to 18 hours and were wrapped in blankets for a period of five days to reduce temperature gradients. In general, a five-day cycle for construction was achieved for the building.

225 West Wacker Drive, Chicago

225 West Wacker Drive in Chicago is a 31-story building which utilizes 14,000 psi (97 MPa) compressive strength concrete in the columns of the basement levels and the first five stories above ground. Concrete strengths change at appropriate floor levels. Concrete strengths used in the building include 14,000, 12,000, 10,000 and 8000 psi (97, 83, 69 and 55 MPa).

Although 225 West Wacker is a relatively short building, high-strength concrete was used in the columns simply because it was still an economical means of satisfying the structural requirements.

Average compressive strengths measured at 56 days for the 14,000 psi(97 MPa) concrete were 16,140 psi (111 MPa) with a coefficient of variation of 5.8%. Maximum strength measured was 18,040 psi (124 MPa). In addition to the use of 14,000 psi (97 MPa) design strength, one experimental column was cast with 17,000 psi (117 MPa) compressive strength concrete. The casting of this column was part of ongoing research to address the practical problems of producing and placing higher strengths concretes in the Chicago area.[27] Measured compressive strengths at 56 days were in excess of 18,000 psi (124 MPa).

Two Union Square, Seattle

The highest strength concrete used in any large scale commercial application is the 19,000 psi (131 MPa) in the 58-story 720-ft (220-m) tall Two Union Square in Seattle.[28] The concrete compressive strength originally specified for the structure was 14,000 psi (97 MPa). However, the designer also wished to achieve a modulus of elasticity of 7.2×10^6 psi (50 GPa). To achieve the modulus of elasticity, testing showed that it was necessary to go to a concrete compressive strength of 19,000 psi (131 MPa). Some strengths in excess of 20,000 psi (138 MPa) were achieved for the concrete used in the structure.

The building frame consists of four 10-ft (3-m) diameter core columns and 14 perimeter columns that range in diameter from 3 to 4 ft (0.9 to 1.2 m). All concrete in all 18 columns including the top stories of the building has a 19,000 psi (131 MPa) compressive strength. The columns consists of reinforcing in the form of a permanent $\frac{5}{8}$ in. (16 mm) thick steel shells surrounding the perimeter. The steel skin is tied to the concrete using shear studs at 1 ft (300 mm) centers. The technique of using steel shells with an interior core of concrete was made to reduce the cost of the structure. Concrete was pumped into the steel shells of the columns from the bottom of each tier. No vibration of the concrete was found to be needed. Extensive quality control was also introduced on this project to ensure that all of the concrete would achieve the specified strengths.

Miglin-Beitler Building, Chicago

If construction is completed, the Miglin-Beitler Tower will establish a new record as the world's tallest building.[29,30] The tip of the tower of the structure will be 1,999 ft 6 in. (609.5 m) above street level. The structural frame consists of a cruciform tube structure to achieve structural efficiency, required dynamic performance, simplicity of construction and unobstructed integration of the structure into the leased office floor space. The structural solution consists of a 62 ft 6 in. (19 m) square concrete core with walls varying in thickness from 36 to 18 in. (0.9 to 0.5 m). On the exterior of the building, eight large columns extend outside the footprint of the building. These columns vary in dimension from 6.5×33 ft (1.9×10.0 m) at the base to 5.5×15 ft (1.7×4.6 m) at the middle to 4.5×30 ft (1.4×9.2 m) at the top. Link beams connect the four corners of the core to the eight fin columns at each core level. Link beams are made of reinforced concrete. In addition, 3 two-story deep outrigger walls located at three different levels in the building further connect the exterior columns to the concrete core. The floor system consists of a conventional structural steel composite system utilizing rolled steel sections, corrugated deck and concrete topping. During erection, the floor system is supported on light steel erection columns. The last structural component of the cruciform tube is the exterior virendeel truss which consists of horizontal spandrels and two vertical colummns at each of the 60 ft (18 m) faces on the four sides of the building. The foundation system of the project consists of caissons varying in diameter from 8 to 10 ft (2.4 to 3.0 m).

The availability of concrete with strengths up to 15,000 psi (103 MPa) was a critical factor in selecting concrete for the primary structural system. The columns of the building will contain concrete with compressive strengths of 14,000, 12,000 and 10,000 psi (97, 83 and 69 MPa). As illustrated in Fig. 9.6, construction of the Miglin-Beitler Tower will add a unique feature to the Chicago skyline. It is also likely to stand as the world's tallest building for a number of years.

Fig. 9.6 Miglin-Beitler Building

Fig. 9.7   Bridges utilizing high-strength concrete

9.6   Bridges

Without doubt, the largest use of high-strength concrete has been in the columns of buildings. However, high-strength concrete is receiving more and more attention for use in bridge structures. Three examples of the usage in North America are illustrated in Fig. 9.7. The tensile strength of high-strength concrete increases with compressive strength. This is beneficial in the design of prestressed concrete members such as bridge girders where the tensile strength may control the design. The reduced creep of high-strength concrete is also beneficial in reducing prestress losses in bridge girders. Consequently, utilization of high-strength concrete results in economies in prestressed concrete girders. Table 9.4 lists a selection of bridges that have utilized concrete strengths of 6000 psi (42 MPa) or greater.[7]

Prestressed concrete girders

Research studies have indicated the potential applications of high-strength concrete in solid section prestressed concrete girders.[31-33] In general, the studies have generally reached the same conclusions. For a given girder size, it is possible to increase the span capability by the utilization of the

Table 9.4   Bridges with high-strength concrete

Bridge	Location	Year	Maximum span ft	m	Maximum design strength psi	MPa
Willows Bridge	Toronto	1967	158	48	6000	41
Houston Ship Canal	Texas	1981	750	229	6000	41
San Diego to Coronado	California	1969	140	43	6000[1]	41[1]
Linn Cove Viaduct	North Carolina	1979	180	55	6000	41
Pasco-Kennewick	Washington	1978	981	299	6000	41
Coweman River Bridges	Washington	–	146	45	7000	48
East Huntington	W. VA to Ohio	1984	900	274	8000	55
Annacis Bridge	British Columbia	1986	1526	465	8000	55
Nitta Highway Bridge	Japan	1968	98	30	8500	59
Kaminoshima Highway Bridge	Japan	1970	282	86	8500	59
Joigny	France	1989	150	46	8700	60
Tower Road	Washington	–	161	49	9000	62
Esker Overhead	British Columbia	1990	164	50	9000	62
Fukamitso Highway Bridge	Japan	1974	85	26	10,000	69
Ootanabe Railway Bridge	Japan	1973	79	24	11,400	79
Akkagawa Railway Bridge	Japan	1976	150	46	11,400	79

1. Lightweight concrete

higher strength concretes. Tower Road Bridge shown in Fig. 9.7 is an example. For fixed girder dimensions and span lengths, it is possible to use fewer girders in a bridge when high-strength concretes are utilized. This also results in a lower unit cost for a given length structure. The utilization of longer span lengths for a multispan structure results in the need for less piers and foundations. This, in turn, reduces the cost of the substructure.

In general, prestressed concrete girders have been produced with compressive strengths in excess of 6000 psi (41 MPa) for many years. Although the design may be based on 6000 psi (41 MPa) at 28 days, strengths required for release of the prestressing result in 28-day strengths well in excess of the minimum specified. However, the higher strengths have not been utilized in design. It should be noted that the State of Washington in the United States has been utilizing high-strength concretes in prestressed concrete girders for many years.

Research has indicated that the advantages of utilizing higher and higher strength concretes do not continue forever. A point is reached at which the higher compressive strengths cannot be utilized.[33] It becomes impossible to induce sufficient prestressing force into the girders using existing prestressing strand dimensions and strengths to take advantage of the higher concrete strengths. This maximum occurs somewhere in the range of 8000 to 12,000 psi (55 to 83 MPa) depending upon the particular situation being analyzed.

East Huntington Cable stayed bridge

The East Huntington Bridge shown in Fig. 9.7 is a segmental prestressed concrete cable stayed bridge over the Ohio River between the states of West Virginia and Ohio.[34] Span lengths are 158, 300, 900 and 608 ft (48, 91, 274, 185 m). In a cable stayed bridge, the superstructure is a long compression member designed to resist the horizontal components of the forces in the cable stays. Also, in long span bridges, the superstructure weight is the predominant load for which the structure must be designed. Consequently, the selection of high-strength concrete results in a lighter structure to resist the longitudinal compressive forces. In the case of the East Huntington Bridge, high-strength concrete was also selected because of its improved durability and higher tensile strengths. Specified compressive strength of the superstructure elements was 8000 psi (55 MPa) at 28 days. Actual strengths averaged 9900 psi (68 MPa) at 28 days and 10,500 psi (72 MPa) at 90 days. In competitive bidding against a steel box girder alternate design, the concrete option was bid at 10 million US dollars less than the steel alternate.

Annacis cable stayed bridge

The Annacis Bridge also named the Alex Fraser Bridge crosses the Fraser River near Vancouver, British Columbia, Canada.[35,36] The cable stayed portion of the structure consists of five continuous spans with a center span of 1,526 ft (465 m). The composite superstructure consists of twin 83-in. (2.1-m) deep I-beams of constant depth, transverse floor beams that taper in depth, and a composite precast concrete deck with a cast-in-place overlay. The precast deck panels are typically 44 ft × 13 ft × 8.5 in. (13.5 × 4.0 × 0.215 m) with a specified compressive strength of 8000 psi (55 MPa) at 56 days. Composite action between the precast concrete elements and the steel framework is achieved through shear studs welded to the floor beam top flanges. Deck panels are integrated with the steel superstructure by cast-in-place strips of concrete. Required strength for the cast-in-place concrete was also specified at 8000 psi (55 MPa) at 56 days. A cost comparison between a composite concrete deck and an orthotropic steel deck indicated that the total cost of the structure with a concrete deck was substantially less.

Joigny Bridge

The Joigny Bridge is an experimental bridge designed to demonstrate the possibility of producing high-strength concrete bridges using existing batching plants and local aggregates.[37] The bridge is a three-span structure with span lengths of 111, 151 and 111 ft (34, 46 and 34 m). In cross section, the bridge is a double-T with a deck width of 49 ft (15 m). The bridge was designed based on a concrete with a characteristic compressive strength of

Fig. 9.8   Glomar Beaufort Sea 1

8700 psi (60 MPa) after 28 days and utilizes external longitudinal prestres-sing. This bridge was part of the French program to introduce high performance concretes into bridge construction.

9.7 Special applications

Offshore structures

Concretes with compressive strengths in excess of 6000 psi (41 MPa) have been used in offshore structures since the 1970s. High-strength concrete is important in offshore structures as a means to reduce self weight while providing strength and durability. Various applications have been de-scribed by Ronneberg[38] and CEB/FIP.[1]

In 1984, the Glomar Beaufort Sea 1, shown in Fig. 9.8, was placed in the Arctic.[39] This exploratory drilling structure contains about 12,000 cu yd (9200 cu m) of high-strength lightweight concrete with unit weights of about 112 lb/ft^3 (1.79 Mg/m^3) and 56-day compressive strengths of 9000 psi (62 MPa). The structure also contains about 6500 cu yd (5000 cu m) of high-strength normal weight concrete with unit weights of about 145 lb/ft^3 (2.32 Mg/m^3) and 56-day compressive strengths of about 10,000 psi (69 MPa).

Miscellaneous

Miscellaneous applications of high-strength concrete have been summa-rized by ACI Committee 363[7] and CEB/FIP.[1] For example:

Precast panels for dam	– 9000 psi (62 MPa)
Prestressed concrete poles	– 10,000 psi (69 MPa)
Grandstand roofs	– 7500 and 8850 psi (52 and 61 MPa)
Marine foundations	– 8000 psi (55 MPa)
Underwater bridge	– 9400 psi (65 MPa)

Grandstand elements	– 8700 psi (60 MPa)
Avalanche shelters	– 10,900 psi (75 MPa)
Piles	– 10,900 psi (75 MPa)

Acknowledgements

The author acknowledges information provided by the designers and contractors for the various structures described in this chapter. Portions of this chapter were written by S.H. Gebler and D.A. Whiting of Construction Technology Laboratories. Photographs were provided by the Portland Cement Association.

References

1   CEB/FIB Working Group on HSC (1990) *High strength concrete – state of the art*. The Institution of Structural Engineers, London.
2   Burnett, I.D. (1989) High-strength concrete in Melbourne, Australia. *Concrete International Design and Construction*. American Concrete Institute, **11**, No. 4, April, 17–25.
3   Holand, I., Helland, S., Jakobsen, B. and Lenschow, R. (1987) *Utilization of high strength concrete*, Proceedings, Symposium in Stavangar, Norway, June 15–18. Tapir, Trondheim, 688 pp.
4   Hester, W.T. (1990) *High-strength concrete, second international symposium*, Publication SP-121. American Concrete Institute, Detroit, 786 pp.
5   Russell, H.G. (1990) Use of high-strength concretes. *Building Research and Practice*, **18**, No. 3, May/June, 146–152.
6   ACI Committee 318 (1989) Building Code Requirements for Reinforced Concrete. American Concrete Institute, Detroit, 353 pp.
7   ACI Committee 363 (1984) State-of-the-art report on high strength concrete. *ACI Journal*, **81**, No. 4, July/August, 364–411.
8   Roller, J.J. and Russell, H.G. (1990) Shear strength of high-strength concrete beams with web reinforcement. *ACI Structural Journal*, **87**, No. 2, March–April, 191–8.
9   Russell, H.G. (1985) High-rise concrete buildings: shrinkage, creep and temperature effects, *Analysis and design of high-rise buildings*, Publication SP-97. American Concrete Institute, Detroit, 125–37.
10 Pauw, A. (1960) Static modulus of elasticity of concrete as affected by density. *Journal of the American Institute*, Proceedings **32**, No. 6, December, 679–787.
11 Martinez, S., Nelson, A.H. and Slate, F.O. (1982) *Spirally-reinforced high strength concrete columns*, Research Report No. 82-10. Department of Structural Engineering, Cornell University.
12 Cook, J.E. (1989) Research and application of 10,000psi (f_c') high-strength concrete, *Concrete International Design & Construction*. American Concrete Institute, **11**, No. 10, October, 67–75.
13 Pfeifer, D.W., Magura, D.D., Russell, H.G. and Corley, W.G. (1971) Time-dependent deformations in a 70-story structure, *Designing for effects of creep, shrinkage and temperature in concrete structures*, Publication SP27-7. American Concrete Institute, Detroit, 159–85.
14 Russell, H.G. and Corley, W.G. (1977) Time-dependent behavior of columns in Water Tower Place, *Douglas McHenry international symposium on concrete and concrete structures*, Publication SP-55-14. American Concrete Institute,

Detroit. Also printed as PCA Research and Development Bulletin RD052.01B, Portland Cement Association, Skokie, Illinois, 10 pp.

15   *Standard test method for compressive strengths of cylindrical concrete specimens*, ASTM C 39. American Society for Testing and Materials, Philadelphia.

16   *Standard practice for capping cylindrical concrete specimens*, ASTM C 617. American Society for Testing and Materials, Philadelphia.

17   Schmidt, W. and Hoffman, E.S. (1975) 9000-psi concrete – why?, why not? *Civil Engineering*, ASCE, **45**, No. 5, May, 52–5.

18   Giraldi, A. (1989) High-strength concrete in Washington, D.C. *Concrete International Design and Construction*. American Concrete Institute, **11**, No. 3, March, 52–5.

19   Russell, H.G. and Larson, S.C. (1989) Thirteen years of deformations in Water Tower Place. *ACI Structural Journal*, **86**, No. 2, March–April, 182–91.

20   Colaco, J.P. (1985) 75-Story Texas Commerce Plaza, Houston – the use of high-strength concrete. *High-Strength Concrete*, Publication SP-87. American Concrete Institute, Detroit, 1–8.

21   Pickard, S.S. (1981) Ruptured composite tube design for Houston's Texas Commerce Tower. *Concrete International Design & Construction*. American Concrete Institute, **3**, No. 7, July, 13–19.

22   (1983) Tower touches few bases. *Engineering News Record*, June 16, 24–25.

23   LeMessurier, W.J. (1982) Toward the ultimate in composite frames, *Building Design and Construction*. Cahners Publication Company, **23**, No. 11, November, 14–21.

24   Case history report – the world's tallest concrete skyscrapers, *Bulletin No. 40*. Concrete Reinforcing Steel Institute, Schaumburg, IL, 8 pp.

25   (1989) Tall concrete buildings come of age, *Engineering News Record*, November 30, 25–27.

26   Page, K.M. (1990) Pumping high-strength on world's tallest concrete building, *Concrete International Design & Construction*. American Concrete Institute, **12**, No. 7, January, 26–8.

27   Moreno, J. (1990) 225 W. Wacker Drive, *Concrete International Design & Construction*. American Concrete Institute, **12**, No. 1, January, 35–9.

28   (1989) Put that in your pipe and cure it. *ENR*, February 16, 44–53.

29   Thornton, C.H. (1990) The world's tallest building – Chicago's Miglin-Beitler Tower. *Engineered Concrete Structures*, **3**, No. 3, Portland Cement Association, December, 1–2.

30   Thornton, C.A., Hungspruke, O. and DeScena, R.P. (1991) Looking down at the Sears Tower. *Modern Steel Construction*. American Institute of Steel Construction, August, 27–30.

31   Jobse, H.J. (1981) Applications of high-strength concrete for highway bridges, Executive Summary, U.S. Department of Transportation, Federal Highway Administration, Washington, DC, Report No. FHWA/RD 81/096, October, 27 pp.

32   Jobse, H.J. and Moustafa, S.E. (1984) Applications of high strength concrete for highway bridges. *Journal of the Prestressed Concrete Institute*, **29**, No. 3, May–June, 44–73.

33   Zia, P., Schemmel, J.J. and Tallman, T.E. (1989) *Structural applications of high strength concrete*, Report No. FHWA/NC/89-006, Center for Transportation Engineering Studies, North Carolina State University, Raleigh, June, 330 pp.

34   (1984) Hybrid girder in cable-stay debut. *Engineering News Record*, November 15, 32–6.

35   Taylor, P.R. and Torrejon, J.E. (1987) Annacis Bridge – design and construction of the cable-stayed span. *Quarterly Journal of the Federation Internationale de la Precontrainte*, 4, 18–23.

36   (1986) Stayed girder reaches a record with simplicity. *Engineering News Record*, May 22.
37   Pliskin, L. and Malier, Y. (1990) The French R&D Project, 'New developments for concrete, the high strength concrete Bridge of Joigny', Preprint No. 89-0586, Transportation Research Board, 69th Annual Meeting, Washington, D.C., January.
38   Ronneberg, H. and Sandvik, M. (1990) High strength concrete for North Sea platforms, *Concrete International Design & Construction*. American Concrete Institute, **12**, No. 1, January, 29–34.
39   Fiorato, A.E., Person, A. and Pfeifer, D.W. (1984) The first large scale use of high-strength lightweight concrete in the Arctic environment, *Second Symposium on Arctic Offshore Drilling Platforms*, Houston, Texas, April.

10 High strength lightweight aggregate concrete

T A Holm and T W Bremner

10.1 Introduction

It may be argued that the first practical use of high strength concrete took place in World War I when the American Emergency Fleet Corporation built lightweight concrete ships with specified compressive strengths of 5000 psi when commercial normal weight concrete strengths of that time were 2000 psi. It was fully recognized by these forward looking engineers that high self-weight was the major impediment in the use of structural concrete.

Concrete density can be reduced in several ways, i.e., lightweight aggregates, cellular foams, high air contents, no fines mixes, etc., but only high quality structural grade lightweight aggregates can develop high strength lightweight aggregate concretes. As such, the letter 'A' in the abbreviation LAC for lightweight aggregate concrete will be dropped and similarly HSLC indicates high strength lightweight aggregate concrete. In a similar fashion, LWA, NWA, NWC and HSNWC represent lightweight aggregate, normal weight aggregate, normal weight concrete and high strength normal weight concrete respectively.

Structural efficiency

The entire hull structure of the USS *Selma* was constructed with HSLC in a shipyard in Mobile, Alabama and launched in 1919. The strength/density (S/D) ratio (structural efficiency) of 50 used in the USS *Selma* (+5000 psi/100 pcf) was extraordinary for that time.[1] Improvements in structural efficiency of concrete since that time are shown schematically in Fig. 10.1, revealing upward trends in the 1950s with introduction of prestressed

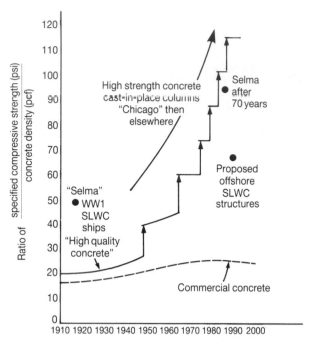

Fig. 10.1   The structural efficiency of concrete. The ratio of specified compressive strength density (psi/pcf) through the recent history of construction (from Holm and Bremner[1])

concrete, followed by production of HSNWC for columns of very tall cast-in-place concrete frame commercial buildings. It would appear that the strength/density ratio for the LC produced in the World War I ship construction program was only exceeded by HSNWC 40 years later.

Compression strength of the LC cores taken at the water line from the USS *Selma* and tested in 1980 were found to be twice the 28 day specified strengths, and from a structural efficiency standpoint are not appreciably different from the HSNWC of today. Analysis of the physical and engineering properties of the HSLC in the ships of World War I, the 104 HSLC World War II ships, as well as numerous recent bridges built, can be found in other reports that amply prove the long-term successful perform-ance of HSLC.[2–4]

Maximum strength 'ceiling'

The strength 'ceiling' of HSLC is reached when further additions of binder materials do not significantly increase strength. Figure 10.2 demonstrates that the compressive strength ceiling for the particular $\frac{3}{4}$ in. (20 mm) top size LWA tested was somewhat more than 8000 psi (55 MPa) at an age of 75 days. When the top size of this aggregate was reduced to $\frac{3}{8}$ in. (10 mm), the strength ceiling significantly increased to more than 10,000 psi

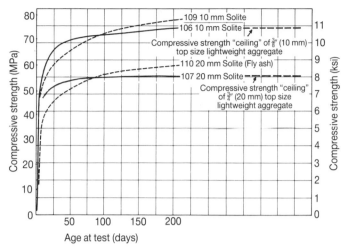

Fig. 10.2 Compressive strength versus age of lightweight concrete (1975–1979 Series)[5]

(69 MPa). Mixes incorporating fly ash demonstrated higher strength ceilings at later ages than non-fly ash control concretes. Results on concretes containing $\frac{1}{2}$ in. top (13 mm) size were intermediate between the $\frac{3}{4}$ in. (20 mm) and $\frac{3}{8}$ in. (10 mm) curves are reported in *Criteria for Designing Lightweight Concrete Bridges*[5] and are not shown for clarity.

Analyzing strength as a function of the quantity of cementitious binder as shown in Fig. 10.3, however, reveals that mixes incorporating binder quantities exceeding an optimum volume are not cost effective.

Pore system of structural lightweight aggregate

Strength ceilings of LWA produced from differing quarries and plants will

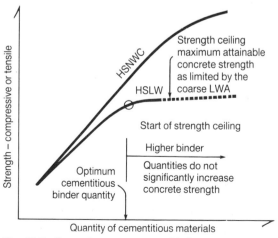

Fig. 10.3 Strength versus concrete binder content

Fig. 10.4   Contact zone – structural lightweight concrete, W P Lane Memorial Bridge over Chesapeake Bay, Annapolis, MD – constructed in 1952[6]

vary considerably. This variation is due to structural characteristics of the pore system developed during the production process. The producer's goal is to manufacture a high quality structural grade LWA which has non-interconnected, essentially spherical, well distributed pores surrounded by a strong, crack free vitrious ceramic matrix. The scanning electron micrograph in Fig. 10.4 demonstrates a well developed pore distribution system of a concrete sample cored from a highly exposed 30 year old bridge deck.[6]

Internal integrity at the contact zone

Micrographs of concretes obtained from mature structural LC ships, marine structures and bridge have consistently revealed minimal microcracking and a limited volume of unhydrated cement grains. The boundary between the cementitious matrix and coarse aggregates is essentially indistinguishable at the 'contact zone' transition between the two phases in all mature HSLCs. The contact zone in LC is enhanced by several factors including: pozzolanic reactivity of the surface of the lightweight aggregate developed during high temperature 2000 °F (1100 °C) production, surface roughness and, most importantly, the opportunity for th two-phase porous system to reach moisture equilibrium without developing the water gain lenses frequently observed under and on the sides of NWA.[7]

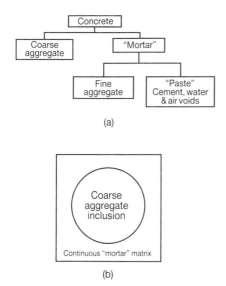

Fig. 10.5   Cement, water, air voids and fine aggregate combine in (a) to form the continuous mortar matrix that surrounds the coarse aggregate inclusion in (b) to produce concrete

Principles of elastic compatibility of a particulate composite

A particular composite is by its very definition heterogeneous, and concrete is perhaps the most heterogeneous of composites with size of inclusions varying from large aggregate down to unhydrated cement grains, and containing voids the size of entrapped air bubbles down to the gel pores in the cement paste. The general understanding of concrete as a particulate composite previously used in the analysis of regular strength LC may be extended to HSLC.[8]

Concrete can be considered as a two-phase composite composed of coarse aggregate particles enveloped in a continuous mortar matrix. This latter phase includes all the other concrete constituents including fine aggregate, mineral admixtures, cement, water and voids from all sources. This division, schematically shown in Fig. 10.5, is visible to the naked eye and may be used to explain important aspects of the strength and durability of concrete.

The elastic mismatch between coarse aggregate particles and the surrounding mortar matrix gives rise to stress concentrations when the composite is subjected to an applied stress. These stress concentrations are superimposed on a system already subjected to internal stresses arising from dissimilar coefficients of thermal expansion of the constituents and from aggregate restraint of matrix volume changes. The latter can be caused by drying shrinkage, thermal shrinkage from hydration temperatures, or changes that result from continued hydration of the cement paste. These inherent stresses are essentially self-induced and may be of a

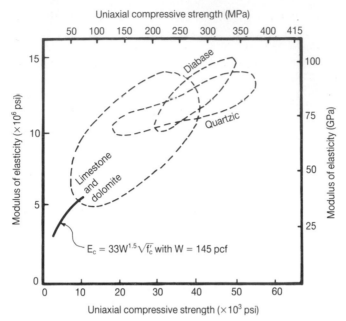

Fig. 10.6 Range of stuffness of concrete caused by variability in the stiffness of the aggregate (adapted from Stagg[9])

magnitude that extensive microcracking may take place before any super-imposed stress is applied.

Natural aggregates have an extremely wide range of elastic moduli resulting from large differences of mineralogy, porosity, flaws, lamina-tions, grain size and bonding. It is not uncommon for a fine-grained diabase rock to have an elastic modulus greater than 13×10^6 psi (90 GPa) while poorly bonded, highly porous natural aggregates have been known to have values lower than 3×10^6 psi (20 GPa). Aggregate description by name of rock is insufficiently precise, as demonstrated in one rock mechanics text which reported a range of elastic modulus of 3 to 10×10^6 psi (20 to 69 GPa) for one rock type.[9]

Figure 10.6 illustrates compressive strength and stiffness characteristics reported for several rock types and compares these wide ranges with the modulus of elasticity of concrete as suggested by equation $E_c = 33\omega^{1.5}\sqrt{f_c'}$ (in psi) of ACI 318–89 Code. The ratio of the coarse aggregate modulus to that of the concrete composite can be shown to be as much as 3, signaling a further difference between the two interacting phases (mortar and coarse aggregate) of as much as 5 to 1. That the strength-making potential of the stone or gravel is normally not fully developed is evident from visual examination of fracture surfaces of concrete cylinders after compression testing. The nature of the fracture surface of concretes is strongly influenced by the degree of heterogeneity between the two phases and the extent to which they are securely bonded together. Shah[10] reported on the profound influence exerted by the contact zone in compressive strength tests on concretes in which aggregate surface area was modified by coatings. The degree of heterogeneity and the behavior of the contact zone between the two phases are the principal reasons for the departure of some concretes from estimates of strength based upon the water-to-cement (w/c) ratio. As has been suggested,[8] undue preoccupation with the matrix w/c ratio may lead to faulty estimates of compressive strength and even greater misunderstanding of concrete's behavior from durability, permeability and tensile type loading conditions.

Obviously the characteristics of the NWA will have a major effect on elastic compatibility. The interaction between the absolute volume percentage of coarse aggregate ($\pm35\%$) and the mortar phase ($\pm65\%$) will result in a concrete with a modulus intermediate between the two fractions. At usual commercial strength levels the elastic mismatch within structural LC is considerably reduced due to the limited range of elastic properties of usual LWA particles.

Elastic matching of components of commercial strength lightweight concrete

Muller-Rochholz measured the elastic modulus of individual particles of LWA and NWA using ultrasonic pulse velocity techniques.[11] This report concluded that the modulus of elasticity of structural grade LWA exceeded values of the cementitious paste fraction, and suggested that instances when LC strength exceeded that of companion NWC at equal binder content, were understandable in light of the relative stress homogeneity.

The *FIP Manual of Lightweight Aggregate Concrete*[12] prepared by the Federation Internationale de la Precontraintc (FIP) reports that the modulus of elasticity of an individual particle of LWA may be estimated by the formula: $E_c = 0.008p^2$ (MPa), where p is the dry particle density. Usual North American structural grade LWA having dry particle densities of 1.2 to 1.5 (1200 to 1500 kg/m^3) would result in a particle modulus of elasticity from 1.7 to 2.6×10^6 psi (11.5 to 18 GPa). At these densities the modulus of elasticity of individual particles of LWA approaches that measured on the mortar fraction of air-entrained commercial strength LC.[8]

The elastic modulus of air-entrained and nonair-entrained mortars is shown as a function of compressive strength in Fig. 10.7. The modulus of a typical individual particle of coarse LWA, as well as a range of values of modulus for stone aggregates, is also shown. These results were obtained by testing concretes and equivalent mortars with the same composition found in concrete, with the exception that the coarse aggregate had been fractioned out.

Mortar matrix mixes were produced to cover usual ranges of cement contents at the same time as companion structural LCs were cast with all other mix constituents kept the same. Data and analysis of these tests are beyond the scope of this chapter.

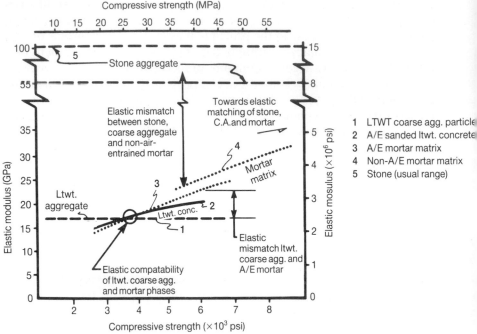

Fig. 10.7 Elastic mismatch in lightweight and normal weight concrete

Sand LC with compressive strength of approximately 4000 psi (28 MPa) made with typical North American structural grade LWAs has values of E_a/E_m approaching unity. From a stress concentration point of view, this combination of constituents would act as a homogeneous material resulting in concrete with minimum stress-induced microcracking. Thus, at ordinary commercial strengths, the elastic match of the two components will be close for air-entrained concrete made with high quality LWA. Matching of the elastic properties of ordinary concrete using a high modulus NWA such as a diabase will be possible only with the ultra-high quality matrix fractions recently developed incorporating superplasticizers and supplementary cementitious materials.

Air entrainment in concrete significantly reduces the stiffness of the mortar fraction and, as shown in Fig. 10.7, results in a convergence of elastic properties of the two phases of sanded structural LC while increasing the degree of elastic mismatch in ordinary concrete. This fact, combined with the slight reduction in mixing water caused by air entrainment, explains why the strength penalty caused by air entrainment is less significant for structural LC than for concretes using highly rigid NWA.

Elastic mismatching of components of high strength lightweight concrete

Combining ultra-high strength, low air content mortar matrix fractions with coarse LWA will produce an elastic mismatch resulting in fracture that starts with transverse splitting of the structural LWA particles. Splitting action stemming from lateral strains is indirectly responsible for the strength ceiling of structural LC observed when improvements in mortar matrix quality result in little or no increase in compressive strength.

In general, for concretes using high quality NWA, elastic compatibility between the two fractions will occur only at extremely high compressive strengths. Ultra-high strength mortar fractions developed by superplasticizers and mineral admixtures will increase the possibility of achieving elastic compatibility at higher compressive strengths when ordinary aggregates are used.

While elastic mismatching plays an important yet incompletely understood role in the compressive strength capabilities of the composite, the influence on other properties (tensile and shrinkage cracking, and particularly limitations on in-service permeability and durability due to microcracking) are far more significant.

10.2 Materials for high strength lightweight aggregate concrete

For the purpose of this chapter, HSLC should have a maximum equilibrium unit weight of 125 pcf (2000 kg/m^3), as defined in ASTM C567, and a specified compressive strength of at least 5000 psi (34.5 MPa). This level on unit weight may be necessary when developing compressive strengths approaching 10,000 psi (70 MPa) while still maintaining benefits of weight reduction. HSLC with compressive strengths ranging from 5000 psi (41 MPa) to 7000 psi (48 MPa) are commercially available in some areas and testing programs on HSLC with ultimate strengths approaching 10,000 psi (70 MPa) are ongoing.

With the exception of LWA, the materials (cement, admixtures, NW fine aggregate) and methods used to produce HSLC are essentially similar to those used in their NWC counterparts. As materials and proportioning methods common to HSNWCs are extensively reported on in other chapters, they will not be addressed except where differences require explanation.

Cementitious materials

Cement

Portland cements used for HSC should conform to the requirements of ASTM C150. Granulated iron blast furnace slags used as a replacement for portland cement should conform to ASTM C989. Compressive data on cements, supplementary cementitious materials, admixtures, etc. are reported on in ACI 363.[13]

Supplementary cementitious materials

Production of ultra-HSLC generally requires the use of supplementary cementitious materials. High quality fly ash meeting the requirements of ASTM C618 will reduce permeability, improve placing qualities, lower heat of hydration, and improve long term strength characteristics. Microsilica will improve compressive strength at all ages and also provide significantly improved resistance to chloride penetration.

Supplementary cementitious materials function very effectively in HSLC, because pozzolanic activity requires the combination of the calcium hydroxide liberated during cement hydration with finely divided silica in the presence of moisture. As shown in Fig. 10.2, mixes incorporating mineral admixtures achieved higher strength ceilings than the control concretes. Favorable hydrating environments will be provided for a longer time due to the internal curing provided by the LWA absorbed moisture, thus promoting increased activity of the pozzolanic materials.

Admixtures

When used in HSLC, admixtures offer reduced water demand, enhance durability, and improved workability in a manner comparable to that of HSNWC. Water reducers, retarders and high range water reducers should conform to ASTM C494 and be dosed according to manufacturers' recommendations.

Air entrainment

LC mixtures, normally contain entrained air. Entrained air serves to increase the cohesiveness of the fresh concrete mix, and to make concrete resistant to the effects of freezing and thawing when in a wet environment. When freezing and thawing is not a consideration, then small amounts of entrained air (3 to 5%) are adequate. Entrained air volumes should meet the requirements of ACI 201 according to the severity of the exposure conditions. While air entrainment will diminish strength making characteristics of the cementitious matrix, it will also lower water and sand volumes necessary to achieve satisfactory workability, with the net effect being only a modest reduction in the strength of HSLC.

Coarse aggregate

HSLCs normally require only coarse LWA. As reported earlier, most, but not all, HSLC mixes require a reduction of the LWA top size, particularly in the 7000 to 10,000 psi (48 to 70 MPa) range. Certain LWAs, however, because of the strength of the vitreous material enveloping the pores, have routinely used the 3/4 to #4 (20 to 5 mm) gradation in production of high strength precast concrete for more than four decades. Most LWA manufacturing plants will limit coarse aggregate to two sizes to minimize production and stockpiling problems, but these plants will entertain other gradations if project volumes warrant.

Gradations of 3/4 to #4 (20 to 5 mm) or 1/2 to #4 (13 to 5 mm) will normally be appropriate for usual size HSLC members while 3/8 to #8 (10 to 5 mm) gradations may be necessary in highly reinforced members to allow adequate placement conditions.

Fine aggregate

HSLC normally incorporates normal weight sand as the fine aggregate fraction. Quality criteria developed for sands used in HSNWC (e.g., FM of about 3.0 for optimum workability and strength, etc) are identical to those used in manufacturing HSLC.

10.3   High strength lightweight concrete laboratory testing programs

Systematic laboratory investigations into physical and engineering properties of HSLC are too numerous for all to be elaborated on here. Most early programs extending strength/density relationships were conducted by LWA manufacturers and innovative precast concrete producers striving for high early release strengths, longer span flexural members, or taller one-piece precast columns.[5] These in-house programs developed functional data directly focused on members supplied to real projects. In general, project lead times were short, practical considerations of shipping and erection immediate, and mixes targeted towards satisfying specific job requirements. This type of research brought about immediate incremental progress but, in general, was not sufficiently comprehensive.

Unfortunately some researchers did not use advanced admixture formulations or supplementary cementitious materials (i.e., high range water reducer, microsilica, fly ash, slag cement) that significantly improve matrix quality and as such provide data of no commercial value. These investigations, as well as others incorporating unrealistic mixtures, inappropriate LWA, or impractical density combinations, are not reported.

Special requirements of offshore concrete structures have now brought about an explosion of practical research into the physical and engineering properties of HSLC. Several large confidential joint industry projects are now becoming publicly available as the sponsors release data according to an agreed upon timetable.[14-16] These monumental studies, comprehensively summarized by Hoff and presented at the November 1991 American Concrete Dallas Symposium on the 'Performance of Structural Lightweight Concrete',[17] are widely preferred throughout this chapter. In addition to providing comprehensive physical property data on HSLC, these programs developed innovative testing methods: revolving disc, tumbler and sliding contact ice abrasion wearing tests, freeze/thaw reistsance to spectral cycles, and freeze bond testing techniques, etc., that measured properties unique to offshore applications in the Arctic.

In addition, Hoff summarizes and reports on four other major investigations that included data on HSLC.[18] These programs included HSLC data in investigations into structural qualities of high strength concrete and include SINTEF/FCB Trondheim, Norway,[19] Hovik, Norway,[20] Gerwick,[21] and National Institute of Standards and Technology, Washington, DC.[22]

Major North American laboratory studies into properties of HSLC include those conducted at or sponsored by Expanded Shale Clay & Slate Institute,[23–25] CANMET,[26–28] Ramakrishnan,[29] Berner,[30] and Luther.[31] Because of their special structure needs, much pioneering into this issue has been conducted by Norwegian sources[32–34] with additional important contributions from other Russian, German, and UK sources.[35–39]

It has been estimated that the cost for these commercially supported research programs into the physical and structural properties of HSLC has exceeded one million dollars.[40] While much research has been already effectively transferred into actual practice on current projects, there remains a formidable task of analyzing, digesting and codifying an immense body of data into design recommendations and code standards.

10.4   Physical properties of high strength lightweight aggregate concrete

Compressive strength

There is a substantial body of information, developed over a long period of time, demonstrating that high strength can be achieved with LC. As early as 1923, Duff Abrams[41] commented, 'The high strength secured from lightweight aggregates consisting of burnt shale, has shown the fallacy of the older views that the strength of the concrete is dependent upon the strength of the aggregate.' Moderately HSLC can be achieved without significantly higher binder contents than used with NWC when the designer selects the appropriate LWA. For every commercially available LWA there is, however, a strength ceiling which is reached when compression strength is limited by crushing the aggregate. It is uneconomical to endeavor to produce HSLC exceeding the strength ceiling by using greater amounts of cement. Aggregate and readymix concrete suppliers are generally aware of the strength making potential of both NWA and LWA and should be consulted early in the design process.

Tensile strength

Tensile strength of HSLC is limited by the fact that approximately 50% of the aggregate volume is pore space. The ACI code (ACI 318–89) requires

LWA producers to supply tensile splitting test data on concrete incorporating their aggregate, allowing structural engineers to modify code equations for shear, torsion and cracking.

High tensile splitting developed on mature specimens of HSLC have shown clearly visible high moisture contents on the split surface, demonstrating that well compacted mixes with high binder content, and particularly those incorporating mineral admixtures (microsilica, fly ash), are essentially impermeable and will release moisture very slowly. High strength specimens drying in laboratory air for over several months were still visibly moist over 90% of the split diameter.[5] The reductions in splitting strength observed in tests on air dried commerical strength LC that are caused by differential dry moisture gradients in the concrete prior to reaching hygro-equilibrium are significantly delayed and diminished in high binder content HSLC.

Elastic properties

Modulus of elasticity

Concrete is a composite material composed of a continuous matrix enveloping particulate inclusions. Stiffness of the composite is related to the stiffness of its constituents in a rather complex way, and it is surprising that the recommended ACI formula has been so effective. One factor affecting stiffness is the variation of aggregate modulus of elasticity within a particular density range. At the same specific gravity, LaRue found the modulus of elasticity of natural aggregates could vary by a factor of as much as three.[42] Concrete strengths also tended towards a maximum where the aggregate modules matched the modulus of the concrete made from them.

Although the ACI 318 formula, $E_c = 33\omega^{1.5}\sqrt{f_c'}$ $(E_c = 0.043\omega^{1.5}\sqrt{f_c'})$ has provided satisfactory results in estimating the elastic modulus of NWC and LC in the usual commercial strength range from 3000 to 5000 psi (20 to 35 MPa), it has not been adequately calibrated to predict the modulus of high strength concretes. Practical modification of the formula was first provided by ACI 213-77[43] to more reasonably estimate the modulus of HSLC:

$$E_c = C\omega^{1.5}\sqrt{f_c'}$$

$C = 31$ for 5000 psi $(C = 0.40$ for 35 MPa)
$C = 29$ for 6000 psi $(C = 0.38$ for 41 MPa)

When designs are controlled by elastic properties (e.g. deflections, buckling, etc), the specific value of E_c should be measured on the proposed concrete mixture in accordance with the procedure of ASTM C 469, *Standard Test Method for Static Modulus of Elasticity and Poisson's Ratio of Concrete in Compression*.

In general, structural grade rotary kiln produced LWAs have a comparable chemical composition and are manufactured under a similar temperature regime. They achieve low density by formation of a vesicular structure in which the vesicles are essentially spherical non-interconnected pores enveloped in a vitreous matrix. It would be expected that, with such similarities, the variability in stiffness of the aggregate would be principally due to the density as determined by the pore volume system. Aggregate density, in turn, is reflected in a reduced concrete unit weight which is accounted for in the first term in the above equation.

As with NWC, increasing matrix stiffness is directly related to matrix strength which, in turn, affects concrete strength. When large percentages of cementitious materials are used, the LC strength ceiling may be reached causing the above equation to overestimate the stiffness of the concrete.

Poisson's ratio

Testing programs investigating the elastic properties of HSLC have reported an average Poisson's ratio of 0.20, with only slight variations due to age, strength level, or aggregates used.[44,45]

Maximum strain capacity

Several methods of determining the 'complete' stress-strain curve of LC have been attempted. At Lehigh University,[15] the concrete cylinders were loaded by a beam in flexure, while at Illinois University, the approach was to load a concrete cylinder completely enclosed within a steel tube of suitable elastic properties.[46]

Despite formidable testing difficulties, both methods secured meaningful data; one of the more complete stress-strain curves obtained by loading the concrete through a properly proportioned beam in flexure is demonstrated in Fig. 10.8.

At failure, HSLC will cause the release of a greater amount of energy stored in the loading frame than an equal strength concrete composed of stiffer NWA. As energy stored in the test frame is proportional to the applied load moving through a deformation inversely proportional to the modulus of elasticity, it is not unusual for a HSLC to release almost 50% greater frame energy. To avoid shock damage to the testing equipment, it is recommended that a lower percentage of maximum usable machine capacity be used when testing HSLC and that suitable precautions are taken for testing technicians as well.[5]

Dimensional stability

Shrinkage

Figure 10.9 demonstrates the shape and ultimate shrinkage strains from

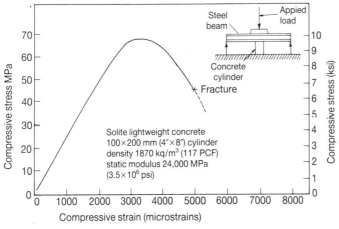

Fig. 10.8   Stress versus strain under uniaxial compression

one extensive testing program that incorporated both HSLC and HSNWC.[5] These results are consistent with data reported in other investigations.[17] Of interest is the relative equality of the maximum shrinkage of the HSNWC mix and the 4500 psi (30 MPa) LC mix introduced for comparative purposes, indicating an apparent trade-off between the rigid skeletal structure of the NWA and the contractive forces developed by the higher binder content. Shrinkage of the 10 mm top size HSLC mix lagged behind early values of the other mixes, equalled them at 90 to 130 days, and reached an ultimate value at one year approximately 14% higher than the reference HSNWC. Shrinkage values of mixes incorporating cement containing interground fly ash averaged somewhat greater than their high strength non-fly ash counterparts.

Shrinkage and density data were measured on $4 \times 4 \times 12$ in. ($100 \times 100 \times 300$ mm) concrete bars fabricated at the same time and from the same mix as the compressive strength cylinders. Curing was provided by damp cloth for 7 days, after which the specimens were stripped from the

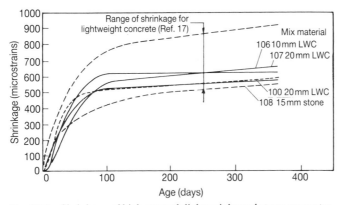

Fig. 10.9   Shrinkage of high strength lightweight and stone concretes

molds. Brass wafers were epoxied at one day to the bar surface at a 10 in. (250 mm) gage distance with mechanical measurement by a Whitemore gage. Reference readings were established 7 days after fabrication after which specimens were allowed to dry in laboratory air, 70 °F (21 °C), (50% ± 5 RH), with no further curing. Shrinkage and weight readings were taken weekly for three months then monthly with results shown to one year. Ten year shrinkage strains were only slightly higher than one year results and will be reported in another publication.[47]

Creep

Rogers reported that the one year creep strains measured on several high strength North Carolina and Virginia LCs were similar to those measured on companion NWC.[48] Greater creep strains measured on HSLC containing both fine and coarse LWA, when compared to reference HSNWC in reports by Reichard[44] and Shideler,[45] could be anticipated because of the larger matrix volume required because of the angular particle shape of the LWA fines.

   While the Prestressed Concrete Institute provides recommendations for increasing stress losses due to creep when using HSLC, it may be advisable to obtain accurate design coefficients for long span HSLC structures by conducting pre-bid laboratory tests in accordance with the procedures of ASTM C512, *Standard Test Method for Creep of Concrete in Compression*.

   As reported in ACI 213,[43] and shown in Fig. 10.10, specific creep values decrease significantly with increasing strength of LC. Additionally, at higher strength levels, the creep strain envelope developed from a wide range of LCs tested converged towards the performance of the reference NWC.

Thermal properties

Accurate physical property input data is essential when considering the thermal response of restrained members in exposed structures. Obvious

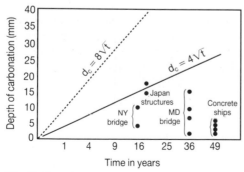

Fig. 10.10   Measured depth of carbonation (mm) of exposed lightweight concrete structures

cases in point include exposed exterior columns of multi-story cast-in-place concrete frames as well as massive offshore concrete structures constructed in temperate zones and then towed to harsh Arctic marine environments. As thermal behavior of concrete is a function of the thermal contributions of the separate components, the low expansions, conductivities and diffusivities of HSLC are predictable in the light of thermal measurement data reported on regular strength LC.[43] A summary of data relating to the coefficient of linear thermal expansion, thermal conductivity, and diffusivity data extracted from the results obtained in the joint industry investigation digested by Hoff[17] is shown below.

- *Coefficient of linear thermal expansion*
  At high moisture contents typical of marine exposed concrete, the coefficient of linear thermal expansion for HSLC ranged from 4 to 7 microstrain/°F (7 to 13 microstrain/°C).[17]
- *Thermal conductivity*
  Over a temperature range of −50 to 70 °F (−46 to 21 °C), thermal conductivity values of the HSLC varied from 5.6 to 7.6 Btu.in/hr.sf. F (0.8 to 0.9 W/m. K) According to density and mix composition factors.[17]
- *Specific heat*
  The specific heat of HSLC tested averaged 0.23 Btu/lb. F (1.0 J/kg K).[17]
- *Thermal diffusivity*
  The thermal diffusivity of the HSLC tested ranged from 0.020 to 0.022 sq ft/hr (18.3 to 20.8 sq cm/hr) for the concrete densities tested.[17]

Durability

Numerous laboratory durability testing programs evaluating freeze/thaw resistance have confirmed field observations of the long-term, proven performance of highly exposed HSLC. The United States Federal Highway Administration closely examined 12 study cases of more than 400 North American LC bridge decks and concluded:

> 'Some have questioned the durability, wear resistance, and long-term freeze/thaw qualities of lightweight concrete. No evidence was found that these properties differ from those of normal weight concrete. In fact, there is evidence that these properties could be better for lightweight concrete, especially if the normal weight concrete is of poor quality. This leads to the suggestion that the designer might consider specifying lightweight concrete if natural aggregates are not of high quality. Although lightweight aggregates vary depending on the raw material source, they are usually of a more consistent quality than some natural aggregates. Specified material tests will provide the necessary quality characteristics.'[4]

These findings supported similar conclusions reported earlier in the LC Bridge Deck Survey of the Expanded Shale Clay and Slate Institute.[24] Reports of the inspections of the durability of the HSLC used in World War I and World War II concrete ships also attested to excellent long-term performance.[2,3,6,7,49]

Freeze/thaw resistance

Most accelerated freeze/thaw testing programs conducted on structural LC have incorporated HSLC on the high end of the compressive strength range of the specimens tested. Investigations in North America[25,50] and in Europe[39,51] researching the influence of entrained air volume, cement content, aggregate moisture content, specimen drying times, and testing environments have arrived at essentially the same conclusion: air entrained LCs properly proportioned with high quality binders provide satisfactory results when tested under usual laboratory freeze/thaw testing programs.

Core samples taken from the hulls of 70 year old LC ships as well as the 30 to 40 year old LC bridge decks have demonstrated concretes with high internal integrity and low levels of microcracking. This proven record of high resistance to weathering and corrosion is due to physical and chemical mechanisms; they include superior resistance to microcracking developed by significantly higher aggregate/matrix contact zone adhesion as well as internal stress reduction due to the elastic matching of coarse aggregate and matrix phases. High ultimate strain capacity is also provided by concrete with a high strength/modulus ratio. In addition, because of elastic compatibility, the stress/strength ratio at which the disruptive disintegration of concrete begins is higher for LC than for equal strength NWC. A well dispersed void system provided by lightweight fine aggregates will assist the entrained air pore system, and may also serve an absorption function by reducing disruptive mechanisms in the matrix phase. Additionally, long-term pozzolanic action is provided by the silica rich expanded aggregate combining with calcium hydroxide liberated during cement hydration. This will reduce permeability and minimize leaching of soluble compounds.

It is widely recognized that while ASTM C666, *Standard Test Method for Resistance of Concrete to Freezing and Thawing*, provides a useful comparative testing procedure, there remains inadequate correlation between accelerated laboratory test results and the observed behavior of mature concretes exposed to natural freezing and thawing. Inadequate laboratory/field correlation observed when testing NWC is compounded when interpreting results from laboratory tests on structural LC prepared with high aggregate moisture contents. A proposed modification to ASTM C666 recommends a 14 day air drying period prior to the first freezing cycle, to improve correlation between laboratory test data and observed field performance.[2]

Durability characteristics of any concrete, both NWC and LC, are decisively influenced by the protective qualities of the paste fraction. It is imperative that the concrete matrix provide high quality, low permeability characteristics in order to protect steel reinforcing from corrosion, which is clearly the dominant form of structural deterioration observed in current construction. The protective quality of the matrix in concretes proportioned primarily for thermal resistance that incorporate high air contents and low cement quantities will be significantly reduced. Very low density, non-structural LC will not provide resistance to the intrusion of chlorides, carbonation, etc comparable to the long-term satisfactory performance demonstrated with high quality, structural grade LC.[2]

For a number of years field exposure testing programs have been conducted by the Canadian Department of Minerals, Energy and Technology (CANMET) on various types of concrete exposed to a cold marine environment at the Treat Island Severe Weather Exposure Station maintained by the U.S. Army Corps of Engineers at Eastport, Maine.[52,53] Concrete specimens placed on a mid-tide wharf experience alternating conditions of sea water immersion followed by cold air exposure at low tide. In typical winters the specimens experience about 100 cycles of freezing and thawing. In 1978, a series of prisms were cast using commercial NWAs with various cement types and including supplementary cementitious materials. W/c ratios of 0.40, 0.50, and 0.60 were used to produce 28 day compressive strengths of 4350, 3770, and 3480 psi (30, 26 and 24 MPa) respectively. In 1980, these mixes were essentially repeated with the exception being that the 1½ in. (40 mm) gravel aggregate was replaced with a 1 in. (25 mm) expanded shale LWA. Fine aggregates used in both 1978 and 1980 were commercially available natural sands. Cement contents for the semi-LC mixtures were 800, 600, and 400 pcy (480, 360, and 240 kg/m^3), which produced compressive strengths of 5220, 4350 and 2755 psi (36, 30 and 19 MPa) respectively. All specimens continue to be evaluated annually for ultrasonic pulse velocity, and resonant frequence as well as visually ratings. Ultrasonic pulse velocities are measured centrally along the long axis of the prisms. Negligible differences exist between the structural LC (8 years) and NWC (10 years) after exposure to twice daily sea water submission and approximately 1000 cycles of freezing and thawing.[2]

Permeability

Conventional strength concrete employs a matrix with a w/c ratio significantly higher than that required for the chemical reaction associated with hydration of portland cement. This excess of uncombined water is free to either move due to an applied hydraulic gradient or, if allowed to evaporate, provide conduits through which gas can either diffuse into the concrete or flow in response to a pressure differential. Concentration gradients of chemicals in liquid form can also diffuse into saturated concrete or can be absorbed by an initially dry concrete.

The above defines a material which is porous, in that it contains both pores and conduits that communicate with a free surface and, as a result, is permeable to liquids and gases. The normal definition of the adjective 'porous' usually contains any, or all, of the following: possessing pores, or containing vessels and conduits, or being permeable to liquids and gases. In the case of concrete of usual strength, all of the above are applicable.

Both HSNWC and HSLC are usually made with water contents that only slightly exceed that required to hydrate the cement. This means that while the hydrated cement paste still contains pores and conduits, these channels are not fully continuous nor do all of them communicate with the surface. High strength concrete normally contains both silica fume and a superplasticizer, which leads to a densification of the cement paste matrix after a relatively short period of moist curing. Slag cement, fly ash, or both, are frequently incorporated into the mix which further densifies the cement paste matrix.

Aggregates inserted into the cementitious matrix are surrounded by the paste and isolated from one another by this essentially impermeable matrix. Microcracks, however, may form in the concrete as a result of the volume changes associated with hydration of the cement paste matrix. Microcracking can also result from stresses that arise in concrete as a result of differing aggregate/matrix thermal expansion coefficients when the concrete is heated or cooled. Aggregates and matrix fractions expand and contract at different rates as the concrete gains and loses water, further increasing prospects for microcracking. Microcracking will also result from the lack of elastic compatibility that exists between the aggregate inclusion and the cement paste matrix when the composite concrete is subjected to an applied stress. It is these crack networks that, more than any other factor, render concrete permeable to gases and liquids. The high stiffness of the matrix fraction of high strength concrete results in a closer elastic match and a lower propensity to form stress-induced microcracks.

For all concrete types, the ratio of aggregate to matrix stiffness starts out at infinity as hydration begins, therefore any volume change due to the hydration process, moisture change, or thermal changes may lead to microcracking in the concrete prior to the superposition of any design loads on the concrete. In most instances where liquid permeability is of concern, or where presence of moisture is associated with deterioration due to corrosion, autogenous healing also is operative. Laboratory testing is, however, not usually conducted on mature concrete and this beneficial effect is not observed. Field and early age laboratory tests on high strength concrete do indicate permeability coefficients substantially lower by several orders of magnitude, however, when compared with commercial concrete. In fact, satisfactory techniques to satisfactorily measure the permeability of high strength concrete in most cases have not been developed.

The role the aggregate plays in the liquid and gas permeability of concrete is minor for both LC and NWC. With both types of aggregate, small cracks formed during the final crushing to size of the aggregate particles may have some small effect. However, few of the cracks would be oriented in the direction of flow, and the matrix essentially seals the cracks at the surface of the aggregates. Scanning electron microscopy studies have indicated that structural grade LWAs have a vesicular structure with essentially no interconnection of pores so that flow of liquid or gas is insignificant. Porous aggregate particles will not be permeable because of the lack of continuous channels through the particles. Permeability investigations conducted on LC and NWCs exposed to the same testing criteria have been reported by Khokrin,[51] Nishi,[54] Keeton,[55] Bamforth,[56] and Bremner.[57] It is of interest that in every case, despite wide variations in concrete strengths, testing media (water, gas and oil) and testing techniques (specimen size, medium pressure and equipment) structural LC had equal or lower permeability than its heavier counterparts. This result has been attributed to the reduction of microcracks in elasticly compatible LC and the enhanced bond and superior contact zone present in structural LC.

Corrosion resistance

Corrosion resistance of HSLC is at least comparable to the performance of HSNWC. Investigations of mature bridges and marine structures report that internal integrity and minimal microcracking have effectively limited rapid intrusion of aggressive forces into the concrete.[2,3,6] Internal integrity effectively limits disruptive effects to diffusion mechanisms which are orders of magnitude slower in their deteriorating actions.

Carbonate resistance

Penetration of the carbonation front into concrete is primarily determined by the vapor diffusion characteristics of the mortar matrix. With high quality matrixes typical for all HSCs, this issue will not be a concern. Field measurements of carbonation depths in LC marine and bridge structures have demonstrated that the rate of carbonation is extremely low, with results shown in Figs. 10.10 and 10.11. Adequate protection against

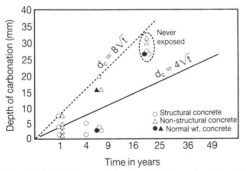

Fig. 10.11   Measured depth of carbonation (mm) of laboratory specimens of lightweight concrete

carbonation depassivating reinforcing steel in the service life of high quality structures is provided by covers recommended by ACI 318.

Abrasion resistance

Investigations into the abrasion resistance of LC using tests conducted in accordance with ASTM C779, *Standard Test Method for Abrasion Resistance of Horizontal Concrete Surfaces*, reported that the

> 'wear trends were similar for both normal weight and lightweight concretes and, because of the small differences in the wear amounts between both types of concretes, it appears that lightweight concrete could be used satisfactorily in abrasive situations where normal weight concrete might normally be specified.'[17]

Bare LC bridge decks exposed to more than 100 million vehicles crossing over a more than 20 year period have demonstrated wear patterns comparable to NWC.

Unit weight

The fresh unit weight of HSLC is a function of mix proportioning, air contents, water demand, and the specific gravity and moisture content of the LWA. Decrease in density of exposed concrete is due to moisture loss which, in turn, is a function of ambient conditions and surface area/volume ratio of the member. Design professionals should specify a maximum fresh unit weight for LC, as limits of acceptability should be controlled at time of placement.

Dead loads used for design should be based upon equilibrium density which, because of the very low permeability of HSLC, may be assumed to be reached after 180 days for moderate sized members. Extensive tests conducted during North American durability studies demonstrated that despite wide initial variations of aggregate moisture content, equilibrium unit weight for commercial strength LC was found to be 3.1 pcf (50 kg/m^3) above the oven dry unit weight. European recommendations for in-service density are similar.[12] For HSLC however, reduction in density from fresh to equilibrium will be significantly less due to the low permeability and high cementitious content which, because of internal curing periods, will serve to hydrate a larger fraction of the mix water. Because of the high cementitious content of HSLC, equilibrium density of non-submerged concrete will be close to the fresh density. Hoff reported that HSLC mixtures containing supplementary cementitious materials exposed to hydrostatic pressures equal to 200 ft. (61 m) of water demonstrated an increase of density of less than 4 pcf (64 kg/m^3).[17]

Ductility

The ductility of concrete structural frames should be analyzed as a composite system – that is, as reinforced concrete. Ahmad's studies indicate that the ACI rectangular stress block is adequate for strength predictions of HSLC beams and that the recommendation of 0.003 as the maximum usable concrete strain is an acceptable lower bound for HSLC members with strengths not exceeding 11,000 psi (76.5 MPa) and p/p_b values less than 0.54.[58] Moreno found that while LC exhibited a distinctive descending portion of the stress-strain curve, it was possible to obtain a flat descending curve with reinforced LC members that were provided with a sufficient amount of confining reinforcement slightly greater than that with NWC.[59] This report also included studies that showed that it was economically feasible to obtain desired ductility when increasing the amounts of steel confinement.

Rabbat *et al.*, came to similar conclusions when analyzing the seismic behavior of LC and NWC columns.[60] This report focused on how properly detailed reinforced concrete columns could provide ductility and maintain strength when subjected to inelastic deformations from moment reversals. These investigations concluded that properly detailed columns made with LC performed as well under moment reversals as NW columns.

10.5 Constructability of high strength lightweight concretes

Proportioning

Proportioning rules used for ordinary concrete mixes also apply to HSLC, with increased attention given to concrete density and the absorption characteristics of the LWA. Most structural grade LC is proportioned by absolute volume methods where the volume of fresh concrete produced is considered equal to the sum of the absolute volumes of cement, aggregates, net water and entrained air. Proportioning by this method requires determination of absorbed and adsorbed moisture contents and the as-batched specific gravity of separate aggregate sizes.

As with HSNWC, air entrainment is required for HSLC for durability and resistance to scaling. With moderate air contents, bleeding and segregation are reduced and mixing water requirements lowered while maintaining equivalent workability. In recognition of HSLC low permeability and the impact of air on strength properties, recommended air contents may be adequate when slightly lower than that required for usual exposed concretes. Because of the elastic matching of the LWA and the cementitious binder paste, strength reduction penalties due to high air contents are somewhat lower for LC than for NWC.[8] Air content of LC is determined in accordance with the procedures of ASTM C173, *Test for Air*

Content of Freshly Mixed Concrete by the Volumetric Method. Volumetric measurements assure reliable results while pressure meters may develop erratic data due to the influence of aggregate porosity.

When absorbed moisture levels are greater than that developed after a one day immersion, the rate of further aggregate absorption will be very low and for all practical purposes HSLC is batched, placed and finished with the same facility as their NWC counterparts. Under these conditions net w/c ratios may be established with a precision comparable to concretes containing NWA. Water absorbed within the LWA particle prior to mixing is not available for establishing the cement paste volume at the time of setting. This absorbed water is available, however, for continued hydration of the cement after external curing has ended.

Mixing, placing, finishing and curing

Properly proportioned HSLC can be delivered, placed and finished with the same facility and with less physical effort than that required for NWC. The most important consideration in handling any type of concrete is to avoid separation of coarse aggregate from the matrix fraction. Avoid excessive vibration as this practice serves to drive the heavier mortar fraction down from the surface of flat work where it is required for finishing. On completion of final finishing, curing operations similar to ordinary concrete should begin as soon as possible. LCs batched with aggregates having high absorptions carry their own internal water supply for curing within the aggregate and, as a result, are more forgiving to poor curing practices or unfavorable ambient conditions. This 'internal curing' water is transferred from the LWA to the matrix phase as evaporation takes place on the surface of the concrete, thus continuously maintaining moisture balance by replacing moisture essential for an extended continuous hydration period determined by ambient conditions and the as-batched moisture content of the LWA.

Pumping

LWA may absorb part of the mixing water when exposed to high pumping pressures. To avoid loss of line workability, it is essential to raise the level of absorption of the LWA prior to pumping. Presoaking is best accomplished at the aggregate production plant where a uniform moisture content is achieved by applying water by spray bars directly to the aggregate moving on belts. This moisture content can be maintained and supplemented at the concrete plant by stockpile hose and sprinkler systems for at least one, but preferably, three days.

Presoaking will significantly reduce LWA rate of absorption, minimizing water transfer from the matrix fraction which can cause slump loss during pumping. Higher moisture contents developed during presoaking will result in an increased specific gravity which, in turn, develops higher fresh concrete density. High water content due to presoaking will eventually diffuse out of the concrete, developing a longer period of internal curing and a larger fresh to equilibrium density differential than that usually associated with LC using aggregates of a lower moisture content. Aggregate suppliers should be consulted for mix design recommendations necessary for consistent pumpability. Mix designs and the physical properties measured on samples of HSLC pumped 830 ft. (250 m) to the 60th floor of the NationsBank project in Charlotte, NC (Fig. 10.12) are shown in Table 10.1.

Laboratory and field control

Changes in LWA moisture content, gradation or specific gravity as well as the ususal job site variation in entrained air suggest frequent fresh concrete

Fig. 10.12

Table 10.1   Mix design and physical properties for concretes pumped 830 ft. (268 m) on Nationsbank Building, Charlotte, NC, 1991

Mix #	1	2*	3
Mix proportions:			
Cement type III (lbs.)	550	650	750
Fly ash (lbs.)	140	140	140
Solite 3/4 to #4 (lbs.)	900	900	900
Sand (lbs.)	1370	1287	1203
Water (gals.)	35.5	36.5	37.2
WRA (oz.)	27.6	31.6	35.6
Superplasticizer	55.2	81.4	80.1
Fresh concrete properties:			
Initial slump (inches)	$2\frac{1}{2}$	2	$2\frac{1}{4}$
Slump after superplasticizer	$5\frac{1}{2}$	$7\frac{1}{2}$	$6\frac{3}{4}$
% air	2.5	2.5	2.3
Unit weight (pcf.)	117.8	118.0	118.0
Compressive strength (psi):			
4 days	4290	5110	5710
7 days	4870	5790	6440
28 days (avg.)	6270	6810	7450
Splitting tensile strength (psi):	520	540	565

* Mix selected and used on project

checks to facilitate adjustments necessary for consistent concrete characteristics. Standardized field tests for consistency, fresh unit weight and entrained air content should be employed to verify field concrete conformance with trial mixes and the project specification. Sampling should be conducted in accordance with ASTM C172, *Standard Practice for Sampling Freshly Mixed Concrete*, and ASTM C173, *Standard Test Method for Air Content*. The *Standard Test Method for Unit Weight of Structural Lightweight Concrete*, ASTM C567, describes methods for the determination of the in-service, equilibrium unit weight of structural LC. In general, when variations in fresh density exceed ±2%, adjustments in batch weights are required to restore specified concrete properties. To avoid adverse effects on durability, strength and workability, air content of HSLC should not vary more than ±1% from specified values.[4]

10.6   Applications of high strength lightweight aggregate concrete

Precast and prestressed structures

HSLC with compressive strength targets ranging from 5000 to 8000 psi (35 to 55 MPa) has been successfully used for almost four decades by North American precast and prestressed concrete producers. Presently there are

Fig. 10.13

ongoing investigations into somewhat longer span lightweight precast concrete bridges that may be feasible from a trcuking/lifting/logistical point of view.

Garage members with 50–63 ft. (15–20 m) spans are generally constructed with double tees composed of sanded LC with air dry density of approximately 115 pcf (1850 kg/m³) (Fig. 10.13).

Weight reduction is primarily for lifting efficiencies and lower transportation costs. One prestressed garage project is of interest from the perspective of the precast producer's quality control adjustments of the mix components, when statistical studies of the plant's first use of HSLC indicated unduly high strengths. The first series of statistical test results with a mix that included 755 pcy (450 kg/m³) of cement yielded seven day strengths of 7450 psi (51 MPa) and 28 day strengths in excess of 9000 psi (62 MPa). Cutting cement back to 705 pcy (429 kg/m³) developed a 28 day strength of 7910 psi (54.5 MPa) and after a final cement reduction to 660 pcy (390 kg/m³), the HSLC developed 7500 psi (52 MPa) compressive strength at 28 days.[5]

Buildings

The 450 ft. (140 m) multi-purpose Federal Post Office and Office Building constructed in 1967 with five post office floors and 27 office tower floors was the first major New York City building application of post-tensioned floor slabs. Concrete tensioning strengths of 3500 psi (24 MPa) were routinely achieved at 3 days for the 30 × 30 ft (9 × 9 m) floor slabs with a design target strength of 6000 psi (41 MPa) at 28 days. Approximately 30,000 cubic yd (23,000 m³) of structural LC were incorporated into the floors, and the cast-in-place architectural envelope serves a structural as

Fig. 10.14

well as aesthetic function. Despite the highly polluted urban atmosphere, the buff colored concrete has maintained its handsome appearance (Fig. 10.14).

The recently completed (1991) North Pier Tower (Chicago) utilized HSLC in the floor slabs as an innovative structural solution to avoid construction problems associated with the load transfer from HSNWC columns through the floor slab system. ACI 318 requires differences in compressive strength between column concrete and the intervening floor slab concrete to be less than a ratio of 1.4. By using HSLC in the slabs with a strength greater than (9000/1.4) = 6430 psi (44 MPa) the floor slabs could be placed using routine placing and techniques thus avoiding scheduling and placing problems associated with the 'mushroom' technique (Fig. 10.15). In this approach, high strength column concrete is overflowed

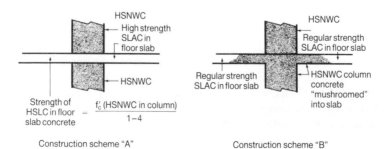

Fig. 10.15   Alternate construction schemes for transfer of HSNWC column loads through floor slabs

from the column and intermingled with the regular strength floor slab concrete. The technique used in the North Pier project avoids delicate timing considerations that are necessary to avoid cold joints.

Bridges

Of the more than 400 LC bridge decks constructed throughout North America, most have been produced with concretes at higher than usual commercial level. The Sebastian Inlet Bridge, which utilizes extra long HSLC drip-in spans during its construction in 1965, is included in one LWA supplier's listing of almost 100 completed bridges. Transportation engineers generally specify higher concrete strengths on bridge decks, primarily to insure high quality mortar fractions (high strength combined with high air content) that will minimize maintenance costs. One state authority has completed more than 20 bridges utilizing HSLC using a target strength of 5200 psi (36 MPa), 6–9% air content, and an air dry density of 115 pcf (1850 kg/m^3). Recent studies have identified tens of thousands of bridges in the United States which are functionally obsolete with low load capacity, unsound concrete, or insufficient number of traffic lanes. To remedy limited lane capacity, Washington, DC engineers have replaced a 4-lane bridge with 5 new lanes providing a 50% increase in one-way, rush hour traffic, without replacing the existing structure, piers, or foundations.

Marine structures

Because many offshore concrete structures will be constructed in shipyards located in lower latitudes and then floated and towed to the project site, there is a special need to reduce weight and improve the structural efficiency of the cast-in-place structure. Because shallow water conditions mandate lower draft structures the submerged density ratio of

$$\frac{HSNWC}{HSLC} \frac{2.50 - 1.00}{2.00 - 1.00} = 1.50$$

which is greater than the air density ratio

$$\frac{2.50}{2.00} = 1.25$$

becomes increasingly important.

These requirements have already been satisfied by several projects placed in the Arctic, e.g., HSLC was used in 1981 in the TARSUIT CAISSON Retained Island project constructed in Vancouver, Canada and transported to the Beaufort Sea on the Alaskan North Slope.[61] This project was followed in 1984 with the HSLC used in the construction of the Concrete Island Drilling System built in Japan and also towed to the

Table 10.2 Physical properties, strength and chloride ion permeability of structural lightweight concrete microsilica[1]

Lightweight concrete mix		K	O	F	C	N+ (With Microsilica)	N−[1] (Without Microsilica)
Fresh concrete properties	Wet density psi (kg/m^3)	116.5 (1870)	120.5 (1930)	117.9 (1870)	116.5 (1870)	118.2 (1890)	116.8 (1870)
	Slump in (mm)	7.5 (190)	8.5 (215)	8.5 (215)	8.8 (225)	8 (205)	6 (150)
	Air content %	5.5	5.2	5.8	5.2	5.8	5.2
Compressive strength at age (days)	2	6360 (43.9)	4850 (33.4)	4180 (28.8)	5920 (40.8)	5800 (40.0)	2890 (19.9)
	7	8270 (57.0)	5740 (39.6)	5310 (36.6)	7470 (51.5)	6990 (48.2)	3960 (27.3)
	28	9648 (66.5)	6840 (47.2)	6050 (41.7)	8600 (59.3)	7460 (51.0)	5050 (34.8)
	90	9855 (68.0)	6550 (45.2)	6225 (42.6)	8990 (62.0)	7525 (51.9)	5270 (36.3)
W/B ratio[2]		.36	.33	.35	.31	.32	.43
Chare passed (coulombs)[3]		260	450	450	220	370	4800

(1) Mix N− identical to N+ with exception of no microsilica/superplasticizer.
(2) Includes water from slurry.
(3) Rapid determination of the chloride permeability of concrete. AASHTO T277. Microsilica Elkem, Pittsburgh, PA.

Beaufort Sea.[62] This project called for compressive strengths of 6500 psi (45 MPa) with density of 115 pcf (1840 kg/m^3). In addition to reducing draft during construction and towing, use of HSLC in offshore gravity based structures can be justified by the improved floating stability as well as the opportunity to carry more topside loads.

Rehabilitation of bridges and parking decks

Numerous opportunities exist for the efficient rehabilitation of existing deteriorated bridges and parking decks. For example, replacing 3 in. of deteriorated NWC with 4 in. of low permeability HSLC will also provide opportunities to improve deficient surface geometry; for example, increasing slopes to drain and improved super elevation on curves.

Cooperative research testing programs with a microsilica producer[1] have demonstrated significant reductions in the permeability of concrete to chloride ions when measured by AASHTO T277, *Rapid Determination of the Chloride Permeability of Concrete*. Table 10.2 reports the physical properties, compressive strength, and shows the dramatic reduction of current passed by LC 'N−' (without microsilica) from 4800 to 370 coulombs when microsilica is added to the concrete.

References

1   Holm, T.A. and Bremner, T.W. (1990) *70 year performance record for high strength structural lightweight concrete*, Proceedings of First Materials Engineering Congress. Materials Engineering Division, ASCE, Denver, Colorado, Aug.

2   Holm, T.A. and Bremner, T.W. (1991) The durability of structural lightweight concrete, ACI SP-126, *Durability of Concrete*. Second International Conference, Montreal, Canada, August.

3   Holm, T.A. (1980) Performance of structural lightweight concrete in a marine environment, ACI Publication SP-65, *Performance of Concrete in a Marine Environment*. International Symposium, St. Andrews-By-The-Sea, Canada, August.

4   (1985) Criteria for designing lightweight concrete bridges, by U.S. Dept. of Transportation, Federal Highway Administration, Report No. FHWA/RD-85/045, August.

5   Holm, T.A. (1980) Physical properties of high strength lightweight aggregate concretes, Second International Congress of Lightweight Concrete, London, UK, April.

6   Holm, T.A., Bremner, T.W. and Newman, J.B. (1984) Lightweight aggregate concrete subject to severe weathering. *Concrete International*, June.

7   Bremner, T.W., Holm, T.A. and deSousa, H. (1984) Aggregate-matrix interaction in concrete subjected to severe exposure, FIP-CPCI International Symposium on Concrete Sea Structures in Arctic Regions. Calgary, Canada, August.

8   Bremner, T.A. and Holm, T.A. (1986) Elastic compatibility and the behavior of concrete. *Journal American Concrete Institute*, March/April.

9   Stagg, K.G. and Zienklewicz, O.C. (1968) *Rock mechanics in engineering practice*. J. Wiley and Sons, New York.

10 Shah, S.P. and Chandra, S. (1968) Critical stress, volume change and microcracking of concrete. *Journal American Concrete Institute*, September.

11 Muller-Rochholz, J. (1979) Determination of the elastic properties of lightweight aggregate by ultrasonic pulse velocity measurements. *International Journal of Lightweight Concrete*, Lancaster, UK, Vol. 1, No. 2.

12 (1983) *FIP manual of lightweight aggregate concrete*, 2nd Edition. Federation Internationale de la Precontrainte/Surrey University Press, Glasgow.

13 ACI 363 (1984) State of the art report on high-strength concrete. *ACI Journal*, July/August.

14 ABAM Engineers, Inc. (1983) Developmental design and testing of high-strength lightweight concretes for marine Arctic structures, Program Phase I, Joint Industry Project Report, AOGA Project No. 198, Federal Way, Washington, May.

15 ABAM Engineers, Inc. (1984) Developmental design and testing of high-strength lightweight concretes for marine Arctic structures, Program Phase II, Joint Industry Project Report, AOGA Project No. 203, Federal Way, Washington, DC, September.

16 ABAM Engineers, Inc. (1986) Developmental design and testing of high-strength lightweight concretes for marine Arctic structures, Program Phase III, Joint Industry Project Report, AOGA Project No. 230, Federal Way, Washington, August.

17 Hoff, G.C. (1991) High strength lightweight concrete for Arctic applications, ACI Symposium on *Performance of lightweight concrete*. Dallas, Texas, November (To be published by ACI).

18 Hoff, G.C. High strength lightweight aggregate concrete – current status and future needs.

19   (1987–1989) High strength concrete, Joint Industry Project, SINTEF, Trondheim, Norway.
20   (1985–1988) Ductility performance of offshore concrete structures, Joint Industry Project, Veritec, Hovik, Norway.
21   (1984–1987) Design of peripheral concrete walls subjected to concentrated ice loading, Joint Industry Study, Ben C. Gerwick, Inc., San Francisco, California.
22   (1987–1988) Punching shear resistance of lightweight concrete offshore structures for the Arctic, Joint Industry Project, National Institute of Standards and Technology, Washington, DC.
23   (1960) Story of the *Selma* – expanded shale concrete endures the ravages of time. Expanded Shale Clay and Slate Institute, Bethesda, Maryland, June, Second Edition.
24   (1960) *Bridge deck survey*. Expanded Shale Clay and Slate Institute, Washington, DC.
25   (1970) *Freeze-thaw durability of structural lightweight concrete*. Lightweight Concrete Information Sheet No. 13, Expanded Shale Clay & Slate Institute, Salt Lake City, Utah.
26   Malhotra, V.M. (1987) CANMET investigations in the development of high-strength lightweight concrete, *Proceedings*, Symposium on the Utilization of High Strength Concrete, Stavanger, Norway, June 15–18, TAPIR, Trondheim, Norway.
27   Malhotra, V.M. (1981) *Mechanical properties and durability of superplasticized semi-lightweight concrete*, SP 68-16. American Concrete Institute.
28   Seabrook, P.I. and Wilson, H.S. (1988) High strength lightweight concrete for use in offshore structures – utilization of fly ash and silica fume. *International Journal of Cement Composites and Lightweight Concrete*, August, Lancaster, UK.
29   Ramakrishnan, V., Bremner, T.W. and Malhotra, V.M. (1991) Fatigue strength and endurance limit of lightweight concrete, ACI Symposium, *Performance of structural lightweight concrete*, Dallas, Texas, November. (To be published by ACI.)
30   Berner, D.E. (1991) High ductility, high strength lightweight aggregate concrete, ACI Symposium on *Performance of structural lightweight concrete*, Dallas, Texas, November. (To be published by ACI.)
31   Luther, M.D. (1991) Lightweight microsilica concrete, ACI Symposium, *Performance of structural lightweight concrete*, Dallas, Texas, November. (To be published by ACI.)
32   Zhang, M. and Gjørv, O.E. (1991) Characteristics of lightweight aggregates for high strength concrete. *ACI Materials Journal*, March/April, Detroit, Michigan.
33   Zhang, M. and Gjørv, O.E. (1991) Mechanical properties of high strength lightweight concrete. *ACI Materials Journal*, May/June, Detroit, Michigan.
34   Zhang, M. and Gjørv, O.E. (1990) Development of high strength lightweight concrete, *High strength concrete* – Second International Symposium, SP-121. American Concrete Institute, Detroit, Michigan.
35   Kudriatsev, A.A. (1973) The modulus of elasticity and the modulus of deformations of structural keramzit concrete, *Structure, strength and deformation of lightweight concrete*, Moscow (in Russian).
36   Dovzhik, V.G. and Dorf, V.A. (1973) The relationship between the compressive strength and modulus of elasticity of keramzit concrete and the strength and deformation properties of its components, *Structure, strength and deformation of lightweight concrete*, Moscow (in Russian).
37   Buzhevick, G.A. and Zhitkevich, R.K. (1973) Investigation of the distribution of average deformations and stresses in the components of high strength

keramzit concrete for various combinations of their moduli of elasticity, *The strength, structure and deformation of lightweight concrete*, Moscow (in Russian).

38 Weigler, H. and Karl, S. (1972) *Stahlleichtbeton*. Bauverlag, Wiesbaden, Germany.

39 Swamy, R.N. and Jiang, E.D. (1991) Pore structure and carbonation of lightweight concrete after ten year exposure, ACI Symposium, *Performance of structural lightweight concrete*, Dallas, Texas, November. (To be published by ACI.)

40 Hoff, G.C. (1991) Private communication, November.

41 Abrams, D. (1923) Influence of aggregates on the durability of concrete, *ASTM Proceedings*, **23**, Part II Technical Papers.

42 LaRue, H.A. (1946) Modulus of elasticity of aggregates and its effect on concrete, *Proceedings*, ASTM **46**.

43 ACI 213 (1989) *Guide for structural lightweight aggregate concrete*, Reported by ACI Committee 213. American Concrete Institute, Detroit, Michigan.

44 Reichard, T.W. (1964) *Creep and drying shrinkage of lightweight and normal-weight concretes*, Monograph No. 74, National Bureau of Standards, Washington, DC, March. (Available from Superintendent of Documents, US Government Printing Office, Washington, DC.)

45 Shideler J.J. (1957) Lightweight-aggregate concrete for structural use. *ACI Journal, Proceedings*, **54**, No. 4, October. Also, *Development Department Bulletin*, No. D17, Portland Cement Association, Skokie, Illinois.

46 Shah, S.P., Naaman, A.E. and Moreno, J. (1983) Effect of confinement on the durability of lightweight concrete. *International Journal of Cement Composites and Lightweight Concrete*, Lancaster, UK, February.

47 Holm, T.A. and Bremner, T.W. Long term shrinkage of structural lightweight concrete, unpublished.

48 Rogers, G.L. (1957) On the creep and shrinkage characteristics of solite concretes, *Proceedings, World Conference on Prestressed Concrete*, July, San Francisco, California.

49 Holm, T.A., Bremner, T.W. and Vaysburd, A. (1988) Carbonation of marine structural lightweight concrete, *Proceedings Second International Conference on Concrete in a Marine Environment*, St. Andrews-By-The-Sea, Canada, SP-109. American Concrete Institute, Detroit, Michigan.

50 Klieger, P. and Hansen, J.A. (1961) Freezing and thawing tests of lightweight aggregate concrete. *Journal American Concrete Institute*, Detroit, Michigan, January.

51 Khokrin, N.K. (1973) The durability of lightweight concrete structural members, SAMARA, USSR (in Russian).

52 Malhotra, V.M., Carette, G.G. and Bremner, T.W. (1987) Durability of concrete containing supplementary cementing materials in marine environment, Katherine and Bryant Mather International Conference, SP-100. American Concrete Institute, Detroit, Michigan.

53 Malhotra, V.M., Carette, G.G. and Bremner, T.W. (1988) Current status of CANMET's studies on the durability of concrete containing supplementary cementing materials in marine environment, *Second International Conference on Concrete in a Marine Environment*, St. Andrews-By-The-Sea, Canada, SP-109, ACI, Detroit, Michigan.

54 Nishi, S., Oshio, A., Sone, T. and Shirokuni, S. (1980) Water tightness of concrete against sea water. Onoda Cement Co., Ltd.

55 Keeton, J.R. (1970) Permeability studies of reinforced thin-shell concrete, Naval Civil Engineering Laboratory, Port Hueneme, California, Technical Report R692 YF 51.42.001, 01.001, August.

56 Bamforth, P.B. (1987) The relationship between permeability coefficient for

concrete obtained using liquid and gas. *Magazine of Concrete Research*, **39**, No. 138, March.

57   Bremner, T.W., Holm, T.A. and McInerney, J.M. (1991) Influence of compressive stress on the permeability of concrete, ACI *Symposium on Performance of Structural Lightweight Concrete*, Dallas, Texas, November. (To be published by ACI.)

58   Ahmad, S.H. and Batts, J. (1991) Flexural behavior of doubly reinforced high strength lightweight concrete beam with web reinforcement. *ACI Structural Journal*, May/June, Detroit, Michigan.

59   Moreno, J. (1986) Lightweight concrete ductility. *Concrete International*, American Concrete Institute, November, Detroit, Michigan.

60   Rabbat, B.G., Daniel, J.I., Weinmann, T.L. and Hanson, N.W. (1986) Seismic behavior of lightweight and normal weight concrete columns. *ACI Journal*, January/February, Detroit, Michigan.

61   (1982) Concrete island towed to Arctic, American Concrete Institute, *Concrete International*, March, Detroit, Michigan.

62   McNarey, J.F., Ono, Y., Okada, T., Imai, M. and Kuroki, (1984) Freeze-thaw durability of high strength lightweight concrete, American Concrete Institute Fall Convention, November, Detroit, Michigan.

11 Applications in Japan and South East Asia

Shigeyoshi Nagataki and Etsuo Sakai

11.1 Introduction

In Japan high strength concrete was first achieved as early as the 1930s. For example, Yoshida reported in 1930 that high strength concrete with 28-day compressive strength of 102 MPa was obtained.[1] This result was obtained by a combination of compression and vibration processes without chemical or mineral admixtures. This isolated development was not followed by systematic development in the production and use of high strength concrete till the mid 1960s.

In 1968, high-strength reinforced mortar piles were developed by a process of autoclave curing with the use of silica powder in cement. The compressive strength of concrete used was 11,300 psi (78 MPa).[2] In 1970, high strength concrete with a compressive strength of 12,700 psi (88 MPa) was developed and used for prestressed concrete piles.[3] This high strength concrete was produced by using superplasticizer and autoclave curing. In the 1970s high strength concrete was utilized in the construction of railway bridges,[4] and since then its use in the construction industry has continued to increase. The growth of the use of high strength concrete is primarily due to the development of naphthalene sulfonate condensed superplasticizer, which was developed in Japan in 1964.[5]

The definition of high strength concrete as defined by the Japan Society of Civil Engineering (JSCE) is different from that of Architectural Institute of Japan (AIJ). In 1980 JSCE published *Proposed Recommendations for the Design and Construction of High Strength Concrete*,[6] which defined high strength concrete as a concrete with a design compressive strength 8500 to 11,500 psi (59 to 79 MPa). On the other hand, AIJ defines high strength concrete as a concrete with strength of 3900 to 5100 psi (27 to 35 MPa) for normal weight concrete, and with strength of 3500 to 3900 psi (24 to 27 MPa) for a lightweight concrete.[7]

For civil engineering structures in Japan, high strengtrh strength concrete is used mainly for bridges, high-rise buildings, and piles.[4] Studies in the use of high strength concrete in super high-rise reinforced concrete buildings are being promoted under the leadership of the Japanese National Project by Ministry of Construction (MOC). Studies related to the development and production of high strength concrete of strengths up to 17,100 psi (118 MPa) are under way and techniques for making these concretes practical for high-rise construction are being investigated.[8]

In South East Asia, studies into the utilization of high strength concrete have just started. However, in some of the countries, especially in Singapore, practical applications of high strength concrete have already commenced in high-rise buildings.

11.2   Methods of strength development

In this section, different methods employed for developing concretes with higher strengths are summarized, with special emphasis on the methods being used or studied in Japan. In recent years, many studies aimed at developing practical methods for producing higher strength concretes have been undertaken, especially in connection with the Japanese National Project related to high-rise buildings. There are a number of techniques for attaining higher strength concretes and they are summarized in Fig. 11.1.[9]

The strength development of concrete primarily depends upon the characteristics of hardened cement paste, aggregates, and the interface boundary between the hardened cement paste and the aggregates. It is well known that the strength of hardened cement paste depends on its degree of porosity, and laws have been proposed to explain the relationship between porosity and strength, e.g., Knudsen's formula.[10] According to these laws, the smaller the pore ratio the greater the strength. Therefore, to achieve higher strengths, it is necessary to reduce capillary pores. This can be done by reducing the water cement ratio which results in an increase in the amount of hydrates generated during the hydration process. Another approach for reducing the capillary pores is to impregnate the concrete with polymers. Polymer impregnated concrete (PIC) is one type of high performance concrete which also has a high strength.

Reduction in water cement (w/c) ratio increases the strength. With the use of superplasticizers, the water cement ratio can be reduced while maintaining the workability of the concrete. For concretes with very low water cement ratios, special techniques such as harmonic vibration and centrifugal compacting process have been developed in Japan and are being used by the construction industry.

A combination of a high dosage of dispersing agents and ultra-fine particles make it possible to maintain the workability of concrete with a low water cement ratio, e.g., Densified Systems of homogeneously

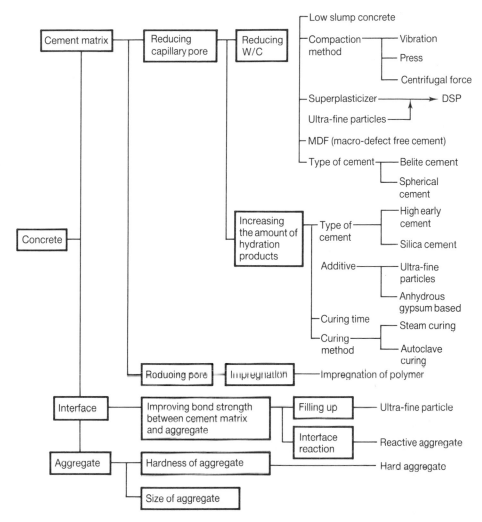

Fig. 11.1   Techniques for attaining high strength

arranged ultra-fine Particles (DSP). In these types of concretes, strengths of over 14,500 psi (100 MPa) are achievable.[11] The microstructure of the hydrate of this concrete is quite different from that of conventional concrete because the ultra-fine particles play a more important role in the reaction than the cement particles. It is reported that most of the cement particles which remain unaffected by the hydration reaction become inner-fillers which makes the hydrate of this concrete different from that of conventional concrete.[12] Another approach for reducing the water cement ratio without losing the workability is by using different kinds of cement such as belite cement[13] or spherical cement.[14] This approach is being investigated in Japan and the results are encouraging.

Increasing the amount of hydrates also results in a reduction of the total pore volume in the concrete. The use of high-early strength cement or silica cement is one method which has been used for some time to increase the proportions of the hydrates. In addition to this, like pozzolanic reactions or latent hydraulicity, the calcium hydroxide generated by hydration of cement or alkaline stimulation generates hydrates in the pores. Mineral additives with ultra-fine particles as the main ingredient are usually used for high-strength concrete. Silica fume, classified and ground blast furnace slag or fly ash,[15] and metakaolin[16] are the most popular mineral additives employed in the production of higher strength concretes. In addition, anhydrous gypsum based additives have also been used to produce ettringite actively,[17] which in turn can also reduce the pore volume.

The amount of hydrates in concrete largely depends on the curing method and the curing time. Steam curing or autoclave curing is effective in promoting the hydration within a certain period, which in turn helps in developing the strength.

The strength of the interface between the aggregate and the paste plays an important role in the achievement of higher strength. The interface strength is affected by the interface structure. Use of calcium hydroxide makes the spaces around the aggregates larger than those in other parts of the concrete continuum, and the greater the amount of calcium hydroxide, the lower the aggregate-paste interfacial bond strength, which in turn reduces the overall strength of the composite.

The replacement of cement by silica fume can increase the strength remarkably. This is attributable to the calcium silicate hydrates generated by both ultra-fine particles such as silica fume and the calcium hydroxide around the aggregates which improve the bond between the aggregates and the cement matrix.[18]

Higher strengths can also be achieved by improving the interfacial bond strength by using cement clinkers that are as reactive as the aggregates.[19] Also use of stronger aggregates with superior surface characteristics improves the strength. Figure 11.2 shows the relationship between water to binder ratio and the compressive strength of DSP mortars with various kinds of aggregates.[12] The results indicate that at a certain water binder ratio, the greater strength of the aggregates results in a greater compressive strength in the mortar, which is remarkable, especially in the higher strength zone. This is also the case with coarse aggregates, and one test result shows that at a constant strength of mortar, the maximum difference between the strengths of the concretes achieved by employing different kinds of coarse aggregates is 5800 psi (40 MPa).[20]

Various methods for examining the quality of aggregates have been proposed. These include, for example, a method for obtaining material factors of aggregates by examining the experimental relation between the strength of mortar and that of concrete for each absolute volume of coarse aggregate.[21] Another method is to determine the strength of coarse

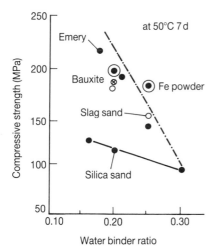

Fig. 11.2   Relation between the water to binder ratio and the compressive strength of DSP mortars with various types of aggregate

aggregates by examining the relation between the strength ratio of concrete to mortar and the reciprocal of the strength of mortar.[22] It is also reported that from the view point of mineralogy or petrology, strength of mineral, fineness of particles, micro cracks within particles, and cleavage should be examined[23] to maximize the intrinsic strength of the concrete composite. Since the quality of aggregates has recently been getting poorer, especially in Japan, it is thought that for producing higher strengths concretes in the future, careful studies of aggregates as well as of cement matrix, will become very important.

11.3 Applications

The applications of high strength concrete in Japan are summarized in Table 11.1. The details of these examples including the methods of strength development employed are described below.

Bridges

High strength concrete has been used for producing prestressed concrete (PC) girders for the purpose of reducing the dead load and for achieving longer span.[8] Japanese Industrial Standard (JIS) requires the compressive strength of concrete used for PC girders to be above 7100 psi (49 MPa) as shown in Table 11.1. Most of the high strength concrete used for railway bridges is usually a superplasticized concrete with a low water cement ratio. Examples include the Dai-ni-Ayaragigawa Bridge, Ootanabe Bridge, Iwahana Bridge, Kazuki Bridge and Akkagawa Bridge. The strengths of concretes used in the highway bridges are in the range of

Table 11.1 Application of high strength concrete in Japan

Concrete constructions and concrete products	Design strength (MPa)	Techniques for high strength	Notes
Railway bridges	59~79	Superplasticizer (+high early cement)	Lightened bridge weight prestressed concrete
Highway bridges	59~69	Superplasticizer (+high early cement)	
Prestressed concrete beams		Superplasticizer (+high early cement)	
for slab bridges	>49		JISA5313
for beam bridges	>49		JISA5316
for light load slab bridges	>49		JISA5319
Diaphram walls	~49	Superplasticizer + low heat cement	Massive concrete
Oil drilling rigs	58~65	Superplasticizer + silica fume	Light weight concrete
Abrasion resistance concrete	39~79	Superplasticizer + silica fume or anhydrous gypsum based additive	Repair of dam (floor)
High rise RC buildings	35~47	Superplasticizer or new type superplasticizer (+ultra-fine particles)	National project 60~112 MPa
Reinforced spun concrete piles	>39	(Superplasticizer)	JISA5310 centrifugal force
Pretensioned spun concrete piles	>49	Superplasticizer	JISA5335 centrifugal force
Posttentioned spun concrete piles	>49	Superplasticizer	JISA5336 centrifugal force
Pretensioned spun high strength concrete piles	>79	Superplasticizer + autoclave curing or anhydrous gypsum based additive	JISA5337 centrifugal force
Steel concrete composite piles	>79	Superplasticizer + autoclave curing or anhydrous gypsum based additive + expansive additives	centrifugal force
Prestressed spun concrete poles	>49 (or 79)	Superplasticizer (+autoclave curing)	JISA5309 centrifugal force
Centrifugal reinforced concrete pipes	49~69	Superplasticizer + anhydrous gypsum based additive or low slump concrete	Centrifugal force Jacked pipe
Railway sleepers	39~49	Superplasticizer (+high early cement)	
Concrete segments	49	Superplasticizer	
Machine beds	79~108	Superplasticizer + silica fume (DSP)	Damping capability

Fig. 11.3 Iwahana Railway Bridge (Sanyo shinkan-Sen)

8400 psi to 9800 psi (59 to 69 MPa). Examples include the Nitto, Kaminoshima, Jodoji and Seto Highway Bridges. In addition to the above examples, concretes with design strengths in the range of 5600 to 6400 psi (39 to 44 MPa) has been utilized in many bridges in the form of prestressed or cast-in-place concrete construction.

The mix proportions of the concrete used for these bridges are listed in Table 11.2. In the Dai-ni-Ayaragigawa Railway Bridge, high strength concrete was employed in order to reduce the dead load. The weight of the main girder with 5600 psi (39 MPa) normal strength concrete would have been 170 tons. This was reduced to 150 tons by the use of 8500 psi (59 MPa) strength concrete. The strength of the concrete with job-site curing resulted in 9000 to 10,400 psi (62 to 72 MPa), with an average of 9400 psi (65 MPa). The production of the PC girders was done in a yard near the construction site, and the concrete used was produced in ready-mix concrete plant about 2 km from the yard.[24]

In the construction of Iwahana Bridge (Fig. 11.3), precast concrete members were primarily used but for the joints cast-in-place concrete was used. Concrete mix information such as cement type, design strength, w/c ratio, dosage of superplasticizer etc. is given in Table 11.2. The w/c ratio for cast-in-place concrete was about 0.29. The average strength attained was 10,700 psi (74 MPa).[25]

In the construction of the Akkagawa Bridge, concrete with a design strength of 11,400 psi (79 MPa) was employed, and it was produced with autoclave curing. The average concrete strength and the standard deviation of the strength after curing, were 14,100 psi (97 MPa) and 580 psi (4 MPa) respectively.[26]

The merit of using high strength concrete for piers was examined.[27] It was reported that by increasing the design strength of concrete from 3900 to 8600 psi (27 to 59 MPa), the cross section of a pier for monorail could be reduced by 36%. It was also estimated that by increasing the design strength from 3900 psi (27 MPa) to 5700 and 11,400 psi (39 and 79 MPa), the cross section of a highway bridge pier could be reduced by 13% and 56% respectively.[27]

Table 11.2 Mix proportions of high strength concrete for railway bridges

Name of bridges	Type of structure	Design strength (MPa)	Type of cement	W/C (%)	s/a (%)	Dosage of SP (CX%)	Average of strength (MPa)	Construction
Da-ni-Ayaragigawa	1 Girder bridge span: 49 m	59	Normal cement	30	40	0.75	65	1973
Iwahana	Span: 45 m	79	High early cement	23	38.5	1.5	83	1974
Akkagawa	Span: 45.9 m	79	Normal cement	30	39.5	1.5	93*	1976

(Slump: 12 + 2.5 cm, G_{max}: 20 mm, * autoclave curing)

Table 11.3 Mix proportions of the main tower for Aomori Oohashi Road Bridge (Type of structure: PC cable-stay bridge, design strength: 59 MPa)

G_{max} (mm)	W/C (%)	S/a (%)	Unit weight (kg/m³)		Target of slump flow	Dosage of new SP (%)	Average of compressive strength (MPa)
			Cement	Water			
25	31.4	39	430	135	40~45 cm	2.2 ~2.5	80
25	35.0	40	386	135	40~45 cm	1.85~2.8	74

Fig. 11.4   Aomori Bay Bridge

The main towers of Aomori Bay Bridge were constructed in 1989 and high strength concrete with a design strength of 8600 psi (59 MPa) was used.[28] Figure 11.4 shows the bridge and Table 11.3 lists the mix proportions of the concrete. In this prestressed concrete cable-stayed highway bridge, high strength concrete was employed to achieve a slender structure with a reduced dead load and attractive aesthetic appearance. The concrete incorporated a new type of superplasticizer. Due to the presence of carboxylate polymer, this new superplasticizer reduced the slump loss which is caused by conventional sulfonated naphthalene formaldehyde condensate or sulfonated melamine condensate superplasticizer. Among these new types of superplasticizer, the one with air-entraining agent is currently being used.

Oil drilling rigs

An oil drilling rig to be used in the Arctic Ocean was made in Japan, using high strength concrete for purposes of greater durability and a lighter dead load.[29] In Table 11.4, some of the mix proportions, related to lightweight high strength concrete with silica fume and superplasticizer are given. Some special methods were introduced into the production of the concrete for this oil drilling rig in order to enhance its freeze-thaw resistance. It should be noted that up to now, only one drilling rig of this type has been made in Japan.

Diaphragm walls

Usually, for diaphragm walls concrete of 3000 to 3500 psi (21 to 24 MPa) is used. However, it is desirable to use higher strength concretes for deep underground large scale walls, which are subjected to higher water and earth pressures. This is very important for Japan, since there is a strong need to develop underground space because of the limited land mass. The

Table 11.4   Mix proportions of high strength concrete using light weight aggregate for oil drilling rigs

W/C (%)	Unit weight (kg/m^3)					Slump (mm)	Air (%)	Compressive strength (MPa, for 28d)
	W	C	Silica fume	Fine aggregate	Coarse aggregate			
30.8	160	520	52	603	493	185	7.7	58
28.5	143	500	75	604	456	170	5.2	65

(with superplasticizer and air entraining agent)

concrete for a diaphragm wall does not need consolidation, because it is cast in bentonite slurry through a tremie pipe. Therefore, the fluidity and the resistance to segregation of concrete are important considerations for designing the concrete mixes.

Table 11.5 shows one of the mix proportions of the concrete used for the diaphragm wall of LNG tanks at the Tokyo Gas Company's Sodegaura Plant.[30] Usually, a diaphragm wall is for temporary use, however in this case the upper part of the diaphragm wall was combined with an inner lining concrete of the side wall to achieve a permanent structure. The inner lining concrete was cast first using the top down lining method in which the excavation was carried out without shoring. The foundation slab was then cast, followed by the construction of the side wall. The thickness of the side wall was 120 cm, and it utilized high strength concrete. Low-heat cement with three binding components was employed in order to solve the problem related to mass concreting. Figure 11.5 shows the uncovered diaphragm wall after excavation.

Abrasion resistant applications

Due to its superior abrasion resistance qualities, concretes with higher strength have been used in the industrial flooring and in the repair of dams.

Fig. 11.5   Diaphragm wall – LNG tank

Table 11.5 Mix proportions of high strength concrete for diaphragm walls

Design strength (MPa)	W/C (%)	s/a (%)	Unit weight (kg/m³)		Dosage of SP (CX%)		Average of compressive strength (MPa)		
			W	C	1	2*	Standard curing (91d)	Strength of core	Standard deviation
49	28.4	38.8	128	450	1.5	1.0	69	69	7

(*: Relayed Addition, Cement: Low heat type; Cement-BFS-FA)

It is known that concrete of over 5700 psi (39 MPa) is very effective as regards to abrasion-resistance.

Table 11.6 lists some of the mix proportions of the concrete used for the repair work on an intake of a dam. The concrete strength for this application varied between 8300 to 8600 psi (57 to 59 MPa). A mineral additive (anhydrous gypsum based additive) was added to the concrete.[31] This additive can reduce the unit cement content, and can also solve the serious problem of thermal cracking in massive concrete. The unit cement content of this concrete was about 508 lbs/cyd (300 kg/m^3).[32] Concrete of about 11,300 psi (78 MPa) with silica fume has seen limited use for abrasion or chemical-resistant floors.

High-rise reinforced concrete buildings

For the high rise construction in Japan, the strengths of the concretes utilized generally range between 5100 psi (35 MPa) to 9000 psi (63 MPa) and chemical additives such as superplasticizer is usually used. Concrete with 17,100 psi (118 MPa) strength is currently being used in a MOC sponsored national project into high-rise construction. All the high-rise buildings in Japan are designed for earthquake excitations. For these special design conditions, the arrangements for the reinforcement in the joints is rather congested, consequently the flow ability and the filling capacity of fresh concrete is very desirable. Slump loss in fresh concrete is a serious problem in relation to flow ability and filling capacity. Recently, a new type of superplasticizer whose main ingredient is carboxylate polymer with a low rate of slump loss, high water reduction and little retardation has been developed. An investigation to establish a standard for new types of superplasticizer for high strength concrete is in progress.

High strength concrete is used in the columns of high rise reinforced concrete buildings to carry the loads economically and to provide relatively more floor space by having smaller sections. Generally, the design strength of the concrete is 5100 to 5900 psi (35 to 41 MPa) for 25-story buildings, and 5900 to 6800 psi (41 to 47 MPa) for 30 to 40-story buildings.[33] A National Project sponsored by the MOC into high-rise buildings using high strength concrete of up to about 17,100 psi (118 MPa) is now in progress. In this project, the employment of DSP materials with ultra-fine particles like silica fume or blast furnace slag and additives like anhydrous gypsum based additive for high strength concrete are being examined. Only one full-scale experiment of construction has been carried out to date.[34]

Table 11.7 shows the mix proportions and the strength test results of high strength concrete with design strength of 5900 psi (41 MPa). This concrete was used for the recently completed 41-storied high rise building. The actual strength achieved in the field was 8100 psi (56 MPa).[36] Table 11.8 shows the mix proportions of the concrete used for a 8-story building.[36] The design strength for this concrete was of about 8600 psi (59 MPa). In this case, precast concrete was used for the columns and

11.6 Mix proportions of abrasion resistance concrete for dams

Execution place	Slump (cm)	W/C (%)	Unit weight (kg/m³)					Compressive strength (MPa, for 28d)
			W	C	Fine aggregate	Coarse aggregate	Additive*	
Slope of dam body	15	36.5	146	400	776	1049	40	59
Flat place of dam body	18	38.0	152	400	769	1040	40	57

(* Anhydrous gypsum based)

Table 11.7 Mix proportions of concrete for high rise RC buildings

Design strength (MPa)	W/C (%)	Unit weight (kg/m³)				Average of comprehensive strength (MPa, for 28d)	Standard deviation (MPa)
		W	C	Fine aggregate	Coarse aggregate		
41	35.5	175	493	673	1009	56	4
35	40.0	175	438	719	1009	50	3
29	44.5	175	393	755	1009	46	3

beams, and cast-in-place high strength concrete for the slabs and the joints of the columns.

The strengths of all these concretes mentioned above are over the upper limit of 5100 psi (35 MPa) which was established by AIJ as a reference strength to distinguish between the normal strength and high strength concrete. Presently there is no standard for selecting materials and mix proportions for the production of high strength concrete. Hence the high-rise reinforced concrete projects are authorized by the Ministry of Construction (MOC) on the basis of design, construction procedure and methodology.

Concrete piles

Concrete strengths for this application range between 5600 psi (39 MPa) to 11,500 psi (79 MPa). The types of concrete pile used in Japan include reinforced spun concrete piles, pretensioned spun concrete piles, post-tensioned spun concrete piles and pretensioned spun high strength concrete piles (Table 11.1). The production and use of Pretensioned High-strength Concrete (PHSC) piles is increasing very fast and presently exceeds 50% of the total market of piles. This is because: (1) A pretensioned high strength concrete pile can withstand higher vertical and horizontal loads compared with other kinds of pile which reduces the number of piles required, resulting in savings, (2) It has higher impact resistance, (3) The production period is shorter. Figure 11.6 shows a prestressed high strength concrete pile being driven. The typical mix proportions for high strength concrete used for concrete piles is given in Table 11.9. The high strength concrete piles are generally made using the autoclave curing method, and sometimes incorporate silica fine powders.[6]

An alternate method for producing high strength concrete piles in Japan, is by the use of ordinary steam curing with an anhydrous gypsum based

Fig. 11.6   PHC piles

Table 11.8 Mix proportions of high strength concrete for high rise RC buildings

Design strength (MPa)	Target strength (MPa)	W/C (%)	s/a (%)	Slump (cm)	Air (%)	Unit weight (kg/m³) C	W	Compressive strength (MPa) Standard curing (28d) Job	Site 56d	Core 56d	Core 28d
59	71	29.1	40.1	25	1.5	550	160	81	70	72	64
								86	87	87	62

Table 11.9 Mix proportions of high strength concrete for concrete piles

Products	Design strength (MPa)	Target strength (MPa)	Type of cement	G_{max}	Slump (mm)	W/C (cm)	s/a (%)	Unit weight (kg/m³) (%)	C	W	Dosage of SP (CX%)
PHC piles*	78	88	Normal	20	7***	32	40		450	142	1.25
PC piles**	57	57	Normal	20	4	36	42		420	152	0.85

(*:Steam curing + autoclave curing, **:Steam curing, ***:By using of pump – 12 cm)

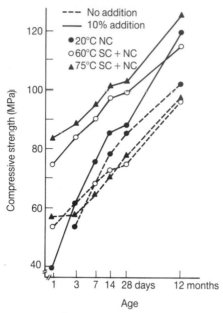

Fig. 11.7   Effect of addition of anhydrous gypsum based additive on compressive strength of concrete

mineral additive as an additional ingredient in concrete. This approach allows the production of larger diameter piles which are becoming popular. The effect of addition of anhydrous gypsum based mineral additive on the compressive strength of concrete is shown in Fig. 11.7. These results are for a mix with the water to cement ratio, the unit cement content and the sand aggregate ratio of 0.29, 470 kg/m^3 and 0.42 respectively. This is one of the mix proportions that is being currently used in Japan for the production of high strength concrete piles. For this mix, the strength is about 11,400 psi (79 MPa) just after steam curing, and the 28-day strength is over 14,500 psi (100 MPa). Figure 11.8 shows the hydrated and unhydrated products of cement with anhydrous gypsum based additive in the concrete. During the curing, anhydrous gypsum is reduced and ettringite is produced actively, which causes a decrease in the total pore volume of hardened paste. The ettringite absorbs a greater amount of water than other hydrates resulting in a low water cement ratio.[18]

One enterprising application of high strength concrete in piles is a composite pile of a steel pipe and concrete. This steel concrete composite pile is made by lining a steel pipe with concrete using centrifugal force. The flexural ductility of this steel concrete composite pile is higher than prestressed high strength concrete piles. An example of mix proportion for the high strength concrete used in the production of composite pile is given in Table 11.10, which shows that the concrete is produced with an expansive additive.[37] The main ingredient of the additive is magnesium oxide or calcium oxide when autoclave curing is employed. On the other

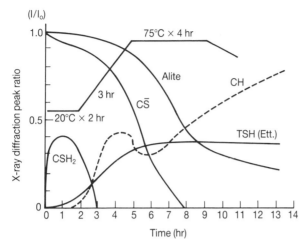

Fig. 11.8 Hydration and unhydration products of cement with anhydrous gypsum based additive

hand, when the composite piles are produced with ordinary steam curing, a combination of anhydrous gypsum based additive and expansive additive is used.[37]

In addition to the piles, high strength concrete is being used for the production of concrete poles. The mix proportions and the manufacturing method are basically similar to those used for the production of the piles.

Other concrete products

The other concrete products in which high strength concrete is being used include hume pipes, sleepers and concrete segments and the mix proportions for the concrete for these applications is given in Table 11.11.

Hume pipes are widely used for sewerage or agricultural canals. Usually, prestressed concrete with an expansive additive added is employed for the production of hume pipes. Because of the jacking method used for driving the concrete pipe, higher strengths of concrete are needed to withstand the static and inertial forces. Considering that longer distance jacking with be desirable in the future, the use of higher strength concretes will increase for this application. In Japan, hume pipes with concrete of up to about

Table 11.10 Mix proportions of high strength concrete for steel concrete composite piles

| G_{max} (mm) | Slump (cm) | W/Binder (%) | s/a (%) | Unit weight (kg/m³) | | | Dosage of SP (Binder X %) |
				W	C	Expansive additive	
20	13	30.7	42	132	404	26	1.5

(Autoclave curing, main component of expansive additive – CaO)

10,000 psi (69 MPa) strengths have been produced with centrifugal compaction and have been used in a number of applications. For these concretes, sometimes only superplasticizer, sometimes a combination of superplasticizer and anhydrous gypsum based additive is used. For the placement of concretes with very low water cement ratio, vibratory centrifugal compaction method is employed.

Prestressed concrete sleepers are used for railway tracks. In addition to the prestressed concrete mentioned in Table 11.1, JIS specifications for various kinds of prestressed concrete products have been established. The specifications require that the design strength of the concrete, and the strength at the time of prestressing must exceed 7100 psi (49 MPa) and 5700 psi (39 MPa) respectively. Because of this requirement, a combination of high-early-strength cement and superplasticizer is usually employed in the manufacturing process.

Concrete shield segments are being used as lining in shield tunneling. High-strength concrete is being used for these segments to make them lighter hence reducing the transportation and erection costs and to make them stronger to better withstand the high earth pressures. Figure 11.9 shows concrete segments ready for use in tunnel construction by the shield tunneling method.

One of the unique applications of high strength concrete is the use of 15,600 psi (108 MPa) concrete with DSP material for the beds of machine tools. Because of the high degree of rigidity and vibration damping properties of higher strength concretes, it is very advantageous for this application.[39] This application is an example of high performance concrete use in which attributes other than strength govern its choice. Figure 11.10 (by Nippei Toyama Corporation) shows a machine tool, the bed for which is made by using concrete with a strength of 15,600 psi (108 MPa).

Fig. 11.9   Concrete segment

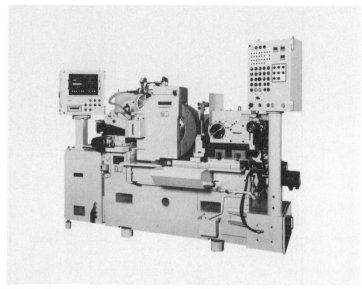

Fig. 11.10   Machine bed (Nippei Toyama Corporation)

Applications in South East Asia

The examples of applications of high strength concrete in the countries of South East Asia are summarized in Table 11.12. The applications of high strength concrete are basically limited high-rise reinforced concrete buildings. The concrete strength usually range from 7200 to 8700 psi (50 to 60 MPa). Figure 11.11 shows a high-rise building in Singapore, which was built by using high strength concrete in the columns from the 3rd underground story to the 7th floor. In the core walls up to the 45th floor, composite structural segments of steel and concrete with strengths of 8700 psi (60 MPa) and 7200 psi (50 MPa) respectively were employed. In addition to Singapore, Korea[39] and Hong Kong, investigations into possible applications of high strength concrete have already been initiated in Thailand.

11.4   Summary and conclusions

There are widespread applications of high performance concretes in Japan and South East Asia. High Performance concrete with very high compressive strength generally called high strength concrete has been used in bridges, oil drilling rigs, diaphragm walls in deep underground tanks etc. For special applications such as industrial flooring and repair of dams, high performance concrete has been used due to its superior abrasion resistance property. For high rise construction, the higher compressive strength of high performance concrete has been utilized in columns to maximize the

Table 11.11 Mix proportions of various concrete products

Products	Design strength (MPa)	Target strength (MPa)	Type of cement	G_{max} (mm)	Slump (cm)	W/C (%)	S/a (%)	Unit weight (kg/m³) W	C	Dosage of SP (CX %)
Centrifugal reinforced concrete pipes	–	54	Normal	20	7	37	43	168	460	0.95
PC railway sleepers	49	64	High early	20	4	32	40	140	440	0.85
PC beams	49	64	High early	20	3	33	35	147	450	0.85
Segments	49	61	Normal	20	5	33	40	146	450	1.05

(Pre-curing for 2~3 h, steam curing for 3~5 h at 65~70°C)

Table 11.12 Application of high strength concrete in South East Asia

Country	Constructions	Design strength (MPa)	Techniques for high strength	Notes
Singapore	Road link	50	Superplasticizer (+ silica fume)	Caisson and bridge
	High rise building	50~60	Superplasticizer or + silica fume	Hoechang building, UOB Plaza, Savu Tower
	Concrete spun piles for wharf	60	Superplasticizer	Expansion of Jurous Port
Korea	High rise building	>49 (for 28d, target strength)	Superplasticizer	Bridges – 40 MPa
Hong Kong	High rise RC building	60 (for 28d)	Superplasticizer + 25% PFA cement	78 floors the expected date of completion: 1992, Aug.

Fig. 11.11   High rise building: Singapore, UOB Plaza

resistance capacity without sacrificing the available floor space. Higher strength concretes have also been used for prestressed concrete piles. Other applications of high performance concrete in Japan and South East Asia include hume pipes for sewerage or agricultural canals, prestressed concrete sleepers, concrete shield segments as lining in shield tunneling, beds for machine tools.

In the last two decades or so, the construction industry in Japan has gained tremendous amount of experience in use of high performance concretes especially high strength concretes. As further development takes place in the high performance concretes, it appears that the use of high performance concretes will continue to increase in Japan and South East Asia.

References

1   Yoshida, T. (1930) *Proceedings Japan Society of Civil Engineering*, **26**, 997.
2   Kokubu, K. and Fukuzawa, K. (1987) *New Concrete Technology*, **8**, 90 pp.
3   Nishi, H., Ooshio, A. and Fukuzawa, K. (1972) *Cement Concrete*, No. 299, 22.
4   Nagataki, S. (1989) *Proceedings MRS International Meeting on Advanced Materials*. Tokyo, **13**, 3.
5   Hattori, K., Yamakawa, K., Tuji, A. and Akashi, T. (1964) *Semento Gijutu Nennpo*, **18**, 200.
6   (1980) Concrete Library, 47 *Japan Society of Civil Engineering*.
7   (1991) JASS-5, *Architectural Institute of Japan*.
8   Masuda, Y. (1990) Text of Concrete Lecture. *Japan Cement Association*, No. 249, 15.
9   Sakai, E. (1991) *Cement Concrete*, No. 535, 25.
10 Knudsen, F.P. (1959) *Journal, American Ceramic Society*, No. 42, 376.
11 Bache, H.H. (1980) WO 80/00959, 15 May.
12 Mino, I. and Sakai, E. (1989) *Proceedings, MRS International Meeting on Advanced Materials*, Tokyo, **3**, 247.
13 Hanehara, T. (1990) *Journal Ready Mixed Concrete*, **9**, 92.
14 Kitamura, M., Hitotsuya, K., Tanaka, I., Take, T. and Suzuki, N. (1991) *Proceedings of 45th General Meeting Cement Association of Japan*, 172.
15 Muguruma, H., Mino, I., Ashida, M. and Sakai, E. (1987) *Proceedings, Symposium on Utilization of High Strength Concrete*, Norway, 63.
16 Bredy, P., Chabamet, M. and Pera, J. (1989) *Proceedings, MRS Symposium on Pore structure and Permeability of Cementitious Materials*, **137**, 431.
17 Kageyama, H., Nakagawa, K. and Nagahuchi, T. (1978) *Journal of the Society of Materials Science*, Japan, **29**, 220.
18 Rosenberg, A.M. and Gaidis, J.M. (1986) Presented to *Second Conference on the Use of Fly Ash, Silica Fume, Slag and Natural Pozzolans in Concrete*, Madrid.
19 Ooshio, A. (1976) *Concrete Journal*, **14**, 34.
20 Kokubu, K. and Hisaka, M. (1990) *Concrete Journal*, **28**(2), 14.
21 Hanehara, T. (1990) *Journal of Ready Mixed Concrete*, **9**, 108.
22 Tanigawa, Y., Nakamura, M., Shibata, T. and Odaka, S. (1991) *Proceedings Annual Meeting, Japan Concrete Association*, No. 209.
23 Sarkar, S.L. and Aitcin, P.C. (1990) *ASTM Special Technical Publication*, No. 1061, 129.
24 Machida, F., Hirose, T., Miyasaka, Y. and Kitta, T. (1974) *Prestressed Concrete*, **16**, 36.
25 Machida, F., Suetugu, H., Yamamoto, T. and Fukumoto, Z. (1975) *Prestressed Concrete*, **17**, 4.
26 Matsumoto, Y., Saito, T., Miura, I. and Mine, Y. (1977) *Proceedings, Japan Society of Civil Engineering*, No. 264, 97.
27 Guide-line of Design and Execution for Constructions with High Strength Concrete, *Japan Cement Association*.
28 Ishibashi, T., Yoshida, H., Ooba, M. and Takeuchi, K. (1990) *Concrete International*, **28**(5), 59.
29 Ono, Y., Suzuki, T., Niwa, M. and Iguro, M. (1984) *Cement Concrete*, No. 450, 8.
30 Okada, T., Imai, M. and Kimura, K. (1987) *Foundation Work*, **111**, 11.
31 Nakajima, K., Matsunami, H., Shimizu, H. and Fukushima, I. (1991) *46th Annual Meeting of Japan Society of Civil Engineers*, V.500.
32 Sugita, H., Nagamatsu, T. and Fufimoto, H. (1989) *Journal Civil Engineering for Electric Power*, No. 223, 63.
33 Masuda, K. (1990) *Concrete International*, **28**(12), 14.

34 Kawai, T., Yamazaki, Y., Imai, M., Tachibana, D. and Inada, Y. (1989) *Cement Concrete*, No. 508, 31.
35 Yagi, S., Tabuchi, H., Hamano, Y., Senda, T. and Itinose, K. (1991) *Cement Concrete*, No. 536, 10.
36 Tomatsuri, K., Kuroha, K. and Iizima, M. (1991) *Cement Concrete*, No. 531, 22.
37 Matsumoto, Y. and Fukuzawa, K. (1980) *Proceedings of Japan Society of Civil Engineering*, No. 301, 125.
38 Nakayama, N. (1982) State of Art of Recent Admixtures. *Japan Concrete Association*, No. 54.
39 Shin, S.W. (1980) *Proceedings of Korea Society of Concrete*, **2**, 165.

Index